# THE RELOADER'S BIBLE

ALSO BY DON GEARY

*Hiking and Backpacking* (coauthor)
*The How to Book of Interior Walls*
*The How to Book of Floors & Ceilings*
*Roofs and Siding*
*How to Build Kitchen Cabinets, Counters and Vanities*
*Interior and Exterior Painting*
*Backpacking* (coauthor)
*The Welder's Bible*
*Step in the Right Direction*
*How to Build Studios and Workshops in Your Home*
*The Outdoorsman's Handbook*
*Maintaining the Exterior of Your Home*
*Plywood Furniture Projects*
*How to Sharpen Anything*
*The Home Brewer's Handbook*
*The Arc Welder's Handbook*

# THE RELOADER'S BIBLE

## The Complete Guide to Making Ammunition at Home

by Don Geary

Prentice Hall Press • New York

*WARNING:* Extreme care should be taken when dealing with explosives or flammable material.

Every effort has been made to ensure the accuracy of the information in this book; however, the Publisher is not responsible for the user's negligence, inadvertent errors, or for the user's failure to follow directions.



Published by Prentice Hall Press
A Division of Simon & Schuster, Inc.
Simon & Schuster Building
Rockefeller Center
1230 Avenue of the Americas
New York, New York 10020

PRENTICE HALL PRESS is a trademark of Simon & Schuster, Inc.

Designed by Jack Meserole/Publishing Synthesis, Ltd.

Library of Congress Cataloging in Publication Data

Geary, Don.
The reloader's bible.

Includes index.
1. Handloading of ammunition. I. Title.
TS538.4.G43 1986 683.4'06 85-28134
ISBN 0-668-06025-5

Manufactured in the United States of America

10 9 8 7 6 5 4 3 2 1

# Contents

# Introduction

Reloading ammunition has a certain appeal for those of us who like to tinker with simple tools and lots of parts. To be sure, quite a bit more is involved than simply assembling a few parts, but reloading ammunition at home can be simple indeed. In fact, almost anyone can start producing quality ammunition on the first try.

I am occasionally asked why I reload my own ammunition. To the uninitiated, all of the hours, hand operations, and outlay of cash (for components) must seem an awful waste of time, but I think the time is well spent. The truth of the matter is that I reload ammunition for a number of reasons, and for what they are worth, I will pass these along to you.

I began reloading about twenty years ago when I discovered it took more than one shotshell to bag one duck. I was relatively new to serious hunting at the time, which meant I could not hit the proverbial side of a barn with a scattergun. It did not take me long to realize it was going to take lots of practice to become a good shot, and with the then two-dollar price tag per box of shotshells, my learning to shoot could easily turn out to be an expensive hobby. So my initial reason for reloading was economy, a reason that is still valid today for me and thousands of other reloaders.

In addition to saving on shooting costs, reloading is the best way to develop high-performance ammunition that is just not available from any other source. To be sure, commercially made ammunition is consistent and generally of high quality, but it can often be improved upon, and, in fact, custom made for a particular firearm.

As a rule, over-the-counter ammunition is available in a rather limited number of bullet weights per caliber. Even though most bullet manufacturers produce a wide range of bullet weights, your local retailer may not stock a very wide selection of any one particular caliber. A good example of this is .30-06 ammunition, which is manufactured in a variety of bullet weights (from 110 to 220 grains), but, since the two most popular bullet weights for hunting are 150 and 180 grain, these weights are quite possibly the only two you will find for sale. The reloader, on the other hand, has a choice of at least twenty different bullet weights—again, from 110 to 220 grains. By varying bullet weight and powder charge according to established reloading data it is possible to develop ammunition for any shooting purpose that performs better, according to your needs.

If your shooting consists of firing your deer rifle fewer than twenty times a year, reloading is probably not for you. But, if you are concerned about your shooting performance, and shoot often, you will want to achieve a level of shooting ability that can only be attained after firing hundreds of rounds per year. To be able to afford to shoot and to develop highly accurate ammunition, you will want to reload your own ammunition. You will find all of the required information in the pages of this book.

Still another reason for reloading is the personal satisfaction of accomplishing a task, really much more than putting primer, powder, and bullet together in a cartridge. While it is entirely possible to develop shooting skills without knowing (or caring) about what happens when you squeeze the trigger, I find it interesting to understand what happens, and I delight in experimenting with different powders, bullets, and primers. This level of understanding can only come from familiarity with reloading and firearms.

This book is designed to explain many of the mysteries of reloading. In the first chapter you will discover the basic tools required for reloading as well as a discussion of more advanced tools and machinery. The second chapter covers all of the components used for reloading both metallic (rifle and pistol) and shotshell ammunition. The third and fourth chapters are devoted to the actual mechanics of reloading, and here you will learn how to customize your ammunition. The fifth chapter covers advanced and specialized reloading techniques. Here you will learn how to make your own bullets from scrap lead, and other handy reloading procedures such as annealing brass casings and reclaiming military cases.

The last three chapters contain reloading data that you can use for reloading pistol, rifle, and shotshell ammunition. Be sure to read the notes about the tables at the beginning. *Follow this data to the letter and observe all cautions.*

It is my sincerest hope that you will find ammunition reloading as rewarding as I have and that you will always exercise due care and respect for components and firearms.

Good shooting!

Don Geary
Salt Lake City, Utah

# HOW TO RELOAD AMMUNITION

# 1 Reloading Tools

Ammunition reloading, like fly fishing, very much falls into the realm of the gadgeteer. There are hundreds of tools and accessories for the handloader, and it is easy for the newcomer to be confused as to which items are really necessary and which are frivolous. In this chapter we will look at the basic tools and accessories required to reload the two types of ammunition—metallic cartridge and shotshell. Along the way we will also look at other tools and accessories that, while not essential, often speed up the process or simply reduce the number of hand operations.

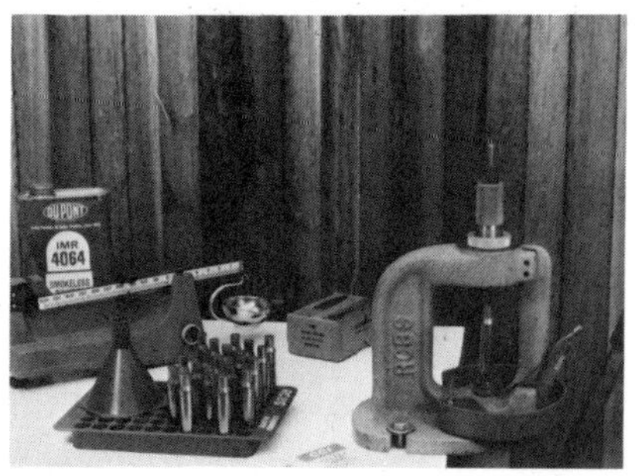


**Only a few specialized tools are required for reloading ammunition at home—reloading press, dies and components, empty brass, primers, powder, and bullets.**

Because of the nature of metallic cartridges—those used for rifle and pistol ammunition—as compared with shotshells, different equipment is required for reloading properly. To simplify reloading equipment choices, I have divided this chapter into two sections. The first covers tools and equipment required for reloading metallic cartridges; the second section covers tools and equipment required for reloading shotshells.

## TOOLS FOR RELOADING METALLIC CARTRIDGES

### Reloading Press

Certainly one of the most important (and expensive) reloading tools required for reloading metallic cartridges is a reloading press. Fortunately for the beginner, all of the reloading presses on the market are well made and should last a lifetime, so there is little chance that your choice will be a poor one.

Although they come in different styles and shapes and have different abilities, all reloading presses are similiar in that they contain a "ram," shell holder, lever, and die screw hole. Generally, reloading presses fall into one of three categories: C-type, H-type, and O-type. The letter designation of the press fairly describes the overall design

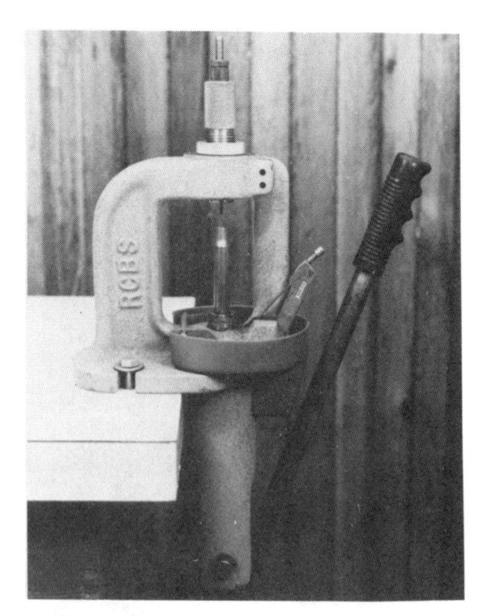

As you can see, the shape of this O-type reloading press resembles the letter "O."

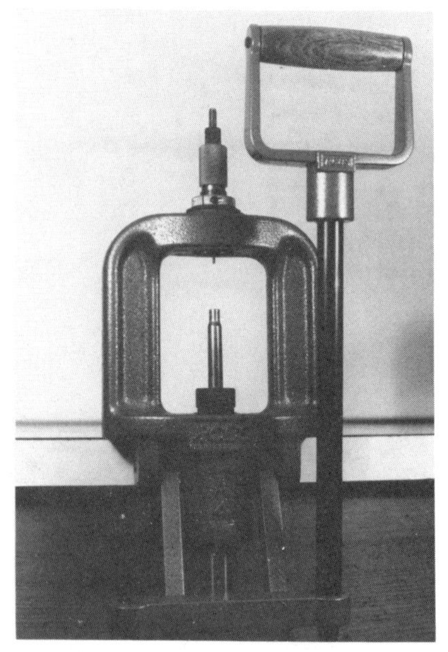

The Big Max reloading press from RCBS offers great strength and has a unique universal shellholder.

of the press. For example, a C-type reloading press resembles the letter C when viewed from the side.

C-type reloading presses are the least expensive of all types and can easily be used for almost all types of metallic cartridge reloading. In my opinion, however, the C-type press is not the best choice, because it is difficult to use for full-length sizing of bottleneck-type rifle cartridges. Nevertheless, if economy is an important factor when you are buying a reloading press, then you will probably purchase a C-type press.

H-type reloading presses are rather specialized reloading machines used for loading a quantity of straight-walled pistol casings quickly. A typical H-type press most commonly has three stations, all of which can be used with a single pull of the lever. In one stroke it is possible to remove a spent primer, seat a new primer, resize the entire cartridge cast, expand the neck, seat a bullet, and crimp the finished cartridge. As you can well imagine, a good H-type press can help you turn out a lot of pistol ammunition with almost lightning speed. As a rule, this type of reloading press is designed for a professional or competitive shooter and, as a result, costs at least $500. The newcomer to reloading will find both the capabilities and price tag of such a machine to be much more than required.

The last type of reloading press, commonly called the O-type press, is undoubtedly the most popular for all types of metallic cartridge reloading. Although there are some variations—such as a turrent top—all O-type reloading presses are basically the same. A brief description of how an O-type press operates may prove helpful in understanding some of the basics and terminology of reloading metallic cartridges with any type of press.

With the press securely mounted to a table or special reloading bench, the proper-size shell holder is first installed into the top of the ram. Next, the proper type and size die is screwed into the top of the press. Now, an empty casing can be inserted into the shell holder and the lever of the press pulled downward. This action raises the casing up and into the die. As the lever is returned to the normal upright position, the casing is withdrawn from the die, and it can then be removed from the shell holder.

Although there are certainly more details about how a reloading press operates in relation to the dies, as you will see in the chapters devoted to reloading specifics, all that is required of a press has been described above. Therefore, if a reloading press can accomplish these tasks—holding the shell securely as it is guided into and out of a die—it will suffice for metallic cartridge reloading.

I have been using a particular reloading press for several months with great satisfaction. The RCBS Big Max™ Reloading Press represents the state of the art in O-type reloading presses. Some of the features I really like are the automatic shell holder, which almost eliminates the need for special-size shell holders, and the general ease of use. The compound leverage of the Big Max makes full-length sizing of bottleneck casings as easy as punching out a spent primer.

## Reloading Dies

After a press, the next most important equipment consideration is a die set. Metallic cartridge reloading requires a set of dies for each caliber. As a rule, straight-walled pistol ammunition (and some rifle ammunition such as 45-70 government, .30 caliber M1, etc.) is reloaded with a three-die set. Rifle ammunition is reloaded with a two-die set. Because of industry-wide standardization, all reloading dies are threaded at ⅞ × 14 inches, so that the dies of any manufacturer will fit into any reloading press.

In a three-die set for reloading pistol ammunition, the first die is used for resizing the empty casing. At the same time it will remove the spent primer with a special punch. A resizing die is simply a hollow metal tube with inside dimensions exactly the same as those of an unfired casing. For example, a .38 special resizing die has inside dimensions of 1.097 inches in length and 0.379 inches in diameter. As a rule, a fired .38 special casing is slightly wider than this—a result of the pressures inside the case upon ignition—and must be brought back to uniformity before reloading.

The second die in a three-die pistol set is used for expanding the rim of the case neck. The end result is often a slight "bell" shaping. This is done to make seating the bullet easier.

**Straight-walled pistol ammunition is commonly reloaded with a three-die set.**

The third die is used for seating the bullet and, in some cases, crimping the neck of the casing tightly against the bullet.

Die sets for reloading rifle ammunition are most commonly composed of two dies. The first die resizes and decaps (punches out the spent primer); the second seats the bullet and (if required) crimps the casing around the bullet. As mentioned earlier, if a rifle casing is straight-walled, the die set for reloading will most commonly have three dies rather than two.

Reloading dies are made from special tool-grade steel and should last for thousands of rounds. The inside of each die is polished to a mirrorlike finish, which reduces the chances of scratching a brass casing. Nevertheless, dies should be disassembled and swabbed out every hundred rounds or so. It is also important to lubricate each case before running it through a die to reduce wear further on these important surfaces.

There are standard reloading dies for all of the standard rifle and pistol calibers. In addition, some companies, notably RCBS®, will make special dies for "wildcat cartridges" (see Glossary). Die sets are made from tool-grade steel and, for extra-long life, can also be ordered in tungsten-carbide steel. Although tungsten-carbide steel dies cost roughly twice as much as tool steel dies, they last almost forever and are therefore a good investment if you load large quantities of ammunition.

If you want the best reloading dies available, consider the so-called competition die sets. While these are the most expensive reloading dies on the market, they also produce the most uniform ammunition. As a rule, however, if your ammunition needs run from hunting to casual target shooting, a standard set of reloading dies should fill your needs.

### Case Preparation Tools

A number of operations must be performed on brass casings before they can be reloaded. While some of these, such as primer removal, primer seating, and resizing, are accomplished with the reloading dies, other case preparation operations must be done manually or with the aid of a special machine. The point is that before you can expect to make uniform and quality ammunition, the cases to be reloaded must be all the same and in good condition.

One case preparation task is to make certain that all of the casings you are working with are the same length; outside diameters are taken care of as the casing is run through the resizing die. Probably the easiest way to check case length is with a *case length gauge.* This handy device is simply a piece of stamped or cast metal inside of which are indicated the proper lengths of a variety of cases. To check a case, insert it into the gauge; almost instantly you will discover if the case is too long, too short, or the proper length.

Unfortunately, case-length gauges tell you not *how* the case in hand differs in length from the standard but only *whether* the case is the right length. A much better way to measure brass casings is with the aid of a *dial caliper.* A dial caliper, which measures both inside and outside dimensions in 0.001-inch increments, is handy for indicating just how much the case in hand differs from the standard for that particular casing.

In the reloading tables, which appear later in this book, you will find case length information for all brass cartridge casings. It is important to keep in mind that each time a cartridge is fired, the casing becomes slightly longer. Brass casing elongation is the result of extreme pressure (up to 50,000 pounds per square inch—PSI) on a ductile metal.

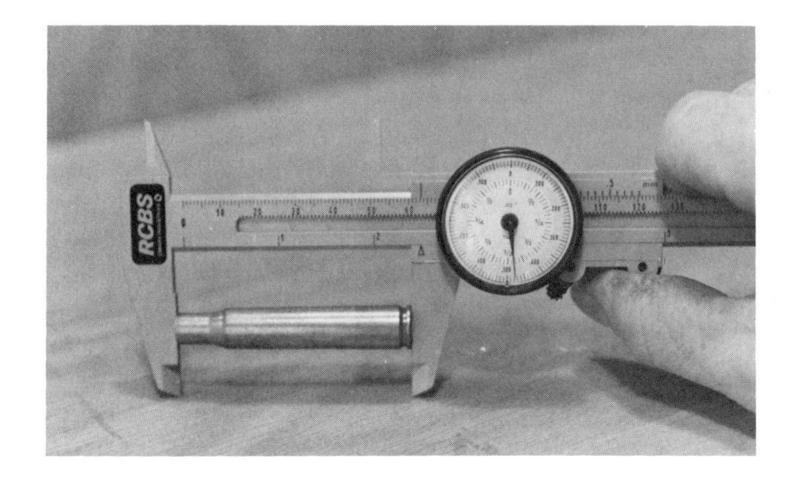


**A dial caliper is handy for checking case lengths and finished cartridge lengths.**

**Brass casings can be reduced in length precisely with a rotary case trimmer; all cartridge cases will elongate in time. (*Photo courtesy of Lyman Company.*)**

While the brass case will expand in all directions, it is limited by the diameter of the chamber and therefore the direction of greatest expansion is forward; the end result is a brass casing that is slightly longer. In time, a brass casing becomes too long to chamber easily—if at all. Such a casing must be trimmed to the standard length before it can be used again.

Undoubtedly the best way to trim metallic cartridge casings is with a special tool designed for the purpose: a *rotary case trimmer.* Each casing is clamped into the trimmer and the cutter bit adjusted until it cuts the case to the proper length. A rotary case trimmer allows you to trim a large quantity of cases to the proper length quickly and efficiently.

Still another way to reduce the length of fired brass casings is to use a special *trim die* and a flat file. A trim die is screwed into the top of the reloading press in exactly the same manner as a reloading die, then an empty casing run into the die. All excess casing that protrudes above the die is then filed off flat with the top of the die. Trim dies are available for all calibers of brass casings and, while slower than the rotary case trimmer, will help you to accomplish the task of trimming brass casings to the proper length.

After a brass casing has been trimmed to the proper length, a tiny burr will develop around the inside or outside edge of the case. Sometimes a burr also develops after several firings of a case even though it has not been trimmed. In any event, the burr must be removed before priming or reloading of the casing. There are two ways of

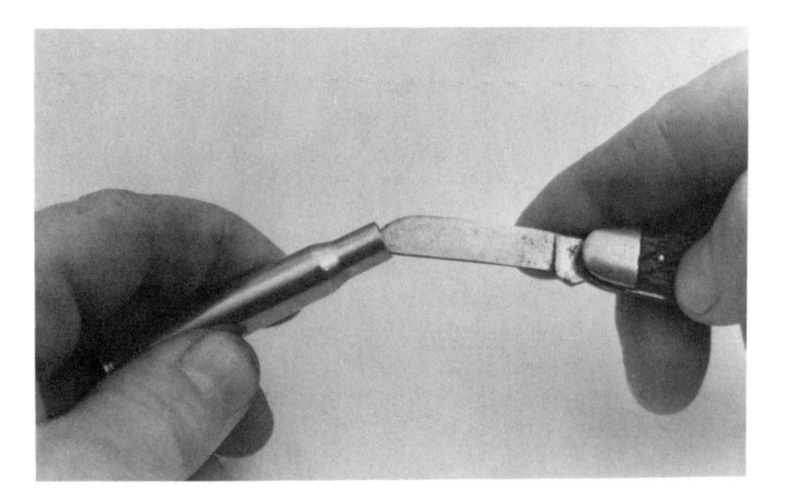

**While a special tool is available, the burrs around a cartridge case mouth can easily be removed with a pocket knife. This makes bullet insertion easier.**

doing this: (1) with a pocket knife, (2) with a special *deburring tool.*

To deburr a brass casing with a pocket knife, simply insert the knife blade into the neck of the shell and, with a slow but firm twist, remove the inside burr. If the burr is on the outside of the case, use the same knife blade to trim this off, but be careful to remove only the burr and not too much of the brass.

A deburring tool makes short work of dressing up the neck edge of any brass casing. The standard deburring tool can be used for any caliber brass casing and will quickly deburr the inside or (by turning the tool end for end) outside with a twist or two.

After you punch the old primer from a fired casing, you will notice a black deposit in the bottom of the primer pocket. This is a residue left after the primer was fired, and should be removed. Remove primer pocket residue with the tip of a small screwdriver filed so it will fit exactly inside the primer pocket or with a special *primer pocket cleaning tool.*

After trimming, deburring, and cleaning the primer pocket on a quantity of brass casings, put these into a *brass polisher* to clean up and polish the casings. Two basic types of brass polishers are available: tumbler polishers and vibrator polishers.

Brass polishers, either type, work by constantly moving the brass through a polishing medium (most commonly

Primer pocket cleaning tool.

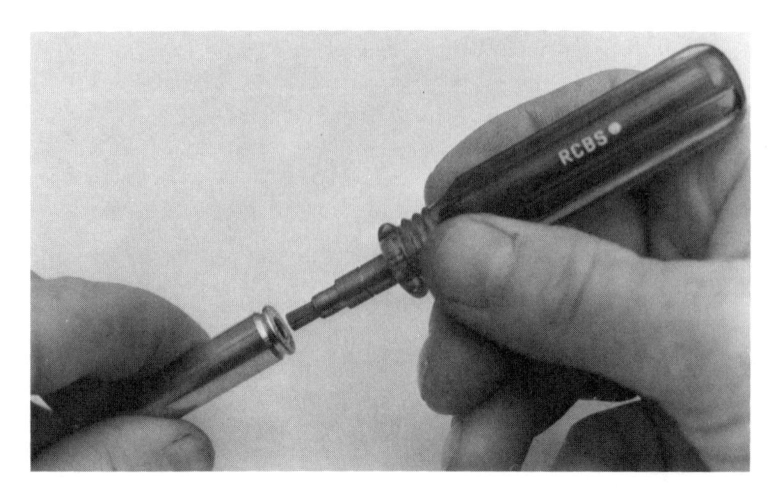

**A few hours in this brass polisher and almost any brass casing will look like new. (*Photo courtesy of Lyman Company.*)**

ground walnut shells and jewelers rouge) until all interior and exterior surfaces have been polished to a mirrorlike finish. Not only does polished brass look much nicer than dirty brass, but the casings won't scratch the interior of your dies. A brass polisher is not a required reloading tool, but you probably will want to use one, especially if you do a lot of reloading.

I prefer the vibrator type of polisher, which tends to be quieter and a bit more efficient than the tumbler type. Brass polishers are generally available in reloading component shops and from mail order companies. If you cannot justify the purchase of a polisher for yourself, consider buying one with a group of friends.

After brass casings are polished, they should be lubricated before sizing. Lubrication on the exterior of a case will make the casing slide more easily into and out of the resizing die. Although resizing lubricant can be applied in a light coat with the fingers, a much more efficient way of doing this is to use a special *case lube pad* designed for the purpose.

To lubricate brass casings with a special pad, begin by placing a small amount of *resizing lubricant* on the surface of the pad and rubbing in until the surface feels almost dry. Next take several clean, empty casings and lay them on top of the pad. Then, with the palm of your hand, gently roll the casings over the pad until each picks up a light coating of lubricant. This small act makes resizing a much easier task.

The only time that brass casings need not be lubricated prior to resizing is when a tungsten carbide resizing die is used. The only suitable lubricant for brass casings is one specifically designed for the purpose. Dry lubricants can be used, but never motor oil or grease, as this may neutralize powder or primer.

Since most resizing problems occur as a result of lack of lubricant, you can virtually eliminate problems by lubricating brass casings before resizing. You should also lubricate the inside neck of brass casings before resizing. This will make the expander ball inside the resizing die work easier. Some companies include brushes with their lube pads. These can be used for cleaning and lubricating the inside of case necks.

One last item handy for case work is a loading block. A loading block simply holds brass casings as they are being charged with powder and while bullets are being inserted. A loading block prevents spilling of powder from the charged cases during the various reloading steps. You can either make a loading block from scrap lumber or buy one for a few dollars.

## Powder-Measuring Tools

Without a doubt, the most important powder-measuring tool for the handloader is a *scale*. It is a fallacy to think for a moment that you can load accurate ammunition without the aid of a reloading scale.

A number of accurate reloading scales are currently on the market. Some are moderately priced and have all the features you need for basic handloading. More expensive models usually have finer adjustments, have greater capacity, and balance quickly.

Purchase the best scale you can afford. A good scale will last a lifetime and will take care of all of your weighing needs easily. Steer clear of low-priced reloading scales from unknown makers. Some of the best scales on the market are available from RCBS™, The Lyman® Company, and Pacific Tool Company.

Since dirt and dust can cause an inaccurate reading, the scale should be protected. Store the scale in a special case or cover it when not in use. Never apply any oil or lubricant to the pivots or bearings on a reloading scale, as this will alter the accuracy of the scale.

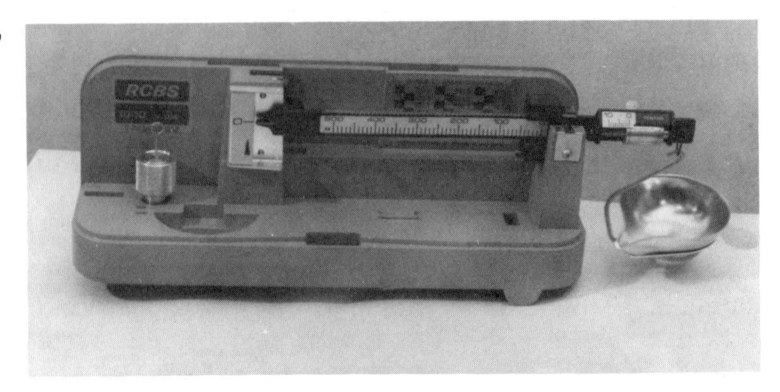


For accurate powder measuring, you will need a reloading scale.

Once you have a reliable scale for measuring powder charges, in grains, then you can begin to think about powder dispensing or measuring tools, which make charging empty cases a simple task.

The simplest way to measure uniform loads of powder is to use some type of dipper. The Lee Precision Company offers a Powder Measure Kit that contains plastic dippers in fifteen different sizes. The dippers are used in conjunction with a slide card that lists most of the powders on the market today. If you wanted to charge pistol ammunition cases with 4.6 grains of Hercules Unique, you would use dipper #5cc.

The Powder Measuring Kit from Lee is handy and quick but does have its limitations, as each dipper holds only a predetermined amount of any particular powder. If, for example, you wanted to load 6 grains of Hercules Unique, there is no dipper available (next size is 6.4

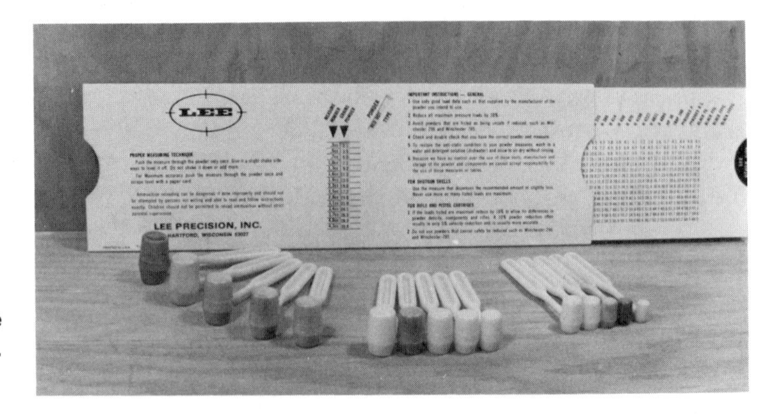


**This powder measuring kit from Lee is inexpensive and easy to use. (*Photo by author.*)**

grains). You must also check the weight of the charge on an accurate reloading scale before you begin reloading and after every ten loads or so to make certain you are charging each case properly.

It is entirely possible to make powder dippers from old pistol and rifle cases. Simply determine how much powder is required for a particular charge, by using an accurate scale, then mark the empty casing accordingly. If you really want to make an effective dipper, cut off the case at the fill line. Then each time you fill the case you will have the approximate charge of powder you require. A bent paper clip handle will make using your homemade dipper easier as well.

A good powder measure is fully adjustable. (*Photo by author.*)

A number of powder dispensers are available for the home reloader. Two made by RCBS™ are worth noting: the Uniflow™ Powder Measure and the Little Dandy® Pistol Powder Measure. The Pacific Tool Company manufactures a Multi-Deluxe Powder Measure and a Pistol Powder Measure. The Lyman Company offers a multipurpose powder measure (No. 55 Powder Measure) suitable for both rifle and pistol ammunition.

The Uniflow powder measure is a bench-mounted powder charging device that will help you to charge a large number of rifle or pistol cases quickly and accurately. With each turn of the handle the Uniflow will measure, within a fraction of a grain, and dispense a charge of powder into a brass casing. The large powder hopper is made from clear plastic so you can see exactly how much powder you have left at all times. The amount of powder to be dispensed must be preadjusted with a special adjustment screw. After checking the charge weight on an accurate scale, the adjustment screw is locked in place so that the charge of powder will remain exactly the same. A periodic check of the powder weight being dispensed (on a reloading scale) is advisable to ensure that all loads are consistent. The RCBS Uniflow™ Powder Measure is handy for large reloading projects for either rifle or pistol ammunition. As a rule, fine-grain pistol powders (such as Hercules Unique) measure and dispense better than long-kernel powders (such as Du Pont IMR 4831).

**The Little Dandy® Pistol Powder Measure by RCBS is very handy for charging pistol cases with powder. Twenty-six different size rotors are available, each dispensing a predetermined amount of a specified powder. (*Photo by author.*)**

The Little Dandy® Pistol Powder Measure from RCBS is a hand-held (or bench-mounted) powder dispenser ideal

**A special reloading funnel will prevent spilled powder on the reloading bench.**

for charging pistol cases with a predetermined amount of powder. Twenty-six different size rotors can be used with the Little Dandy, each predrilled to dispense a specific amount of a particular pistol powder. Unfortunately, the rotors are not adjustable, as the Uniflow powder measure is, so you must more or less settle on a specific charge of powder. Nevertheless, the RCBS Little Dandy™ Pistol Powder Measure is the most reliable, fastest, and easiest powder dispenser for charging a large quantity of pistol cases. Be sure to check the charge weight before and during use for best results.

If you really want the ultimate in accurate powder charges, you will want to use a powder scale for checking each charge. This means presetting your scale to a predetermined grain reading, then adding powder until the scale balances. One aid worth mentioning here is a powder trickler that dispenses small amounts of powder with the twist of a handle. The several different versions of a powder trickler or dribbler available all work about the same way. These small-quantity powder dispensers are handy for precise measuring of powder charges.

One last aid is a special powder funnel, which helps you eliminate a lot of spilled powder and makes charging empty cases easy.

## Priming Tools

If you already own a reloading press and set of reloading dies, you will not require any special tools for priming brass cases prior to reloading. Since most reloading presses

**A powder trickler is a real aid when measuring exact powder charges.**

prime empty cases quite well, a tool that has only one function—to prime cases—may seem like an unnecessary additional expense. In time, however, you may want to look at some of the special priming tools on the market. These tools make priming cases easier, especially if you have a large number of cases to prime at one time, and you will actually do a better job of seating each primer.

When a new primer is inserted and pressed into the clean primer pocket, it should be seated firmly to the bottom of the pocket. This puts the legs of the primer anvil in firm contact with the bottom of the primer pocket and at the same time forces the primer anvil deeper into the explosive primer mix. When using a reloading press, with hundreds of pounds of leverage, it is often impossible to "feel" when the primer is seated properly; an improperly seated primer often causes a misfire when the round is fired. For this reason, special tools that do nothing more than seat primers correctly are a good investment for the home reloader concerned about quality ammunition. There are two basic types of priming tools: bench-mounted and hand-held.

**This automatic priming tool from RCBS will enable you to prime a quantity of brass casings quickly. (*Photo by author.*)**

Bench-mounted priming tools are really quite simple, often consisting of a shell holder (which must be matched to the shell being primed), a primer cup, and a lever that seats the primer when the lever is lowered or raised. Several different kinds of bench mounted priming tools are available; some hold and dispense primers automatically, while others require a new primer be inserted for each casing.

The other type of priming tool is hand-held and works in a similar manner. Instead of a large lever, however, hand pressure is used for seating each primer. Hand-held priming tools are handy for priming a large quantity of cases quickly and do not require the user to sit at the reloading bench during use.

## MISCELLANEOUS RELOADING TOOLS AND ACCESSORIES

One type of accessory you will always have need for is empty boxes for loaded ammunition. These are available in a variety of colors and sizes. They make handy carrying containers for loaded ammunition and last almost forever, because they are made from resilient plastic. *Be sure to label each box of ammunition* so you can be certain of the bullet weight, powder, primer, and date of making. This information will aid you in making the best ammunition possible for your specific needs.

As I mentioned in the beginning of this chapter, reloading has great appeal to those of us who are fond of gadgets. If you walk into any shop that sells reloading supplies you will be amazed at the variety of tools and accessories for the reloader. As a rule, you need nothing more than a reloading press, set of dies, and scale (plus, of course, empty shells, powder, primer, and bullets) to reload quality ammunition. By the same reasoning, you need nothing more than a hook, line, and bait to catch fish. But many gadgets, tools, and aids can make reloading simpler, faster, and more accurate for the do-it-yourself ammo maker. I leave it to you to seek out equipment you think you need and perhaps even equipment you don't need. In your search you will discover that many manufacturers of reloading equipment offer package deals on equipment. This often is the least expensive way of acquiring basic reloading equipment.

There is one last group of reloading machines that are growing in popularity and which are worth mentioning—progressive reloading presses for rifle and pistol ammunition.

Progressive reloading presses—such as the 4 × 4™ from RCBS and the Dillon RL-550—can help you to make quality rifle and pistol ammunition very fast. These machines will turn out from 350 to 500 rounds of ammunition per hour. This capability will be appreciated by those who reload ammunition for automatic weapons—such as .223, 9mm, and .45 ACP. Both progressive reloaders sell for around $200 and $225, respectively, without dies so they are really for the serious or professional reloader. Nevertheless these machines have a number of features that are worth mentioning.

The key to speed in progressive reloaders is a revolving base which holds casings in various stages. For example (on the Dillon RL-550), the first station resizes the case, deprimes it, and automatically seats a new primer to the correct depth.

The second station bells the mouth of the case and dispenses a predetermined charge of powder. The third station seats a bullet to the proper depth. The fourth and last station crimps the bullet into the case mouth then kicks the finished cartridge down a chute into a collection box. Several hand operations are required prior to pulling the handle down: a spent casing must be inserted into the first station, a bullet must be placed into a cartridge case mouth in station three, and the shell holder plate must be indexed to the next station.

Progressive reloading machines, while not for everyone, do offer capabilities not found on single-stage reloaders. While the 4 × 4® from RCBS is available at most stores that sell reloading supplies, the Dillon RL-550 is available only from the manufacturer—Dillon Precision Products, Inc., 7442 East Butherus Drive, Scottsdale, Arizona 85260.

**The Dillon RL-550 progressive reloader is capable of producing up to 500 rounds per hour. (*Photo courtesy of Dillon Precision Products, Inc.*)**

## TOOLS FOR RELOADING SHOTSHELLS

The shotshell reloader plays by a slightly different set of rules than the metallic cartridge reloader. Although the theory of reloading is basically the same—a primer is used to ignite powder, which in turn sends the round (shot) down the barrel—the tools for arriving at this end are different.

At first, reloading shotshells might seem like a poor investment in time and energy, especially when you consider that a box of shotshells can be purchased for around five dollars or roughly twenty cents each. In truth, if you only shoot a few boxes of shotshells a year, you will remain well ahead of the game by buying rather than reloading shotshells. It is when you shoot often that shotshell reloading really pays off. As a guideline, if you shoot ten or more boxes of shotshells per year then you will be able to realize substantial savings in shooting costs over a period of several years.

Another good reason for reloading shotshells is the variety of loads that you can produce. While the manufactur-

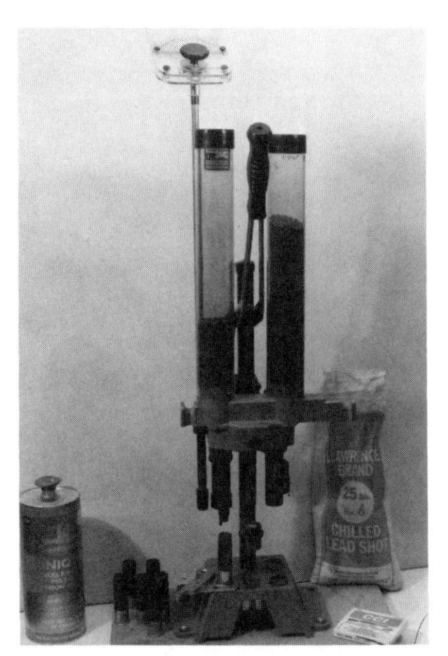


**You will need a special shotshell reloading press to reload shotshells.**

ers of shotshells offer a wide selection of shotshells—target, field, waterfowl loads, for example—it is possible and in some cases desirable to develop specialized shotshell loads for the types of shooting you do. In short, you will be more in control of the shotshell ammunition you reload.

All shotshell reloading is done on a special shotshell reloading press. A typical version has two plastic tubes or bottles on top, one for powder, another for shot. A horizontal bar rides beneath these reservoirs and contains special bushings for measuring and dispensing powder and shot. Since these bushings are preset to drop a specific amount of powder or shot, you can only load what these bushings will allow. Of course, shotshell reloaders have interchangeable bushings, by changing them, you can vary powder and shot charges. It is important, from the standpoint of safe and reliable shotshells, to use the proper size bushings for the particular ammunition you are reloading.

Since no measuring of charges of powder or shot is required (the metering of shot and powder being done automatically by the bushings), shotshell reloading tends to go quicker than metallic cartridge reloading. But this does not mean that less care is required.

A quality shotshell reloader has a base that holds a shotshell in position while it goes through the various stages. These include depriming, priming, resizing, charging with powder, insertion of wad, wad pressure, shot charge, and crimping.

The simplest of shotshell reloaders have shell holders at strategic locations around the base of the reloader, and each shell must be physically moved from one stage to the next. On more expensive shotshell reloaders, often called "progressive" or "automatic" reloaders, the base revolves and each shotshell is moved as a result. The more expensive shotshell reloaders are capable of producing as many as 1800 perfect shotshells an hour (with three operators).

As a rule, the more expensive shotshell reloaders are intended for the professional shooter, trap and skeet, for example, rather than the hunter who simply wants to save on ammunition costs. Chances are good, I suspect, that you are in the second category and you will probably be content with a less expensive shotshell reloader. Of course, one good way to get the use of a high quality shotshell reloader is to purchase one with a group of friends or a club and share the machine.

Some worthwhile features are an automatic primer feed tube, easily replaceable (and obtainable) shot and powder bushings, and large reservoirs for shot and powder. Other features, such as potential for changing from one gauge to another, may not be important now, but if you later acquire a 20 gauge shotgun, you will find this feature handy.

Shotshell reloaders from Ponsness/Warren, Inc., Pacific, Mayville Engineering Company, Inc. (MEC), and Lee are all reliable and will help you to make factory-quality shotshells at a fraction of the factory price.

Now that we have looked at the components, machines, tools, and accessories that make reloading possible, let's get into the nuts and bolts of reloading.

# 2 Casings, Primers, Powders, Bullets, and Other Reloading Components

The handloader has a broad selection of reloading components to choose from. With a little background information you can learn to choose your own components. You can then custom-make ammunition for any type of shooting that is, as a rule, more versatile than commercial loads.

All ammunition reloading must follow specific guidelines as to which components can be used for any given load. For example, many different reloading powders are available, some for shotguns, others for pistols or rifles. To ensure predictable results, it is essential to (1) match the powder to the cartridge, and (2) use the right amount of powder. You will find the reloading tables in the second half of this book very helpful. There you will find specific information about type and amount of powder, bullet de-

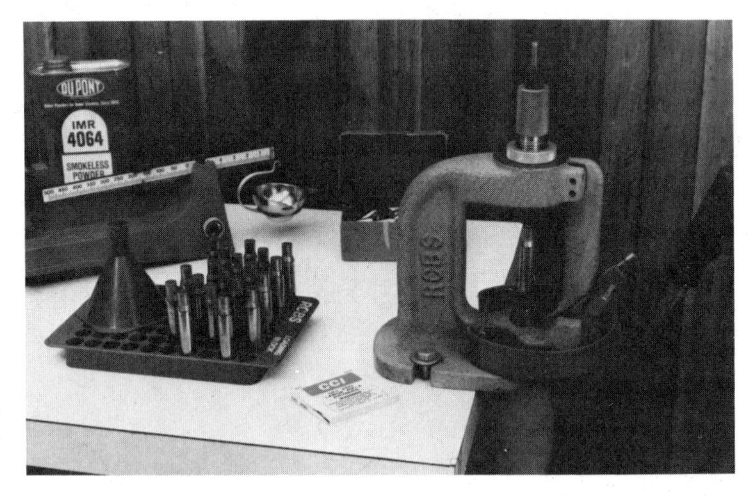


A wide range of reloading components are available to the handloader.

sign and weight, primers, and velocities of the various combinations.

As you already know, there are two basic types of reloading: metallic cartridge and shotshell. Similar components are used for both types, but there are some differences. We'll look at metallic cartridges first.

## METALLIC CARTRIDGE COMPONENTS

With the exception of rimfire ammunition, which cannot be reloaded, all centerfire metallic ammunition comprises four basic parts: brass casing, primer, powder, and bullet. Let's look at each of these parts carefully and, in so doing, learn what each contributes to any given round of ammunition.

### Brass Casings

The most expensive part of any piece of centerfire ammunition is the brass casing. By reloading empty cases, the handloader dramatically reduces the overall costs of making ammunition. Through careful reloading practices, it is possible to reuse an empty cartridge case as many as twenty-five times or more.

The brass used to manufacture rifle and pistol cases is composed, most commonly, of 70 percent copper and 30 percent zinc. In a typical cartridge-manufacturing plant, sheets of high-quality brass are fed into a machine that punches out "cups" of brass about half the size of the finished cartridge. Later steps form the cup into a specified size and shape, .30-30 or .30-06, for example.

All metallic cartridges fall into one of four basic styles or designs: rimmed, semi-rimmed, rimless, and belted. One other design, rebated, is used only on the .284 Winchester and is therefore of limited application. The foregoing terms really describe the base of the cartridge case, and you should know that case body variations also exist.

**Case base types include (l to r) rimmed, semi-rimmed, rimless, and belted.**

A metallic cartridge case may be straight-walled, bottlenecked, or tapered, depending on the cartridge design. You will no doubt learn, through your own reloading experiences, that theories abound as to cartridge design and potential accuracy.

The rimmed cartridge case is one of the earliest of all

**This cutaway (l) of a .30-06 cartridge shows what's inside the same cartridge on the right.**

cartridge designs and is still used today. The edge of the rim is wider than the case. One prime example of this cartridge is the .30-30 Winchester.

A semi-rimmed cartridge case is similar to a rimmed cartridge, except that the rim on the former is only slightly wider than the case body and that an extraction groove is also present. The .220 Swift and .225 Winchester are the only two cartridges with the semi-rimmed case.

The rimless cartridge case is the most common cartridge base design in existence in the world today. It is probably fair to say that all current military ammunition (worldwide) and most sporting ammunition have a rimless case. Some good examples of rimless cases include .30-06, .308, .223, 9mm, and .45 automatic.

Belted cartridge cases are most commonly magnum types such as 7mm Remington magnum, .300 H & H magnum, and .458 Winchester magnum. In some cases a belt is added for cosmetic rather than functional reasons.

The primary purpose of any cartridge case is to hold the powder, primer, and bullet in a self-contained, portable, and secure container, not unlike an egg. Before cartridges were used for shooting, each shot required loading of powder and bullet into the firearm—flintlocks and precussion firearms being examples of precartridge day weapons.

Another function of a cartridge case is to center the round in the chamber of the firearm; this is accomplished by the design of the case itself. For example, a rimmed cartridge case is held in the chamber by the rim, a rimless case is centered by the taper and shoulder of the case, a belted case is held in place by the belt, and a semi-rimmed case is held by the taper and shoulder of the case.

Each time a cartridge is fired in a firearm, the casing expands slightly to the chamber walls. This enables the bullet to exit and prevents any gases from escaping rearward. The chamber pressure for a typical rifle round is between 40,000 and 55,000 CUP (copper units of pressure, roughly comparable to pounds per square inch). After the bullet is on its way down the barrel, the brass casing shrinks back to almost normal dimensions again, at which time it can be extracted and ejected from the firearm. As you can imagine, this kind of pressure puts stress on the brass casing; in time, the cartridge will become slightly longer and in extreme cases fracture.

For obvious reasons, inspect all brass casings before any attempt is made at reloading. Discard any cases that are cracked, split, or otherwise damaged. At the very least, a damaged case cannot be depended upon to provide an accurate round.

**Primers**

A primer in a modern cartridge is often likened to a spark plug in an automobile engine. Just as a spark plug provides the flash to ignite fuel, so does a primer ignite the powder. When a firing pin strikes a primer, a hot flash is produced that flows through the flash hole in the back end of the case and into the powder, igniting it and forcing the bullet down the barrel. Modern primers are rather complex little pieces of metal and explosive mixture, and their development during the Civil War period resulted in ammunition as we know it today.

Primers used for reloading in this country are commonly called "Boxer" type, after their British inventor, Col. Edward M. Boxer. A typical version consists of a small metal cup partially filled with lead styphnate (the explosive mixture), then covered with a piece of foil. In addition, another piece of metal—the anvil—is inserted into the primer and sits against the priming mix. When the primer is seated in the primer hole, the anvil bears against the priming mix and provides the necessary support for crushing the priming mix, thus igniting it, when struck by the firing pin.

It should also be mentioned that an American ordinance officer, Hiram Berdan, also invented a primer around the same time. The Berdan primer is still in use today, largely for military ammunition. Almost all British and European sporting and military ammunition in use today uses the Berdan primer as well.

Since the likelihood of the reader coming up with empty military ammunition is great—especially .30-06, .45 automatic, 7.62 NATO, and 5.56mm—a brief description of the Berdan priming system should prove helpful.

The standard Berdan primer consists of a primer cup partially filled with standard, noncorrosive priming mix, lead styphnate, covered with foil for protection. There is

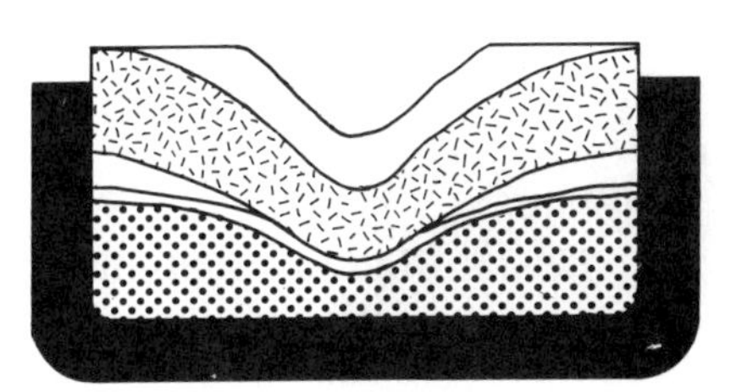

**Boxer primer construction.**

no anvil, as with the standard Boxer-type primer. Instead, the primer pocket in the brass casing has a bump or protrusion, that serves the same purpose, and two or more off-center flash holes. When the Berdan primer is struck by a firing pin, the priming mix is crushed against the small projection and ignites.

Berdan primers are less expensive to make (fewer parts) than Boxer primers and work just as well. From the standpoint of reloading, however, Berdan primers, with their offset flash holes, require a special Berdan primer removal tool. Berdan primers cannot be removed with a standard decapping die, the type used for removing all Boxer-type primers. In addition to removing the Berdan primer, the projection type anvil, part of primer pocket design, must be flattened using another special tool commonly called a "primer pocket swager." More details about Berdan primer removal can be found in Chapter 5.

There are three major groupings of primers for metallic cartridge reloading: standard, magnum, and benchrest. These are further broken down into small or large pistol and small or large rifle.

As a rule, Standard primers from any primer manufacturer are suitable for most types of rifle and pistol ammunition reloading. Standard primers are engineered to provide optimum ignition of fast and medium burning powders, such as Hercules Unique.

Magnum primers are designed to be used with slower-burning powders—Du Pont IMR 4831, for example—as these powders need more of an initial blast to ignite the powder. Magnum primers can be interchanged with standard primers, but all primer manufacturers recommend reducing powder charges by one to two grains during tests. If pressure indications permit, then you may load up to the maximum powder charge. The important thing to remember about magnum primers is that they tend to give more complete ignition of the powder charge, so pressure will almost certainly increase. Some experts suggest that if a round is to be fired in temperatures of zero or below (Fahrenheit), the magnum-type primers are more reliable.



Benchrest primers are relatively new on the reloading scene and are generally interchangeable with standard and magnum primers of the same type, that is, small rifle and large rifle. Although the priming mix and metal parts are the same as for standard and magnum primers, benchrest primers are extremely uniform and ideally suited for benchrest shooting. These primers cost a bit more than the other types, but many benchrest shooters feel this is a small price to pay for uniformity in primers.

## Powders

If primers can be compared to spark plugs in an internal combustion engine, then modern smokeless powders can be compared to gasoline. The development of the powders we now use for handloading has taken centuries. The invention of explosive powder is generally credited to the Chinese and probably took place about 600 years ago. It was not until fairly recently—around 1885—that real improvements began to be made. Up until that point, black powder was the only powder available, and it had (and still has) a number of shortcomings.

Smokeless powders may be grouped into one of two possible categories: single- or double-based. Generally, single-based smokeless powders contain nitrocellulose, while double-based powders also contain a significant amount of nitroglycerine. The Hercules Powder Manufacturing Company produces most of the double-based powders in use today. The Du Pont Company produces single-based powders (except for 700-X, a double-based powder).

All smokeless powders burn at a predetermined rate based on the size, shape, and makeup of the individual pieces. At one end of the scale, we have fast-burning powders, such as Hercules Bullseye, suitable for handgun loads. At the other end we have very slow-burning powders, such as Du Point 4831, used for rifle loads. In between, there are a number of other powders that burn at various rates and are suitable for pistol, shotshell, and rifle. It is important to choose the right powder (with its predetermined burning rate) for any particular load. The reloading tables later in this book specifically list powder types and charge weights, but here the more popular powders in use today from both Hercules and Du Pont will be considered briefly. The powders are discussed from the fastest to the slowest burning rates.

**Hercules Bullseye**—The fastest-burning powder available to the handloader. Primarily a handgun powder, especially suitable for .38 special target loads, although it can also be used for some 12-gauge 2¾ inch target shotshell loads.

**Hercules Red Dot**—Primarily a shotshell powder that has long been used for light to medium target loads. Can also be used for a variety of pistol loads.

**Du Pont Hi-Skor 700-X**—A good target shotshell powder that can also be used for light handgun loads.

**Du Pont Hi-Skor 800-X**—A fairly new target shotshell powder that can also be used for medium handgun loads.

**Hercules Green Dot**—Similar to Red Dot but slightly slower burning. Suitable for shotshell and medium handgun loads.

**Du Pont PB**—A shotshell and handgun powder.

**Du Pont 4756**—Used for 3 inch magnum shotshell loads and handgun loads.

**Hercules Unique**—One of the most versatile handgun load powders; also can be used for shotshells and reduced rifle loads.

**Du Pont IMR 7625**—A shotshell and pistol powder for medium loads.

**Hercules Herco**—One of the oldest powders for high velocity shotshell loads and an ideal powder for 9mm luger and other high-performance handgun loads. Herco is similar to Unique, but not as versatile.



**Hercules Blue Dot**—The newest magnum shotshell powder from Hercules that can also be used for magnum handgun loads.

**Hercules 2400**—Primarily a magnum handgun powder but can also be used for some rifle loads, especially the .22 Hornet and other small-capacity rifle cases.

**Du Pont SR 4759**—One of the best choices for reduced loads in rifle loads.

**Du Pont IMR 4227**—A fairly fast rifle powder quite suitable for small-capacity rifle loads such as .22 Hornet.

**Du Pont IMR 4198**—A versatile rifle powder that can be used for a wide range of loads. A good choice for straight-walled rifle cases such as 45-70 government.

**Hercules Reloader 7**—Can be used for a wide range of rifle loads.

**Du Pont IMR 3031**—Useful for a wide range of rifle loads.

**Du Pont IMR 4895**—One of the most versatile rifle powders on the market today.

**Du Pont IMR 4064**—Similar to Du Pont 4895 but slower burning. Fairly versatile across a broad range of rifle loads and one of the most popular rifle powders in use today.

**Du Pont IMR 4320**—Similar to Du Pont 4064 but slightly slower burning.

**Du Pont IMR 4350**—One of the slowest burning powders from Du Pont and very popular, especially for 6mm loads.

**Du Pont IMR 4831**—Another slow-burning powder and usable for a wide range of .30 caliber loads.

**Du Pont IMR 7828**—Newest powder introduction from Du Pont developed for large-capacity rifle cartridges such as .264 Winchester and 7mm Remington magnum.

Modern smokeless powders should always be treated with due respect. Since the potential for mishap always exists, be sure to follow safe practices for handling and storage. Always observe the following ten safety rules.

1. Exercise care and common sense at all times
2. Store powder only in the original container and in a cool, dry place.
3. Keep all powders locked up, away from the reach of children.
4. *Never* use a powder unless you are certain of its type.
5. *Never* smoke when handling powder.
6. Never mix powders.
7. Don't keep more powder in an open container than you can immediately use.
8. Work up all new loads starting 10 percent below the recommended charge weight.
9. Keep powder away from heat, electrical circuits, and open flame.
10. Keep children, animals, and anyone whose judgment and common sense cannot be trusted out of the reloading room. These materials are extremely dangerous in the wrong hands.

More detailed information about safe handling and storage of modern smokeless powders is available from the Sporting Arms and Ammunition Manufacturers' Institute, Inc., P.O. Box 218, Wallingford, CT 06492.

**Bullets**

Part of the reloading experience must involve bullet design, weight, and function. If you are just a beginner, this may not sound like something you need to take an interest in. But, as you will see, these are considerations that you begin to face when you start thinking about why you reload in the first place.

The design of any bullet indicates how that bullet is likely to perform after ignition. Bullets are described according to shape of base and point type. Bases are either flat (F) or boattail (BT). Imagine, for a moment, the pressure inside a brass casing as the powder is ignited by the primer. When a standard .38 special round goes off,

**Flat-base bullet (l) and boattail bullet designs.**

the pressure inside will be around 18,000 CUP (copper units of pressure, roughly equal to pounds per square inch) and almost all of this pressure is exerted on the base of the bullet. If the bullet is of the flat base variety, as all handgun and many rifle bullets are, the effect will be different than if a boattail (BT) bullet were being fired. One of the main differences between these two bullet designs is the drag factor. A flat-base bullet, because it has a base that fills the bore, drags more than a boattail design.

Theories abound as to which is the better bullet design—flat-base or boattail. Some bullet manufacturers, such as Hornady Manufacturing Company, produce mostly flat-base bullets, while others, such as Sierra Bullet Company, produce mostly boattail design bullets.

Whichever side of the argument you subscribe to, some interesting facts must be taken into consideration. According to the Sierra Bullet Company, all National Match records for 200, 300, 600, and 1000 yards are held with

**A Sierra .30 caliber 180-grain soft-point flat-base bullet in flight captured by a spark photograph. Turbulence caused by the partial vacuum at the base can be clearly seen. (*Photo courtesy of Sierra Bullets.*)**

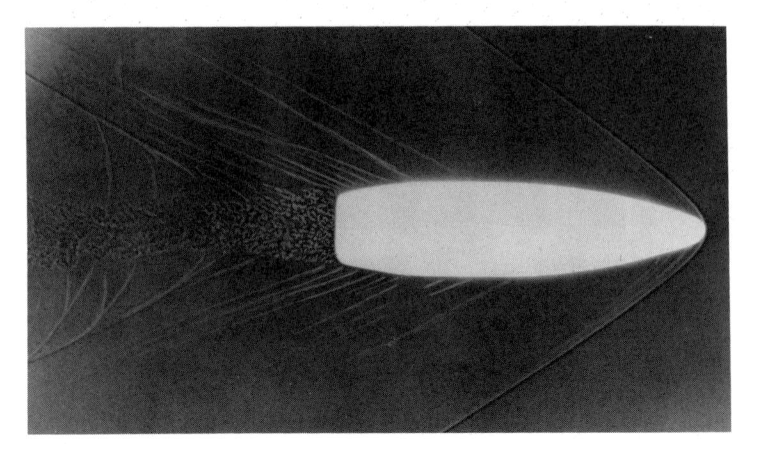

**A Sierra .30 caliber 180-grain Spitzer boattail bullet in flight captured by a spark photograph. Turbulence caused by the partial vacuum at the base can be clearly seen. (*Photo courtesy Sierra Bullets.*)**

boattail bullets. All 300-meter International free rifle records are held with boattail bullets. Almost all benchrest records are held with boattail bullets. All .30 caliber and 7.62mm Army Arsenal Match ammunition is loaded with boattail bullets. All 7.62mm NATO and 5.56mm ball ammunition used by United States armed forces is loaded with boattail bullets. It would seem that the boattail bullet design is the best choice for rifle bullets!

Bullet weight is another important factor for the reloader. Most shooters will agree that the selection of over-the-counter ammunition is rather limited when it comes to bullet weight. For example, if you were to go down to the local sporting goods store to buy a box of .30-06 ammunition today, you would probably have a choice of only two or possibly three different bullet weights—180, 165, and possibly 200 grains. But, if you went down to the same store (assuming reloading components are also sold) you might find as many as twenty or more different bullet weights for the same .30-06, from 100 to 220 grains. If nothing else, this wider choice in bullet weights means that the .30-06 rifle can be used for all kinds of shooting. This selection of bullet weights applies to all calibers of rifle and pistol ammunition. Since bullet weight is a matter of personal choice, you must experiment with different weight (and design) bullets to arrive at your favorite for any given situation.

The last important factor when choosing a bullet is the

**Just a few of the different weight and style bullets available for .38 special and .357 magnum ammunition.**

shape of the tip. There are four basic choices at this time: hollow-point (HP), spitzer (SP), flat-nose (FN), and round-nose (RN). Each of these bullet tip designs is illustrated on page 31. Since the ultimate use of the bullet dictates which type is best, bullet-tip choice is once again a personal matter. Nevertheless, some guidelines are worth mentioning.

Hollow-point bullets expand upon impact through the hydraulic action of the liquid in an animal's flesh and are generally considered to be effective for many types of hunting. Where deep penetration of the bullet is a concern, however, in hunting big game such as elk, for example, hollow-point bullets are a poor choice.

Spitzer bullet-tips are generally believed to expand as well as round-nose designs and offer better accuracy.

An interesting development in bullet tip design has come from the Nosler Bullet Company recently—Nosler Ballistic Tip Bullets. These new bullets are constructed with a hollow cavity in the nose into which a polycarbonate tip is inserted. The polycarbonate tip (and overall boattail design) offers the best possible ballistic shape, which translates into the flattest possible trajectory, maximum retained velocity, and maximum retained energy at any given range. In addition to these strong points, the polycarbonate insert is highly resistant to damage or deformation during normal reloading and handling. Lead tipped bullets, on the other hand, can easily be damaged or deformed and, as a result, accuracy often suffers.

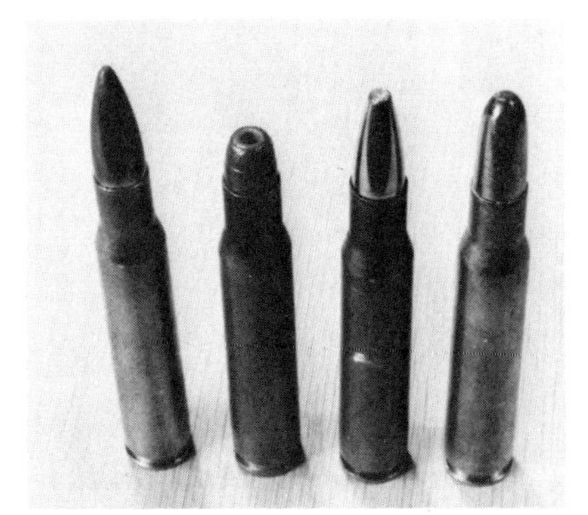

**Four basic bullet type styles (l to r): Spitzer, hollow-point, flat-nose, and round-nose.**

At this time, Nosler Ballistic Tip Bullets are available in a variety of weights for the following calibers: .277 (.270), .284 (7mm), and .308 (.30 caliber). In addition, Nosler indicates that the loading data for these new bullets are identical to the corresponding bullet in the standard solid base bullet line, as listed in the reloading tables later in this book.

Round-nose and flat-nose bullets are generally used in firearms with tubular magazines.

Experimentation with different bullet designs is the best course of action for the reloader. In this way you will discover which types of bullets are best for certain shooting situations.

Up until this point I have been discussing bullets manufactured by the various bullet makers. With few exceptions, these bullets are jacketed and costly. You should also know that with a few simple tools it is also possible to cast your own bullets from lead. This area offers substantial savings for the reloader, and you will probably want to try your hand at casting at some point in the future. Detailed information about bullet casting can be found in Chapter 5.

## SHOTSHELL COMPONENTS

The shotshell reloader has a wide selection of reloading components to choose from. If you shoot your scattergun often, you will be able to save quite a lot of money rather quickly. Of possible greater importance, you will also be able to turn out shotshell ammunition as good as or better than any manufacturer by reloading empty shotshell casings.

Shotshell reloading differs from metallic cartridge reloading only in components. The theory of how a shotshell works is basically the same. A primer is used to ignite the powder charge and a quantity of projectiles are then sent on their way down the barrel.

Since a number of the powders used for handgun and rifle can also be used for shotshell reloading, we will not duplicate the section on powder descriptions here. Instead, you can turn back to the earlier section in this chapter to see which powders are good for reloading shotshells. You will also find powder suggestions in the reloading tables for shotshells (see p. 233). It is important to follow these suggestions as to brand, type, and charge weight to the letter.

Shotshells differ from metallic cartridges in casings, primers, wads, and shot.

Shotshell casings, sometimes called "hulls," are made from plastic or paper. In times gone by, shotshell casings were made from brass. I am afraid that the paper shotshell case is following in the footsteps of the brass casing. At this point, only the Federal Cartridge Company offers a paper-case shotshell, Federal Champion® Target Loads and Federal Special Target Loads. For one such as myself who is fond of paper shotshells, the passing of paper shotshell hulls is a sad event.

Plastic shotshell hulls have many advantages over paper. For one thing, plastic shells can be used as many as thirty times; paper shells, on the other hand, will deteriorate before ten firings. The most popular plastic shotshell cases today are *compression formed;* a prime example is the "AA"® shotshell made by Olin Winchester-Western Division. Considerable pressure and heat are used to form a special plastic into the shape of the shotshell. Special dies are used to form the case with primer flash hole, primer pocket, and base wad. Next, a brass cup is attached to the base of the shotshell to make extraction and ejection of the fired shell easy.

Plastic shotshell cases offer better and more consistent chamber pressures than paper cases. What this means for the reloader is that plastic cases perform better overall than paper cases. In older pump action shotguns paper shotshells chamber and eject easily while plastic shells often jam. Paper shotshells have a slightly lower chamber pressure which results in less recoil and are therefore easier on the shooters' shoulder when many shots are fired—such as during competitive shooting or in the field. Paper shotshells reload easier than plastic shotshells. Lastly, I have been reloading paper shotshells for twenty years and find them more aesthetically pleasing than plastic shells.

**Brass, plastic, and paper shotshells (l to r). Of the three, plastic is most widely available.**

Primers used for shotshells are larger and of a different shape than those used for metallic cartridge reloading. As a rule, shotshell primers are called upon to provide a hot flash that ignites practically all of the powder at once; therefore, a special priming mix is commonly used for

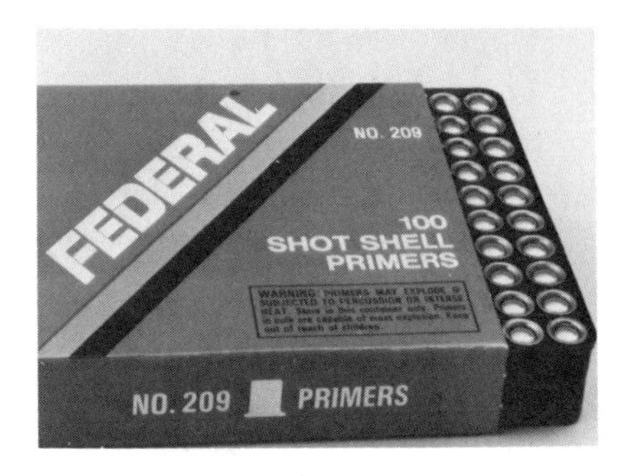


**Shotshell primers are much larger than primers used for metallic cartridge reloading.**

shotshell primers. There are several brands and types of primers for shotshells. It is important to use only the recommended primer when loading to obtain predictable results. Consult the reloading tables for shotshells before you begin any reloading session.

All shotshells must have a wad that separates the powder from the shot. Fifteen years ago the shotshell reloader had to piece together several wads by hand before inserting them into the shotshell over the powder to arrive at the desired wad column length. Now, however, reloaders simply use one-piece plastic wads that not only separate the powder from the shot but also (because of their shape) actually protect the shot on its ride down the barrel of the shotgun. One-piece plastic shotshell wads also increase chamber pressures somewhat. The development of the one-piece shotshell wad is probably one of the biggest developments in shotshells in the past half-century.

Several companies manufacture one-piece shotshell wads; once again, theories abound as to which design is best overall. In the reloading tables later in this book, you will find data for Remington, ACTIV, and Federal wads only, because I believe these are suitable for all kinds of shotshell reloading. Just keep in mind that other brands and designs are available.

Only one type of shot is available to the handloader: lead shot. Shotshells loaded with steel shot are commercially available, but the components for loading steel shot are not. As a result, all manufacturers of shotshell reloading components recommend that only lead shot be loaded. If you want or require steel shot (some Federal lands require steel shot only), then you must buy factory-loaded steel shotshells.

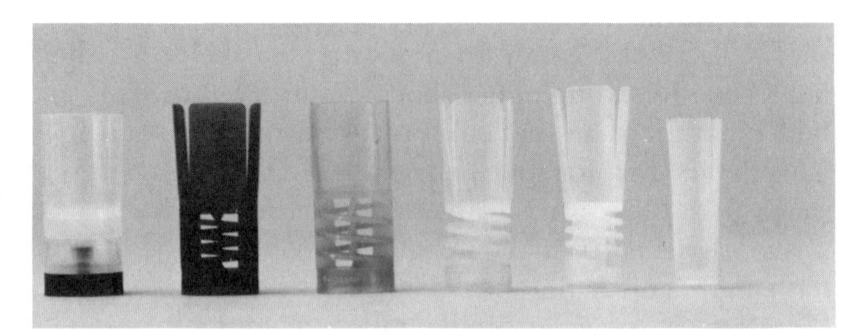

**A wide selection of wads is available and it is important to choose the right wad for the shotshell you are reloading. Check the reloading tables for specific information.**

As you shop for lead shot, most commonly sold in twenty-five-pound canvas sacks, you will come across a variety of terms such as "chilled" shot and "extra-hard" shot. Basically all lead shot is the same, although some lead (because of the introduction of antimony and tin) will be harder than others. However, lead shot is lead shot, and inexpensive lead shot is the best deal of all.

Lead shot ranges in size from "dust" (approximately 4565 per pound, 0.04 inch in diameter) to 000 buckshot (99 per pound. 0.36 inch in diameter). The more popular sizes of lead shot are (in increasing sizes) 8, 7½, 6, 4, 2, and 00 buckshot. The choice of lead shot will depend largely on the purpose of the loaded shotshells. It is possible to shoot (and break) clay pigeons with 7½ shot, but you would certainly not want to use this size for goose shooting—where size 2 or BB sizes would be an infinitely better-size lead shot choice. A Shot Chart in the Appendix (see p. 262) can be used as a guide for choosing shot.

# 3 How to Reload Rifle and Pistol Ammunition

Look for signs of deformity, damaged base, cracks (neck), and holes as in these cartridges.

Until now we have looked at the tools and components required for all types of reloading. In this chapter you will learn the basics of metallic cartridge reloading. As you may recall, metallic cartridges are divided into two major groups—bottleneck shaped and straight-walled—and although the reloading procedures are similar for each type, there are additional steps for reloading straight-walled cases. For this reason, we will cover these two different types of cartridge reloading individually.

## RELOADING RIFLE AMMUNITION

If you are just starting out in reloading, chances are you have a pile of empty brass that you have been saving for years. The first step in the reloading process is to inspect each casing carefully. Look for cracks, dents, holes, scratches, excessive dirt or grime, and anything else that makes the empty casing unsuitable for reloading. Discard any brass casings of questionable quality.

### Cleaning Brass

During your inspection of the brass casings, keep a water-dampened rag close at hand. If necessary, wipe the exteriors of lightly soiled cases. If especially dirty cases are present, you may want to clean them with boiling water and dish detergent.

The best way to clean brass is with some type of cartridge brass polisher, either the tumbler or vibrator type. Since you will probably not have one of these handy machines, your next-best alternative is to boil the cartridges with detergent on top of the kitchen stove. Place a quantity of dirty brass in a large pot with some dish detergent and bring to a boil. Agitate often with a suitable kitchen tool to help dislodge unwanted material on both the inside and outside of the casings. Next, rinse the cases. This is most easily accomplished by dumping the contents into a kitchen colander and rinsing under a strong stream of hot tap water. Shake the colander often to help wash away the dirt and grime. Lastly, place the clean brass casings on an old towel to drain and dry. After cleaning, inspect the brass once again and discard any unsuitable casings. Next, turn your attention to the mouth of each case.

As a bullet is fired and rockets away from the brass casing, a tiny burr commonly forms around the case mouth. Since this burr may hamper seating of the new bullet, it should be removed. To do this, you can use a pocket knife blade or a special deburring tool designed for the purpose. A quick twist of the knife or tool is all that is usually required. Remember that your intention is simply to remove the burr (check both inside and outside of each case) and not any of the case metal. If you remove too much brass from around the case mouth, the brass will become thinner here and, as a result, weaker.

As you are inspecting the cases, separate them into groups of the same caliber that contain the same headstamp. For example, make piles of brass casings from R-P

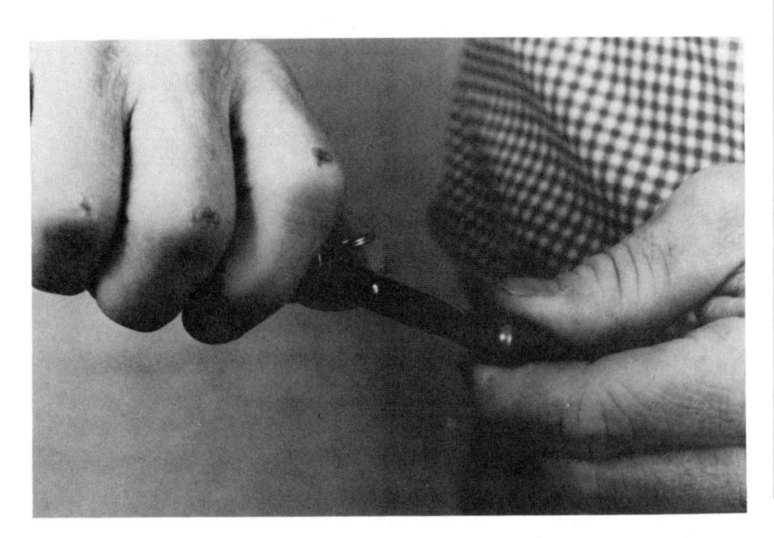

**Use the blade of a pocket knife to remove burrs around case mouth.**

**SAFETY**

**Before getting into the mechanics of metallic cartridge reloading, I would like to take a few moments to review safety on the reloading bench.**

**It is important to remember that while modern primers and smokeless powders are safer to handle than gasoline and other volatile substances, there is always a danger present. For this reason, make a habit of not smoking while reloading. Also, *never* have any kind of open flame in the reloading area, and this includes kerosene heaters or any other heat source. From the very beginning, establish a set routine of reloading and follow this each time you sit down to load some ammunition. Remove all unnecessary reloading components from the reloading bench. Do not reload in a casual or hurried manner, as this will increase the chances of mistakes. Keep all primers, powder, and loaded ammunition locked away from the reach of children.**

It is important to sort brass casings by make. All of these have different head stamps.

(Remington-Peters), W-W (Winchester-Western), Frontier Cartridge Company, military, and so forth. The main reason for sorting your empty brass casings by maker is to achieve a certain degree of consistency. All brass casings are similar, but there are often differences between manufacturers and even between lots of one manufacturer. The brass may differ as to hardness, thickness, and internal capacity. This will cause bullets to shoot at slightly different points of impact. By sorting, reloading, and shooting ammunition with the same headstamp, you increase the chances of consistent rounds.

Once sorted by headstamp, the empty brass casings require lubrication to make them slide easily into and out of the resizing die. The easiest way to accomplish this is to use a special lube pad, like the one illustrated below. It is important that only specially designed bullet-resizing lubricant be used, *never* motor oil or grease. As a rule, all petroleum-based lubricants except those produced specifically for resizing, will neutralize powder and primer, thus making a totally useless piece of ammunition.

## Lubrication

To use a lube pad, begin by placing a few drops of resizing liquid on the pad. Next rub this in with the fingers until the pad seems almost dry. Then, take about six empty casings and lay them flat on the pad. Roll them over the pad with your palm, until each picks up a light coating of the lubricant. Then, with the special neck-lubricating brush (which

Use a bullet lube pad to apply a light coating of special lubrication to casings before sizing.

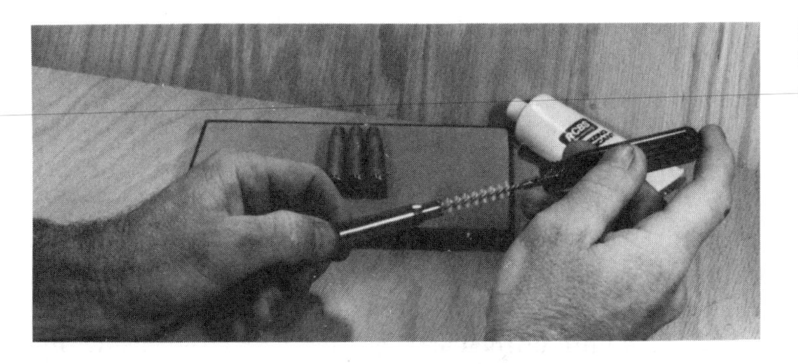

Lubricate the necks of brass casings with a special brush and lubricant.

comes with a lube pad), lubricate the inside of the cartridge neck. This will make the expander ball (on the resizing die) slide easily into and out of the casing.

Since 90 percent of all resizing problems are a result of improper or insufficient lubrication, take the time to lubricate your empty casings properly. Remember that you want lubrication only on the exterior walls of the cartridge and inside the case mouth. Do not lubricate the outside of the cartridge neck around the shoulder area, because this will usually cause dents as the cartridge is resized, a result of pressure in the die forcing the excess oil against the case.

## Resizing

Now the lubed cases are ready to be resized. Begin by first screwing the resizing die into the top of the reloading press. Install the proper size shell holder into the top of the press ram and raise this up to the top of its stroke. Screw the resizing die down until the bottom of the die just touches the shell holder. Then lower the ram slightly and turn the die down an additional one-eighth turn. Next, screw the lock ring down so that the die will remain in this position during use.

After adjusting, secure the sizing die with locking ring as shown.

Generally, there are two resizing procedures for the home reloader, full-length and partial. As a rule, a cartridge case is full-length resized when the finished piece of ammunition may be used in any rifle. Partial resizing, on the other hand, involves resizing just the neck of the cartridge. Partial resizing is commonly used by competitive rifle shooters who use the same firearm for all ammunition being reloaded. Given the choice, it is better to full-length resize brass casings, especially when the finished ammunition will be used for hunting and/or in automatic weapons.

To full-length resize a brass casing, begin by inserting a clean, lubed case into the shell holder on the reloading press ram. Slowly lower the press handle, thus raising the casing up and into the resizing die. Lower the press handle until it "bumps" at the bottom of the stroke. At about this time, the spent primer will be pushed out of the casing. Raise the press handle up to the original position. You should feel a slight resistance as the expander ball passes out of the mouth of the cartridge. Now take the casing out of the shell holder and look over the primer pocket carefully.

Usually a primer pocket will contain a bit of black residue that originated when the primer was fired. Since this residue may prevent proper seating of the new primer, remove it. Use either the tip of a small screwdriver or a special primer-pocket cleaning tool. Usually a few turns of the tool are all that is required for cleaning a primer pocket.

**Priming**

Now place the empty cartridge casing back into the shell holder on the press ram. Raise the ram up until the mouth of the shell comes in contact with the bottom of the resizing die. Then take a new primer and place it into the priming arm on the press (anvil side up) and push the priming arm into the slot on the ram. Now lower the ram until the primer is seated into the clean primer pocket. Since the primer needs to be seated fully, a slight pressure on the lever handle may be required. It is important, however, that not too much pressure be exerted, because this would damage the primer. Since seating primers is done more by "feel" than by eye, it may take several tries to learn the technique on a reloading press. In time, you may want to invest in a special priming tool that makes seating primers easier.

**Sizing die also contains a primer punch pin to remove spent primer.**

After you have seated the new primer, lower the ram and take the shell out of the holder. Inspect the new primer to see that it is just slightly below the surface of the cartridge head. Primers that stick up above the casing are not seated fully. Primers that appear flattened have been mashed and should be removed and replaced with a new primer.

## Charging with Powder

The next step in reloading is to charge the primed case with powder. It is extremely important at this point that you consult the reloading tables in this book to determine not only the type of powder required for the bullet you are reloading but the exact quantity as well. Remember too that each powder manufacturer publishes reloading tables, in most cases annually. *Use the most current reloading data, from a reliable source, that you can get your hands on.*

Before you begin to weigh powder charges, you should first balance your scale at zero. Then set the scale at the desired charge and add powder until the scale balances.

Whenever you are reloading with a new powder (one that you have never used before) or new type of bullet, you should not load the maximum charges shown. When more than one load is listed, use the lowest powder weight first; if these rounds perform well, you can increase the powder charge next time you reload. If only one charge is shown in the table, it is generally acceptable to reduce the amount of powder required by 10 percent the first time around. For example, if loading a rifle cartridge, the tables may list 47, 49, and 51 grains of a particular powder. The first time you load, you should use 47-grain charge, then after test firing—assuming pressures are not excessive, as evidenced by a flattened primer—next time you reload you may use the 49-grain charge, and so on, until you are reloading the maximum powder charge.

After you have checked the table and weighed the proper charge, you can dump the powder into the primed

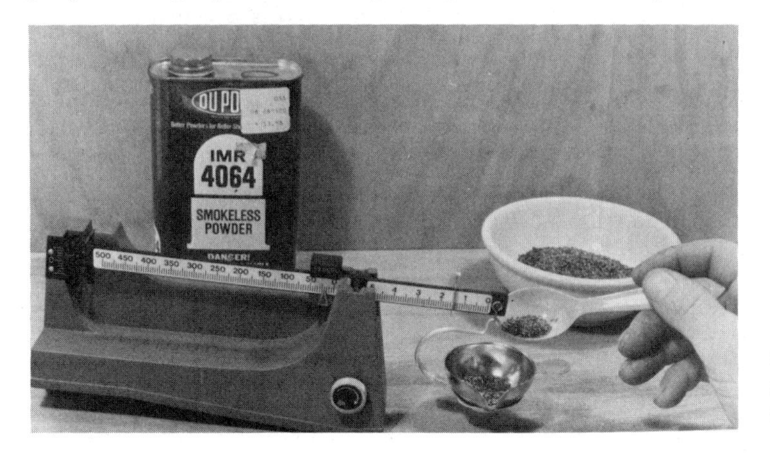


**Use a measuring spoon to add powder and balance your scale. Preset the scale to the desired weight.**

**To avoid spills, use a special reloader's funnel when charging casings with powder.**

casing. A special funnel makes adding powder to a casing very simple. After the casing has been charged with powder, place a new bullet (of the type and weight indicated in the tables) into the case mouth. Set the charged case into a reloading block while you install the seating die in the press.

It should be mentioned at this time that a number of powder measuring and dispensing tools are available for the reloader. These tools meter a specific amount of powder and can greatly speed up this reloading step. In addition, powder charges—the amount of powder for a given load—tend to be more consistent than if each charge is measured separately.

## Seating Bullets

Seating bullets is the last reloading step for most bottleneck casings. As a rule, the case is not crimped, as straight-walled cases are, unless the bullet will be used in a tubular-type magazine.

Begin by removing the resizing die from the press and screwing the seater die partway into the press. Next, place a sized but uncharged casing into the shell holder and run the ram up to the top of its stroke. Screw the seating die down further until you feel it come in contact with the empty shell. When this happens, back the seating die out one-eighth turn and lock the die in place using the locking ring. Now back the seating screw almost all of the way out of the die. Raise the lever handle, thus lowering the ram and empty cartridge, and place a bullet into the case mouth. Lower the press handle again and run the empty casing, with bullet installed, up to the top of the stroke. Now you actually begin to seat the bullet by screwing the seating stem down until it comes in contact with the bullet. Lower the ram and check to see how far the bullet has been seated. Continue doing this until the bullet has been seated to the proper depth. Consult the reloading tables to learn the maximum overall length for the cartridge you are reloading. Once you have achieved the proper setting on the seating screw, lock it in place with the locking nut.

**After all cases have been charged with powder, tip the loading block and check that the powder level is the same in all cases. This will prevent dangerous double charges of powder.**

It is important to remember that a bullet is not necessarily seated to the depth of the *cannelure,* the indented lines midway between the base and tip of the bullet. The proper seating depth of a bullet will result in a finished overall length that is given in the reloading tables. Use this information, given for all calibers, as a basis for your bullet seating.

The empty casing with bullet that you have been working with is of course useless as a piece of ammunition but is handy for helping you to set up the seating die for future use. Mark the shell in such a manner, by drilling holes around the sides, for example, that you will never use it. It is a good idea to keep this so-called "dummy shell" in the box of dies for this caliber. Then, the next time you are adjusting the seating die, use the dummy.

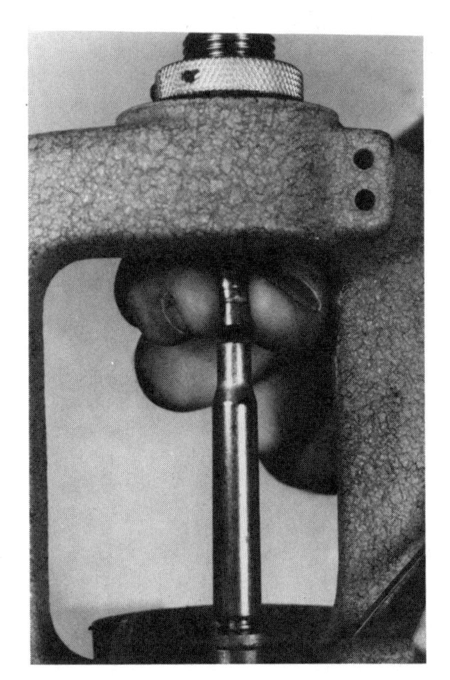

**Place a new bullet into the case mouth just prior to seating.**

After the seating die has been properly adjusted and locked in place, place the fully charged casing with bullet in the shell holder and run it up into the die. Raise the press handle and take the finished shell out and inspect it carefully. What you should see is a clean casing, with both primer and bullet seated to the proper depth. The overall length of the cartridge should be equal to the measurement given in the reloading table for this particular round.

If the finished cartridge is dented, deformed, or heavily scratched, something may be wrong with the dies and they should be checked. Bullets *must* be seated to the proper depth. A bullet seated too deeply may fire properly but will not be accurate. Bullets not seated deeply enough will be difficult or impossible to chamber in your rifle. Keep in mind that the intention of reloading is to produce better-than-factory ammunition and that this is possible only if you perform all reloading steps with care.

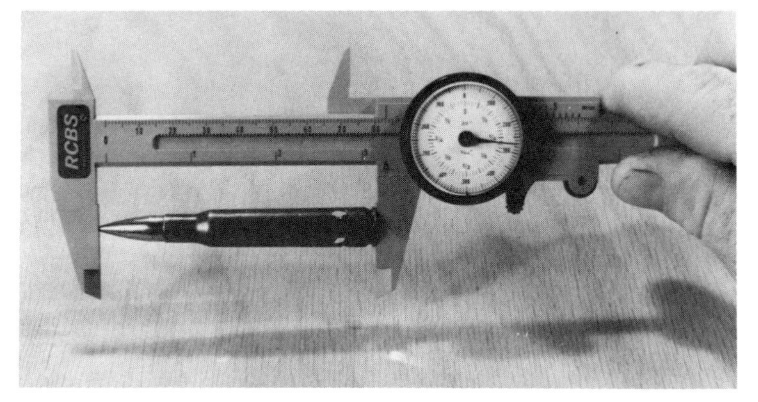


**Check overall length of "dummy shell" with a dial caliper—note holes in side of dummy. Maximum overall length (OAL) is listed for each cartridge in the reloading tables.**

## RELOADING PISTOL AMMUNITION

Before we actually begin discussing the procedure for reloading pistol ammunition, it should be pointed out that by "pistol ammunition" I mean brass casings that have straight sides, rather than a bottleneck shape.

The main distinction between rifle and pistol ammunition is that the former requires a two-die reloading set, while the latter has a three-die set. Generally, pistol ammunition is loaded in the same manner as rifle ammunition, except that since there is an extra die—the expander die—there is an extra step in the reloading process.

There are, to be sure, a number of differences between rifle and pistol ammunition. As a rule, the powders used for reloading pistol ammunition burn much more quickly than the powders used for rifle ammunition. Pistol ammunition generates lower internal chamber pressures and lower velocities. For example, a typical .38 special load will have a chamber pressure of about 18,900 CUP and a muzzle velocity of from 700 to 900 FPS. A standard rifle load such as a .30-06 will have a chamber pressure of about 50,000 CUP (maximum) and a muzzle velocity of from 1900 to 2700 FPS. Another thing to keep in mind is that rifle primers (small, large, magnum, or benchrest) cannot be used for pistol loads, nor are any pistol primers (small, large, or magnum) usable for rifle-cartridge reloading.

Ammunition for pistols differs in other ways as well. Ammunition for revolvers will generally function if the

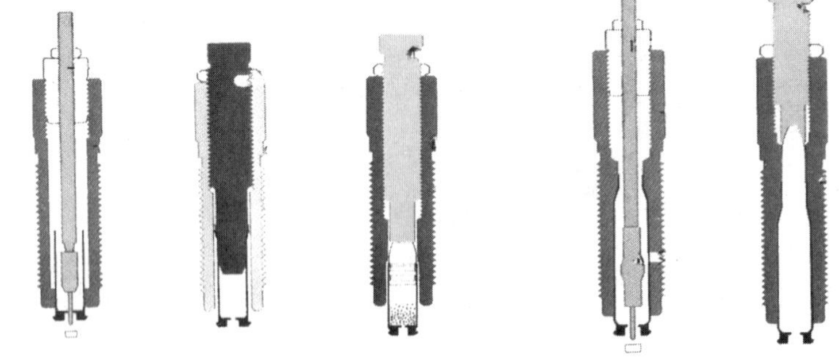

**Straight-walled cases are reloaded with a three-die set (l), while bottle neck casings require only a two-die set (r). (*Courtesy of Lyman Company.*)**

load can be inserted in the cylinder and does not protrude (which would prevent the cylinder from revolving). Ammunition for automatic pistols, on the other hand, will commonly have a minimum and maximum length so that the cartridges will feed and chamber properly. Powder charges for revolvers can vary quite a bit, but automatic pistol ammunition must have rather specific charges of powder to operate the automatic mechanism properly.

Although the bullets used in revolvers can be made from soft lead and have a variety of shapes, such as wadcutter, hollow-point, and flat-nose, automatic pistols generally require full metal-jacketed (often called 'hard ball') bullets. The reason for full metal-jacketing is to ensure proper feeding and chambering of the ammunition.

Full metal jacket (FMJ) bullets (sometimes called "hard ball") are often necessary when reloading automatic pistol ammunition.

One other area where revolver and automatic pistol ammunition differ is in case design. Most are straight-walled, but that is about the only similarity. Revolver cases commonly have a thick rim at the base of the casing and are usually longer than automatic casings. It is this rim on revolver casings that provides the headspace for the cartridge. The bullet is crimped in the casing on most revolver ammunition.

Automatic ammunition, on the other hand, is generally of the rimless case design. This means the cartridge is held in the chamber by the forward edge of the casing rather than by the rim. The rim of the casing on automatic ammunition is used for extraction and ejection of the fired casing. Because the leading edge of the casing is used for headspacing, it obviously cannot be crimped around the bullet and still be expected to function properly.

The rimless case (l) has a rim the same diameter as the casing, while the rimmed case (r) has a rim which is much wider than the case itself.

The equipment required for loading pistol ammunition is basically the same as for rifle ammunition in that you need a set of reloading dies (for the caliber you are reloading) and a reloading press. An accurate scale is handy to have for checking loads but less crucial for pistol ammunition, because there are some very good alternatives.

Probably the least expensive way to charge up pistol ammunition cases with powder is with the Powder Measure Kit by Lee. The current version of this powder-measuring kit comes with thirteen dippers ranging in size from .3 cc (cubic centimeters) to 4.3 cc. To simplify use, the kit also comes with a slide card that lists ninety-five different powders, including black powders. To use, simply locate the powder you want and look down the scale to determine which dipper to use. Although this kit has some limitations, it will enable you to load pistol ammunition with reasonable accuracy.

Other alternatives to a scale for charging pistol ammunition include a variety of powder metering and dispensing tools. Since all of the major reloading equipment manufacturers offer some type of powder measuring device, the reloader will have little problem finding a useful powder measure.

**Inspection**

The first step in reloading pistol ammunition is to inspect your empty casings. Quickly look over each case for signs of dents, cracks, and excessive dirt or grime. While small dents can usually be pushed out with the sizing die and dirty cases can be cleaned, cracks indicate a breakdown of the brass case; these cases should be discarded.

If your empty cases are "once fired," chances are they will not be overly dirty. In most cases, a simple wipe with a damp cloth will clean them sufficiently for reloading. If, however, your empty brass came from an unknown source (such as picked up at the range), they may need a more thorough cleaning. If this is the case, refer to the discussion of cleaning rifle casings to learn how to clean empty pistol cases.

**Resizing**

Once the casings have been inspected and cleaned (if required) the next step is to remove the primers. This is commonly referred to as "decapping." Begin by screwing the sizer die into the top of your reloading press. This die should be clearly inscribed by the manufacturer with the word "sizer." Additionally, a small pin should protrude from the mouth of the die. This pin will punch out the spent primer when the press is operated.

Since the function of the sizer die is important, it will be helpful to know how this die is positioned in the reloading press. With a shell holder in place, raise the press ram to the top of its stroke. Next, screw the sizer die down into the top of the press until the mouth of the die comes in contact with the shell holder. Lower the ram slightly and give the die an additional one-eighth turn downward.

Now raise the ram once again and you should feel a slight "bump" as the linkage of the press passes over center. Next, lock the die in position in the top of the press with the special lock ring.

Lower the ram to the bottom of its stroke. You are now ready for the next step—resizing and decapping. Begin by inserting a clean, empty, lubricated casing into the shell holder on top of the ram. Empty casings can be quickly lubricated with a special lube pad designed for the purpose. The only time brass casings are not lubricated is when a special tungsten-carbide die is being used for resizing.

**The decapping pin can be seen here as a casing is about to be resized.**

Pull the press handle downward slowly until the empty casing passes completely into the resizing die. At the top of the stroke you will feel a slight "bump" of the press linkage; at about the same time, the spent primer should be punched out. When you hear the primer drop, this indicates that the casing has been both resized and decapped. Raise the press handle to the top of its stroke and remove the casing. It should look almost the same except for the absence of a primer.

## Cleaning the Primer Pocket

The next step in the reloading process for pistol ammunition is to clean the primer pocket. As a primer is fired, it will leave a black residue around the flash hole in the primer pocket; this must be removed before a new primer is inserted. The easiest way to clean a primer pocket is with a special tool designed for the purpose. If you do not have such a tool, the primer pocket can be cleaned with the tip of a small screwdriver, but you should exercise great care. For best results, the tip of the screwdriver must be smaller than the primer pocket hole. Usually, a quick twist is all that is required for cleaning. Do not remove any of the brass around the rim of the primer pocket hole.

## Case Neck Expansion

After the primer pocket has been cleaned, install the expander die in the reloading press. The die is screwed down until the mouth of the die just contacts the shell holder on top of the press ram, which should be at the top of its stroke for positioning the die. Lower the ram slightly and give the expander die an additional one-eighth turn. Next lock the die in place with the special locking ring. At this time the expander shaft, which screws into the top of the die, should be screwed upwards at least halfway out of the die. The next step is to correctly position this expander rod.

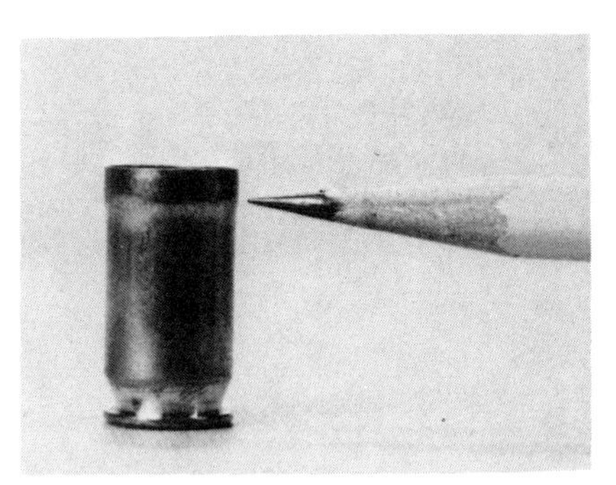

**The expander die flares a case mouth slightly to facilitate insertion of a bullet prior to seating. This case mouth has been flared excessively for illustration purposes.**

The purpose of the expander rod is to flare the mouth of the empty casing. This is done so that seating the bullet, after the powder has been added, will be easy to accomplish. Your intention is to expand the case mouth only slightly. It may take several attempts to do this correctly.

Begin by installing an empty case into the shell holder and running the press ram to the top of its position. This will run the casing into the die. Next, screw the expander rod down into the die until it meets some resistance. This indicates that the expander rod has made contact with the mouth of the case. Lower the ram and look at the case mouth. If it is flared slightly, the expander rod is in the proper position. Adjust the expander rod accordingly if necessary. Keep in mind that if the case mouth is not expanded enough, inserting a bullet will be difficult and trying to seat such a bullet often results in a crushed casing. A case neck that is expanded too much, on the other hand, will often crack or be difficult to crimp. Once you have adjusted the expander rod properly, lock it in position with the locking nut. This ensures that all casing mouths are expanded to the same dimensions.

## Priming

While the empty casing is still in the shell holder after neck expansion, prime it. With the ram at the top of its stroke, place a new primer into the primer arm, open-end-up, and push the primer arm into the primer arm slot of the ram. It will be necessary to hold the primer arm in this position while the ram is lowered. As the ram moves downward, it will come in contact with the primer arm, and the new primer will be inserted into the primer pocket. Gentle but firm pressure is required to seat the new primer properly. Next, raise the ram so that the primer arm can swing out and away from the ram; then lower the ram and remove the primed casing.

Look over the newly primed casing carefully. If you cannot remove the casing from the shell holder, it is probably because the primer was not seated deeply enough. If this is the case, raise the ram, push the primer arm forward, and bring the ram downward again until you "feel" the primer seat more fully. If, on the other hand, the casing comes out the shell holder easily and the face of the primer appears to be just below the surface of the shell base, chances are very good that the primer was seated correctly.

### Charging with Powder

Now charge the primed casing with powder. It is extremely important that you consult the reloading tables for the proper charge weight for the bullet you are reloading. It is also recommended that when more than one charge is listed you use the lower of the two on your first reloading.

Charge the casing with the proper amount of powder. If you are charging more than one case, after all are charged, check visually to see that all casings appear to have about the same amount of powder. If you are using a scale, weigh each charge carefully. If you are using some other means of charging the cases—dippers or some kind of powder measuring and dispensing tool—check the charge on an accurate scale if available.

The last operation is to seat the bullet and crimp the case mouth, if required. Remember that most revolver cases are crimped but rimless (automatic) cases are not.

Before a bullet is seated and crimped, install the bullet seating die properly in the reloading press. To do this begin by screwing the seating die into the reloading press. Run the ram up to the top of its stroke and screw the die down until the mouth contacts the shell holder. Lock the die in position with the special locking ring. Then unscrew the seater plug five or six turns. Next, insert a "dummy" cartridge (or factory cartridge) into the shell holder. A "dummy" cartridge is one that contains no powder but has the bullet seated to the proper depth and is crimped. A dummy cartridge is very handy for setting up the seating/crimp die, and I keep one in every box of reloading dies I own.

Run the press ram up to the top of its stroke, at which time it will pass into the seater die. Now screw the seater plug down until you feel it come in contact with the top of the bullet. Lock the seater plug in this position with the special locking ring. Lower the press ram and remove the dummy cartridge. Next, insert a primed and powder-charged case into the shell holder. Place a new bullet into

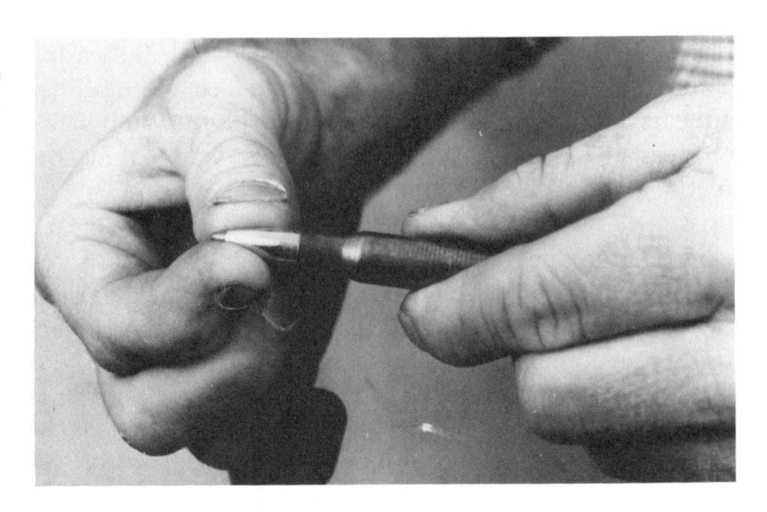

**After the bullet has been seated and crimped, try to twist the bullet. If you can, the crimp is not sufficiently tight.**

**Label and box *all* reloaded ammunition. Include primer type, grains powder, bullet type and weight, and date of reloading.**

the case mouth and run the ram up to the top of its stroke. This will run the casing up into the die and, at the same time, seat and crimp the bullet. Lower the press handle, remove the bullet, and inspect it carefully. The newly reloaded round of pistol ammunition should be the same length as the dummy or factory round. The walls of the cartridge should be uniform and the bullet should be held tightly in the neck of the casing: try to twist the bullet with your fingers. If all is well, the round is ready to be test-fired in a suitable place.

It makes sense to box and label reloaded ammunition so you can develop ammunition that works well in a given firearm. The label should include date; bullet type and manufacturer; powder type and weight; primer type and brand; and velocity. In time you will have fun comparing different loads.

For simplicity, we have followed only one piece of ammunition as it went through the various reloading steps. In actual practice, you will probably run a quantity of cases through each of the reloading steps—resizing and decapping, expanding and priming, charging with powder, seating and crimping the bullet. This greatly reduces the amount of time required to load each round.

# 4 How to Reload Shotshells

There is a vast difference between the average shotshell and the typical metallic cartridge. Internal pressures, as a shotshell is fired, rarely rise above 10,000 PSI, while almost any metallic cartridge has a chamber pressure from 30,000 to 50,000 CUP. At least part of this difference is caused by the different powders for reloading shotshells.

As a rule, shotshell powders burn very rapidly and are flakelike in consistency. Rifle powders, on the other hand, are commonly rod or spherical in shape and are very slow-burning. For more detailed information about the different types of powders used for all reloading, flip back to the second chapter, where you will find a thorough discussion of the popular powders used today.

Because of the major differences in burning rates between powders, it is extremely important to consult the reloading tables later in this book for specific recommendations as to powder type and charge weight. Do not experiment with powders—substituting one powder for another, for example—but instead follow powder recommendations exactly. Also strictly follow other component recommendations as to primer type and brand, wad, and charge of shot. By following the recommended loads in the reloading tables, you can reasonably expect to turn out dependable, quality ammunition.

## SHOTSHELL RELOADING MACHINES

The tools required for reloading shotshells are surprisingly simple and few in number. In fact, almost all shotshell reloading is accomplished on a special machine that performs the various reloading steps—priming, powder charging, wad and shot insertion, then crimping—with an empty casing.

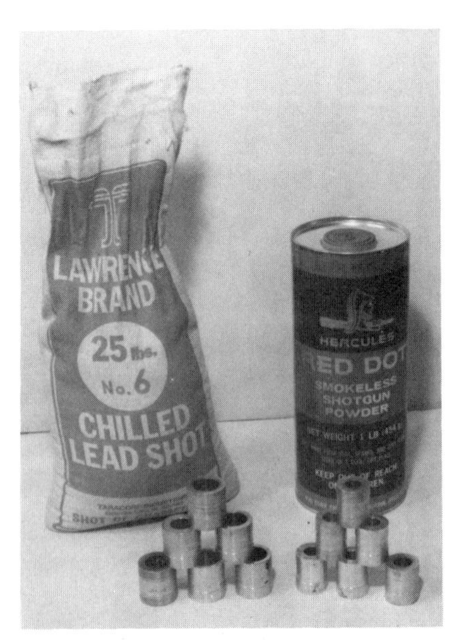


**A selection of shot (l) and powder bushings (r) are necessary for different types of shotshell loads.**

The typical shotshell reloader is easily identified by the two plastic tubes or bottles on top. One tube holds and dispenses powder; the other, shot. The amount of each charge is governed by special bushings that allow a specified amount of shot or powder to drop at one time. Since the better shotshell reloaders have replaceable bushings, it is possible to vary the amount of powder and/or shot to arrive at different types of shotshell loads, such as ⅞-, 1-, 1⅛-, 1¼-, 1⅜-, 1½-, and 1⅝-ounce, simply by changing the bushings for shot. Variations for brand and grain charge of powder are accomplished by inserting suitably sized powder bushings as well.

Shotshell reloading machines are most commonly set up to load only one gauge of shotshell at a time, 12 gauge, for example. The better reloading machines can be altered to load other gauges, 16, 20, 28, and 410, for example, by changing not only the bushing for powder and shot but also the primer punch and sizer, wad insert, and crimping mechanism. The lower-priced shotshell reloaders generally require that the change from one gauge to another be done at the factory, if at all.

Since the amount and brand of powder, primer, wad, and shot charge are all clearly given in the reloading tables, it is generally a simple matter to reload shotshells, providing you have both the bushings and components required. It is not a good idea to substitute any components for those called for in the tables; the results will be unpredictable.

If you are planning to purchase a shotshell reloader, shop wisely. Look at as many shotshell reloaders as you can before you actually purchase your machine. Keep in mind that the more features a machine has, the higher the selling price and the more you may be paying for features that you don't need, such as an attachment that kicks finished shells out of the machine and down a special chute.

It may be helpful to mention a few features that are worthwhile to have included on a shotshell reloading press: a full-length sizing ring or die, adjustable and replaceable; automatic primer feed; replaceable wad-guide fingers; a variety of powder and shot bushings, easily replaced in the machine; spent primer tray or container; large shot and powder reservoirs; adjustable-wad pressure ram. In addition to these features, the shotshell reloader you are considering should be easy to operate and well made. If the finish on metal parts is good, the machine workmanship will probably also be of high quality.

At the time of this writing, the leading shotshell reloader manufacturers include Ponsness/Warren, Pacific, Mec, and Lee. These companies manufacture the best shotshell reloading machines available today. It is always wise, however, to shop around to learn if there are any worthwhile new entries on the market.

## STEP-BY-STEP SHOTSHELL RELOADING

### Inspecting Empty Shells

The first step in reloading shotshells is to gather together empty shotshells for inspection. Look over each empty carefully for cracks and splits and discard any that are in other-than-first-class condition. Pay particular attention to the area just above the brass base; shotshells tend to develop burn holes in this area, most usually in the form of tiny black pinholes.

### Cleaning Shotshell Casings

If the empty shotshells are dirty, they should be cleaned. There are several ways of doing this. If the empty shotshells are paper, about the only way these can be cleaned is with a damp cloth; then they must be allowed to dry thoroughly before reloading. I have found that paper shotshell empty casings dry perfectly if left in a warm area, such as near your furnace, for several days. After cleaning and drying, empty paper shotshells benefit from a light application of beeswax to the exterior. After this treatment, the empty paper hulls will be as good as new and ready for reloading.

Chances are good that most of your empty shotshell casings are plastic, and these are easier to clean than the old paper casings. There are two different types of plastic shotshell hulls: one-piece and multipiece base wad types. One-piece plastic shells, such as Winchester-Western AA Plus and Remington-Peters' Blue Magic, have an internal base wad that is actually part of the plastic hull. To clean these, simply dump a number of them into a pot of hot

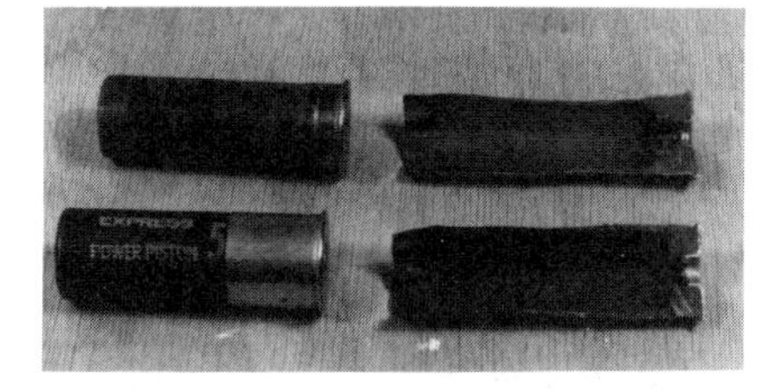

**Plastic-base wad shotshell (l) differs from fiber-base wad shotshell (r).**

water and dish soap, agitate for several minutes, drain, rinse and allow to dry thoroughly before reloading.

If you also have dirty multipiece plastic shotshells, such as Winchester-Western Duck & Pheasant and Federal field loads, these must be cleaned the same as the older paper shotshells. The reason is that these multipiece shotshells have a paper-base wad that will absorb water if immersed. To clean, wipe with a damp cloth and allow to dry thoroughly before reloading.

## Sorting Casings by Manufacturer

The next step in the process is to sort your empty casings by maker and type. Since primers, powders, and wads differ from one manufacturer to another, and since the reloading tables are arranged according to manufacturer, you must segregate your empties along these lines.

This is a good time to explain the difference between the so-called high-brass and low-brass shotshell casings. As a rule, shotshells with a high-brass casing—the actual brass around the base of the shell—have a low internal base wad. Shotshell casings with low brass around the base of the shell have an internal base wad that is high by comparison. High-base, low-brass shotshells are commonly used for target shooting, and the high internal base wad permits lower charges of powder. Low-base, high-brass shotshells permit more powder to be dropped into the casing and are therefore more powerful. These are commonly used for hunting rather than for target shooting.

**Low-brass, high-base wad (l) requires different components than high-brass, low-base wad shotshell (r).**

After your empty shotshell casings have been cleaned, dried, and separated by maker and type, you can begin reloading. Start off by picking a group of same-make empties and consulting the reloading tables for the proper components for this particular type. You must also decide how much of a shot charge you would like for these particular shotshells. As you can see in the tables, the range includes ¾-, ⅞-, 1-, 1$\frac{1}{16}$-, 1⅛-, 1$\frac{3}{16}$-, 1¼-, 1⅜-, and 1½-ounces of shot (your choice of shot size). Once you have decided what amount of shot you want, it is a simple matter to determine the proper-size shot bushing and the corresponding powder bushing—the latter based, once again, on the tables and in conjunction with the bushings for your particular reloader.

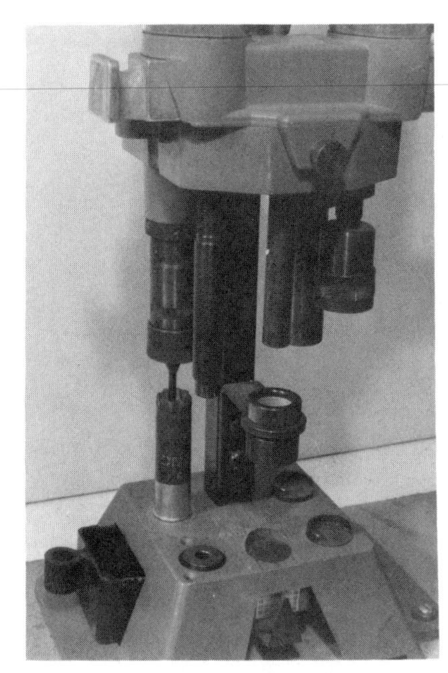

**The first step is to remove old primer and resize shotshell casing.**

Install the proper bushings for powder and shot in your shotshell reloader. Since the installation of bushings differs from one manufacturer to another consult your owner's manual for instructions for your own make. Once this has been accomplished properly, you can begin the actual reloading.

There are some differences in operation between one shotshell reloader and another, but all the following tasks must be accomplished in order to make quality shotshell ammunition.

## Primer Removal

Remove the spent primer. This is commonly done by positioning the empty shotshell below the location of the primer punch tool and pulling the handle downward. This causes the primer punch to move downward into the empty casing. When the punch reaches the bottom of the shell, the spent primer is forced out.

## Resizing

Resize the empty casing. This is often done at the same time that the old primer is removed. A resizing ring is forced over the empty casing, and because this ring is exactly the proper size of an unfired casing, it will force the empty casing back to its original external dimension. This is done for the same reasons that metallic cartridges are resized: upon firing, the casing expands slightly to fill the chamber, then shrinks slightly so it can be extracted easily.

If you are firing shotshells in only one shotgun, the resizing step can usually be eliminated without any chambering problems. This is because the fired shell will be just slightly less in external dimensions than the chamber of a given shotgun. If, on the other hand, the shotshells will be used in any shotgun, it is better to full-length resize each casing, ensuring that the reloaded shotshells chamber properly in all shotguns of the same gauge.

**Priming**

After resizing, install a fresh primer into the casing. Follow the recommendations in the reloading tables, since there are differences among primer dimensions as well as in degree and intensity of flash, so one brand and type of primer will not fit all empty shotshells.

When seating primers in shotshells, exercise the same care taken with seating primers in metallic cartridges. Lower the empty casing down onto the new primer, and, when you begin to feel some resistance, gradually ease the new primer into place. Then remove the primed casing and look it over carefully. A properly seated primer should be set flush in the brass base of the shotshell. It is important that just enough pressure be exerted to seat the primer properly. Too much force will result in the primer being seated too deeply, which often results in a misfire. Primers not seated deeply enough will also result in a misfire.

It takes a bit of practice to learn the feel of a properly seated primer in any type of reloading. If you are new to reloading, keep this in mind, and remember that in time you will learn how much pressure to exert to seat a primer properly.

**Powder Charging**

After priming, charge the casing with powder. By this time, you should have already determined the type and amount of powder to be used and installed the proper-size powder bushing.

Depending on the type of shotshell reloader you are using, you will either pull a lever or push a charge bar to dispense the proper amount of powder into the primed shotshell. This is all that is required unless you are using one of the least-expensive shotshell reloaders. If so, the powder must be measured before loading.

**Wad Insertion**

The next component to be loaded into the shotshell hull is the wad. Since almost all wads used today are of one-piece

There are many shotshell powders on the market and it is important to choose the right one for the type of shotshells you are reloading. Consult the data tables in this book for specific information.

construction, this is a simple matter. In most cases, the wad is simply placed into the wad column station and the lever is pulled downward, causing a special ram to push the wad past the spring fingers (which hold the mouth of the shell open) and into the shell. At the time of inserting the wad, it is common for the wad ramrod to exert a predetermined amount of pressure over the wad. This wad pressure is necessary to compact the powder and give predictable results. In some cases, an average amount of wad pressure exerted is preset at the factory, but in other machines the wad pressure can be set by the user.

Some reduced shotshell loads require the insertion of a 0.135-inch-thick card wad inside a one-piece wad (20-gauge card wads are used inside 12-gauge cup wads). The reason for this is to take up some of the space in the wad cup so that a reduced charge of shot will fit inside and still crimp well. Few loads call for this addition, but it is worthwhile to know how this is done: you may find a notation in the reloading tables later in this book.

A new one-piece wad goes into the charged and primed shotshell casing.

## Shot

After the wad has been inserted and the proper amount of wad pressure applied, drop the shot into the shell. In most cases this is done simply by sliding the charge bar to the left. This action drops a predetermined amount of shot into the shell. It should also be pointed out that this is generally done while the wad-pressure ram (which is also the drop tube for both powder and, in turn, shot) is still inside the shell. After you hear the shot drop into place, raise the

**On this shotshell reloader, you must pull the charge bar to drop shot or powder into the casing.**

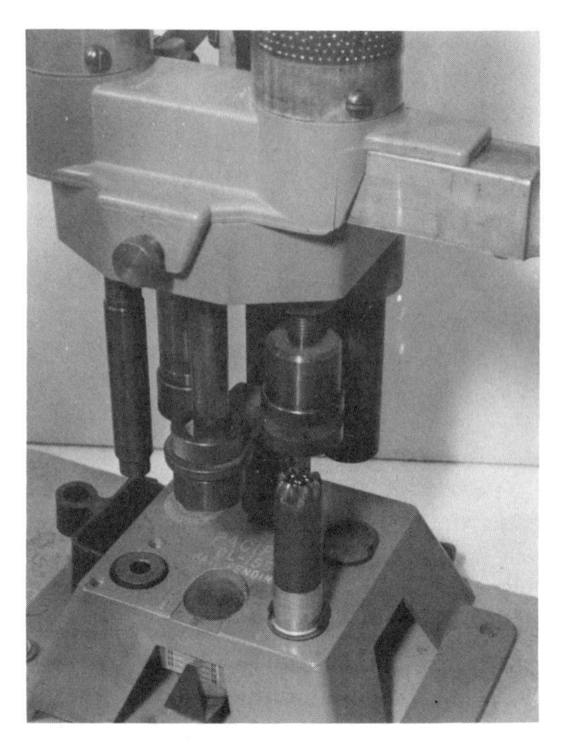

**Started crimp on loaded shotshell.**

reloader handle and move the loaded shell to the next station on the press.

**Crimping**

The last step is to crimp the loaded shell. Depending on the type of shotshell reloading press you are using, this may be done in one or two steps. Plastic shotshells require two crimping steps—crimp starter and final crimp—while paper shells require only final crimping.

The reason for crimp-starting with plastic shells is that the plastic tends to be too stiff to be crimped fully in one operation. The crimp starter simply bends the top of the shell over, in either six- or eight-star pattern, so that the final crimp will be well-formed to begin with.

After the crimp has been started, move the shell to the next and final station for the final crimp. Next, the handle of the reloader is brought down, forcing the final crimp die over the shell and crimping. The final-crimp die fits over the entire shell during crimping to prevent bulging as pressure is exerted for the crimp. Now raise the handle, remove the finished shell, and inspect it carefully.

A perfect reloaded shotshell has straight sides, and the crimp should be tight. The permissible recess for the crimp is about 1/16-inch deep. If there is a hole in the center of the crimp where the folds meet, this should be smaller than the size of the shot inside. The face of the crimp must be flat and tight; a crown or point usually indicates insufficient crimp pressure. It is entirely possible that your first reloaded shotshell will have all of these characteristics, especially when a modern shotshell reloading press is used. If you find that your reloaded shotshell is not quite up to standard, try a few more loads and you will discover that a perfect shotshell is relatively easy to produce, providing the reloading machine is adjusted properly according to the instruction manual that came with it.

**A perfect crimp is possible with the modern shotshell reloaders available today.**

## Boxing the Shells

Box and label the ammunition. The label should contain all important information, such as date, powder (type and amount), primer, shot (size and amount), chamber pressure, and velocity.

## Cleanup

After you have finished reloading shotshells, clean up the area. Drain powder and shot into their original containers. You should also put any loose primers back into their storage box. Get into the habit of cleaning up after each reloading session and you will greatly reduce the chances of an unfortunate mishap. Store powder and primers according to the guidelines given in the second chapter.

## Problems

Most of the problems in reloading shotshells are a result of an improperly adjusted reloading press or substitution of reloading components. In addition, a few problems relate to the shotshell casing.

Assuming that you purchase a new shotshell reloading press, the setup of the machine should be straightforward. In most cases the machine will have been adjusted (wad pressure, crimp, and so forth) at the factory, and all that should be required is to bolt the machine to a sturdy bench. If you find that the machine does not operate properly, consult the owner's manual or a salesperson at the place of purchase.

The reloading tables later in this book have been compiled by the various manufacturers of powder and primers. Since these data list specific components, you should use *only* these. If a given load calls for a Remington #57 primer, do not attempt to substitute another brand or type of primer: the second choice will probably not fit or work properly. The same applies to powder (brand and type) and other components.

The selection of wads for shotshell reloading is also greatly simplified by following the recommendations in the reloading tables. In fact, you should use *only* the recommended wad for any given load. Keep in mind that the reloading data tables were compiled by experts working under laboratory conditions, and they specify the best possible combination of components.

As a rule, a poor crimp is often a result of substituting one wad for another. If a load calls for a Remington RXP wad, you cannot use any other Remington wad (such as RP, R12H, R12L, or SP)—as these will have different shot capacities. The results of substitution are always much inferior to what you can expect when the proper wad is used. If you discover that your shotshell reloaded ammunition is of poor quality, doublecheck all components and amounts of powder and shot; the problem may lie here.

One problem with empty shotshell cases, particularly the older paper hulls, is the phenomenon known as the "dished head." This problem is easy to see when inspecting fired shells: the head area around the primer is pushed in and has a dished appearance. This happens either because the primer is of the wrong type or because the primer hole is larger than normal (this can happen if a primer is seated improperly or after repeated reloadings). When there is a space around the primer, this permits gas to escape rearward, to the face of the bolt. The end result is that the brass is pressed forward, forming a dish-shaped indentation. As a rule, dish-shaped brass should be discarded. Efforts to repair such a casing generally do not work and the problem will recur.



Another problem with fired casings has to do with the original crimp pattern. Most reloaders have a crimp starter that is either six- or eight-point. In some cases, the manufacturer will supply additional crimp-starter dies for both types. If your reloads do not crimp well, the reason may very well be that you are trying to crimp a shell that originally had six points with an eight-point crimp starter.

While we are on the subject of crimping shotshells, you should know that at one time all shotshells were roll-crimped (as well as being made from paper), rather than star-crimped. It is impossible to put a star crimp on an old shotshell casing that was originally roll-crimped. Some manufacturers offer a roll-crimp device, however, so if you have a quantity of roll-crimped empty shotshells, this may be a worthwhile investment. Roll-crimped empties are most commonly paper but plastic ones are also encountered, especially shotgun slugs.

**From left to right: roll crimp, six-point crimp, and eight-point crimp.**

If you have a problem chambering your reloaded shotshells, this may be due to improper or insufficient sizing of the empty casing during reloading. Remember that a shotshell is actually resized twice during reloading: the first time during decapping and then again during final crimp. Check to see that in both of these operations the case is full-length resized. If it is not, you may have to adjust your shotshell reloader slightly. Consult the owner's manual for your machine or a knowledgeable source to learn how to do this correctly. If the problem persists, check your shotgun chambering mechanism: this too may be at fault. Simply try chambering a few factory-loaded shotshells. If the problem still exists, it is in the shotgun and not the ammunition. Take the shotgun to a qualified gunsmith for repairs.

Shotshell reloading is undoubtedly the easiest of all reloading for the do-it-yourselfer. The shotshell reloader machines available today make shotshell reloading so simple a matter that almost anyone can master it on the first try. Just be sure to follow any special instructions for setup and use by the manufacturer and use only reloading data from this book or other reliable sources.

# 5 Specialized Techniques for the Home Reloader

Based on what you have learned up until this point, you can make quality ammunition with just a few tools and reloading components. In addition, you will save substantially on your annual ammunition costs even if you shoot more often. In this chapter, you will see how it is possible to save even more by casting you own lead bullets, using military brass casings, and making a variety of brass casings from the standard .30-06 casing. You will also learn a number of shortcuts to save you time and help you get the most out of reloading.

## CASTING BULLETS

The shooter or hunter who loads his own ammunition from commercially made components can save substantial amounts of money. When we start thinking in terms of making our own bullets or, as some experts prefer, our "projectiles," the overall cost of ammunition is reduced dramatically. For example, when I load lead cast bullets for my .38 special, I estimate that it costs me about two cents a round. When I use factory-made bullets such as jacketed hollow-points for the same round, the cost-per-round is about ten cents. Just in case you are wondering, a box of factory ammunition for this same firearm currently sells for around twelve dollars, or about twenty-four cents per round.

Since economy is almost as important to me as accuracy, I can honestly say that I very rarely use commercially made bullets for my .38 special. Over the years I have discovered that the performance of home-cast lead bullets is entirely suitable for my needs, which, I might add, include plinking, serious target punching, and varmit hunting. When you consider that it costs me about two dollars per hundred rounds using home-cast lead bullets, these rounds enable me to shoot frequently for next-to-nothing.

Before we actually get into the mechanics of casting lead bullets, we need first to talk a bit about the lead, or different types of lead, really, available to the home reloader. These include tire wheelweights, printer's linotype, plumber's lead, and scrap lead from almost any source.

Over the years I have used all of these lead materials for casting bullets and found only slight differences between any two kinds of lead. As a rule, tire wheelweights work as well as linotype, although the weights tend to require a bit more work before the molten lead can be poured into the casting mold. This amounts simply to skimming off the dross and impurities which float on top of the molten lead after the weights have been melted. Linotype is a slightly harder lead, due to a higher tin content in the material. As a result of this extra tin—approximately 10 percent—bullets cast from linotype are harder than pure lead bullets and tend to be a brighter silver color.

As a rule, almost any lead you can find is suitable for casting bullets. In time you will develop a fondness for one or more types because of specific characteristics, such as hardness, shootability, and ease of reloading, but in the

**Many types of lead are usable for casting bullets.**

long run you will be reasonably content with almost any available lead. I find that lead that costs nothing is much more desirable than lead that I have to pay for.

Before you can start casting your own lead bullets, you will need some specialized equipment. A large selection of tools and equipment designed for do-it-yourself bullet casting is available, but you can easily get by with just a few items.

Probably the most important piece of casting equipment is a bullet mold. In its simplest form, the typical bullet mold consists of a two-section block of aluminum, in the center of which is a cavity, of a specific caliber and weight. Molten lead is poured into the top of this mold and in a matter of moments solidifies into the shape of the mold. The actual process of using a bullet mold is a bit more sophisticated, but this is the basic technique. You should also know that multicavity bullet molds are available in a wide range of calibers. These molds will enable you to make a large number of cast bullets quickly.

When bullets are being cast, some means of heating the lead is required. A number of electric melting pots have been designed to make this a simple matter, and in time, if you do a lot of casting, you will probably want to invest in such a unit. The typical version will have a large capacity—around twenty pounds—and a thermostatically controlled heat source. Newer versions are capable of dispensing small amounts of molten lead from the bottom of the furnace. As you can imagine, this makes for quick and efficient casting of bullets. A good lead furnace with these qualities costs around $180.

**A single-cavity bullet mold will enable you to cast your own lead bullets quickly and at a very low cost.**

Moving down the list of efficient bullet-casting furnaces, we find the simple cast-iron pot and dipper. Chances are very good, unless you are really serious about casting bullets, that this will be the type of lead melting setup you will use at first. This type of melting pot is designed to be used on your heat source, most commonly the kitchen range. Kits containing a melting pot, dipper, mold, and special handles are currently available for around $60.

If economy if really your game, you can melt lead in any sturdy pot, although cast iron is usually the favorite. I have had years of success with a discarded kitchen pot and cover. It is generally agreed that an aluminum pot is unsuitable for a number of reasons. The first and probably most important is that aluminum has a melting point not much more than the temperature of molten lead, and there have been instances where the bottom of an aluminum pot has melted, spilling the molten lead. Another good reason for not using an aluminum pot is that even if it does not melt and fail, some of the aluminum will mix with the molten lead and create an undesirable lead aluminum alloy.

Some means of dipping and dispensing the molten lead must also be used. I have found the best tool to be a dipper designed for the purpose. These are available from most shops that sell reloading supplies.

To review the basic casting equipment for the home reloader quickly, all that is really required is a bullet mold (in the caliber and bullet design that suits your needs), a melting pot, and a dipper for pouring the molten lead.

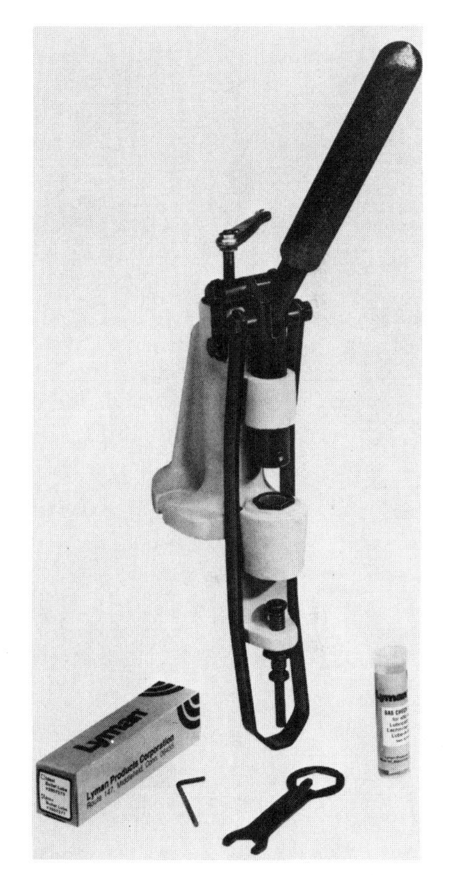

**A bullet sizer/lubricator is necessary for accurate cast bullets. In one stroke this machine will lubricate and size your cast bullets. It can also be used for attaching gas check to base of lead bullets. (*Photograph courtesy of Lyman Company.*)**

## BULLET-SIZING EQUIPMENT

One of the most important pieces of related bullet casting equipment, from the standpoint of making consistent bullets, is a special device called a bullet sizer/lubricator. I highly recommend that you pick up such a piece of equipment.

Basically, a bullet sizer/lubricator forces a cast bullet through a special die that makes the bullet uniform in diameter. This does a lot to improve the ballistics of each round you fire. Sizing dies are available in all calibers and are fairly easy to install in the machine.

In addition to sizing, this machine also performs another important function, that of lubricating each bullet. A lead cast bullet will shed lead inside of a rifle or pistol barrel unless it is lubricated. A bullet sizer/lubricator packs special lubrication into the grooves around the base of the bullet and virtually eliminates lead fouling in the firearm bore.

The last function of a sizer/lubricator is a means of installing or attaching gas checks to the back of each cast bullet. Gas checks are simply copper or brass cups attached to the bottom of a lead-cast bullet during the sizing operation. A gas check prevents hot high-pressure gases from deforming or melting the base of the lead bullet and thus helps the cast bullet to remain fairly uniform. Accuracy is better overall.

Now let's cover the mechanics of casting lead bullets.

## CASTING LEAD BULLETS

Begin by finding an area where you can work undisturbed and safely. You will need a heat source, at least ten pounds of lead, a melting pot (cast iron is preferred), bullet mold, lead dipper (once again, the cast-iron variety is preferred), heavy leather gloves, clear face shield or other eye protection, and a 12-inch-long wooden dowel (about 1 inch in diameter). In addition, I like to use an old cookie sheet to work over. This way, as I remove each cast bullet from the mold, it can be dropped onto the cookie sheet rather than on the counter.

Place a few pounds of lead into the melting pot and place the pot on the heat source. Covering the pot will help the lead to melt more quickly. The liquidus temperature for lead is around 700 degrees Fahrenheit, so you must turn the heat source up fairly high initially. Once the lead has melted, you can usually lower the temperature and still keep the lead in a molten state. Keep in mind that the less lead in the pot, the higher the temperature of the molten lead and the greater the chance of oxidation. For this reason it is necessary to add a few "cold" pieces of lead every five to ten minutes. This helps keep the overall temperature of the molten lead constant.

Once the lead has reached the molten stage, skim the dross off the top. Dross appears as a dull gray surface covering and can be removed easily with a large spoon. As the dross is removed, the molten lead will appear as a bright, silvery liquid.

When I cast lead bullets, I like to have the molten lead to my right front while I hold the mold block in my left hand. I dip some molten lead with my right hand (gloves on) and pour into the top of the mold. The dipper is then placed back into the melting pot. Next I take the wooden dowel and tap the sprue cutter—the metal strip on the top of the molding block through which lead is poured into the molt cavity—so that the mold can be opened. The sprue cutter should be rotated horizontally about 180 degrees. Then I give a tap to the side of the mold and the newly cast bullet drops out. Once the bullet drops onto the cookie sheet I close the mold and tap the sprue cutter back into position with the wooden dowel. Then I am ready to pour more lead and repeat the process.

Casting lead bullets in this manner, you will find that a certain rhythm develops as you cast five to ten new bullets a minute with a single-cavity mold. You will quickly see the appeal of a multicavity mold where four bullets are produced each time.

Some pointers may be valuable at this time. You will save time if you pour just enough molten lead into the cavity for each round. Excess lead is useless as it is simply trimmed off with the sprue cutter or runs down the side of the mold. This trimmed lead is commonly put back into the melting pot periodically.

Make certain that you close the mold tightly each time, or the cast bullets will not be clearly defined. Tap the sprue cutter back into position with the wooden dowel while at the same time closing the handles of the mold.

Adjust the temperature of the heat source if you begin to notice that the molten lead is too hot or not hot enough. A special thermometer is a good investment for keeping the molten lead at the perfect casting temperature, below 750 degrees Fahrenheit. If the temperature of the molten lead rises much above this, the lead will oxidize and the

**When removing the dross, wear heavy leather gloves and eye protection. You should also wear a heavy shirt and pants to protect your body from accidental spills. Ventilation is important, as arsenic is commonly used in the production of lead and is given off as a gas when the lead is melted. Exercise due care when working with molten lead: remember, it is at a temperature of around 700 degrees Fahrenheit.**

overall quality of the cast bullets will be poor. As a rule, when the cast bullets appear grainy—coarse textured rather than smooth—the overall temperature of the lead is too high. If the texture of the molten lead is chunky and does not pour well, the temperature of the molten lead is too low. Make adjustments accordingly throughout the casting, but keep in mind that adjustments do not produce instant results.

When you are finished casting bullets, pour the extra molten lead into special ingot molds. As an alternative to this you can pour the molten lead into an old cupcake tin. Next time you cast bullets you will have small blocks of clean lead to work with.

After casting, the next step—once the bullets have cooled sufficiently—is to size and lubricate the new bullets. While it is entirely possible to use freshly cast bullets as they come out of the mold, you will not be entirely satisfied with their accuracy—or with the lead buildup they cause in the firearm. Lubrication and sizing will generally make better bullets.

## LUBRICATION AND SIZING OF CAST BULLETS

After lead bullets have been cast they must next be lubricated and, for the utmost in accuracy, sized to the proper diameter. It is possible to lubricate cast bullets by hand, but the work will go much quicker and be better accomplished on a special bullet sizer/lubricator machine.

A sizer/lubricator machine forces a cast bullet into a cylindrical die of a specific diameter to make it uniform. This action increases the accuracy of the bullet in flight. Inside the die are several holes through which pass a special lubricant. The lubricant, most commonly a preformed stick of Alox, is held inside the machine under pressure. To maintain consistent pressure on the lubricant, a lever or ratchet-type handle is pulled periodically. As the cast bullet is run into the sizing die, and as the grooves on the side of the bullet pass by the holes in the die, lubricant is forced into them. The Alox lubricant is similar in consistency to axle grease and will not affect the reloading powder.

The sizing die, which must be matched to the caliber of bullet you are sizing, consists of two parts: top punch and base punch. The top punch may be almost any style but should be closely matched to the nose design of the bullet being sized. The base punch is nothing more than a cylinder made of hard steel with dimensions exactly that of the bullet being sized.

If you do not own a cast-bullet sizer/lubricator machine, you can lubricate your cast bullets by hand. While you will reduce leading in the barrel by lubricating, you will still sacrifice accuracy by not sizing. There are two ways of lubricating by hand. The first and simplest is to buy a can of spray lubricant, available at most reloading supply shops, and give your cast bullets a thorough coating. If you do this, first arrange the bullets in a manner that will allow the most efficient use of the spray and coat all sides of the cast bullets. Allow to dry. The cast bullets with sprayed lubricant are now ready for reloading.

The second way of lubricating bullets by hand takes more time, but some think that it is a much more efficient way to coat the bullets with lubricant. Begin by melting a stick of Alox bullet lubricant over moderate heat in an old pan or metal pie plate. Next, stand a batch of cast bullets in the pan so that the melted Alox level is up to the top groove on all of the bullets. Let the pan and cast bullets cool until the lubricant has solidified fully, and then cut each of the bullets out of the lubricant.

The easiest way to get all of the bullets out of the solid lubricant is to use a suitable size cartridge case from which the base has been removed. This gadget is called a "cake cutter," and though it is commercially available, the homemade version works well. As the cutter is pressed down over each bullet, the bullet is separated from the lubricant and the bullet forced up through the top of the casing.

Whichever means you use to lubricate your cast bullets, strive to coat only the grooves in the bullet and not the nose or base. When sizing in a machine, remember that the bullet nose is upwards. Match the nose punch to the nose of the bullet to prevent deformity of the nose.

The reason for striving to lubricate only the grooves of bullets and not the nose or base, is to make handling the lubed bullets easier. Alox is rather sticky and ammunition which is coated with this material may malfunction. A

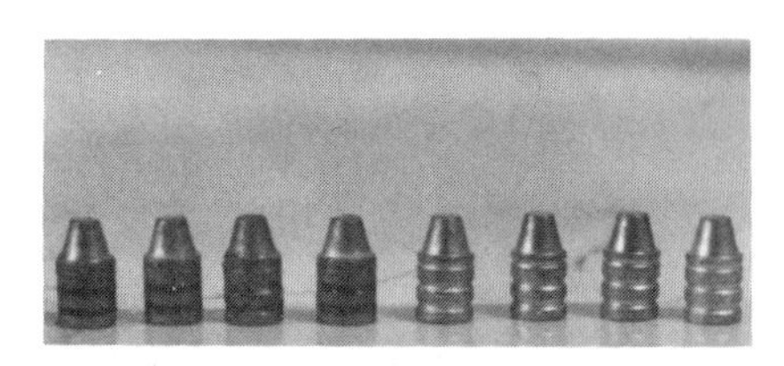

**The four lubed and sized cast bullets on the left will shoot much more accurately than the four cast bullets on the right.**

small amount of Alox on the base of the bullet will have no affect on the powder inside the cartridge.

Another way to make bullets from lead involves the use of special lead wire. The wire is fed through a special die that forces it into the required diameter, then cuts the bullet off at the required length. This way of making bullets from lead, called swaging, is widely used by commercial bullet manufacturers. After swaging, the bullets are commonly sized and lubricated as described above. Bullet swaging equipment and lead wire are available from reloading equipment suppliers. As a rule, swaged bullets perform better than cast bullets, because the wire used in the swaging, subjected to extreme pressure, tends to produce a more uniform lead bullet.

Cast or swaged lead bullets can be used in either rifle or handgun, but serious shooting often requires that the lead bullets be jacketed. For this, you will need special dies for your metallic cartridge reloading press. It is also helpful if your reloading press is a heavier version of the standard O-frame design, such as the RCBS Big Max,™ rather than the lightweight C-type design.

Since the process of making jacketed bullets at home is lengthy and involved, you must turn to other sources for more information. Suffice it to say that while it is possible to make accurate jacketed bullets at home, the equipment is expensive and much time is involved. Unless you are planning practically to go into the business of making bullets, the price of the equipment may be hard to justify.

## MILITARY BRASS CASINGS

Sooner or later you will discover that surplus military brass is available for a very low cost, and then you will probably want to try your hand at reloading this in addition to using commercially made brass casings. Currently available are military casings for .30-06, .308, .223, .38 special, and .45 automatic. A listing in a current magazine offers such military brass for about $25 per 1,000. Unprimed brass from almost any manufacturer, Remington, for example, costs about $80 dollars per 1,000.

Although you can realize substantial savings by using military brass casings for reloading, more work is involved in preparing the empty casings before they can be reloaded successfully. Since military ammunition is produced with Berdan primers, the procedure for removing the spent primer is different than that used for casings from commercial sources. The Berdan primer, as mentioned in Chapter 2, does not have an internal anvil as Boxer primers do, but instead relies on a small protrusion inside the bottom of the primer cup. A conventional decapping die cannot be used for removing Berdan primers; a special Berdan primer removal tool must be used. This tool is available from reloading supply shops.

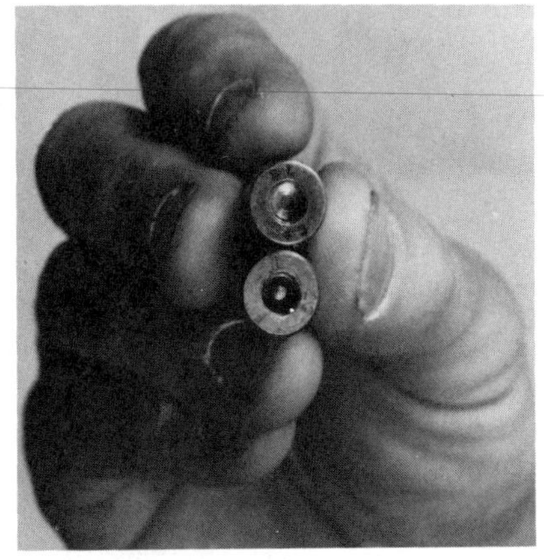

**The military casing on left uses a Berdan primer while the casing on the right uses a standard Boxer-type primer. Note anvil in military casing.**

An alternative method of removing Berdan primers involves using a close-fitting punch. First fill the military casing with water and start the punch into the mouth of the casing. A sharp blow to the punch with a hammer will then cause the Berdan primer, as well as the water, to be forced out. Once the primer has been forced out, you will notice a tiny bump in the bottom center of the primer pocket and two flash holes off to the side of this.

Both methods of Berdan primer removal are time-consuming but effective. After the primer has been removed, the next step is to remove the primer crimp around the edge of the primer pocket. If this is not done, new primers will not fit into the primer pocket but will instead be crushed. This is most easily accomplished with a pocket knife or other suitable tool. Remember that your intention is simply to remove the crimp and not any more of the brass than necessary. Primer pocket swager dies are available that will remove the crimp quickly and easily. This die is used in a conventional reloading press.

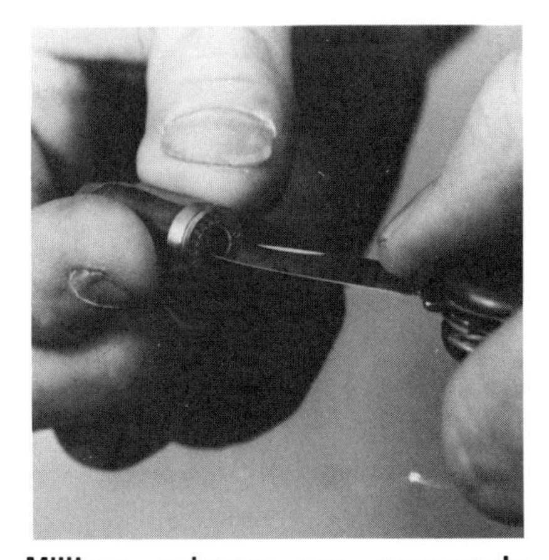

**Military primers are commonly crimped into place. This crimp must be removed with a pocket knife (or special tool) before a new primer can be installed.**

## ANNEALING

Each time a cartridge is fired, the brass casing expands to the chamber walls, preventing the rearward escape of pressurized gas and allowing the bullet to exit down the barrel. In about 0.001 second, the case shrinks back to almost normal dimensions, which makes extraction possible. In addition to lengthening the neck of the casing, repeated firings, under normal pressure and heat, tend to work-harden the case neck much faster than the thicker parts of the casing. The best way to restore the normal ductileness of the brass is to anneal the neck of the casing. There are two ways of neck-annealing a brass casing: torch method and immersion method.

The torch method of brass-casing annealing involves standing an empty casing in a pan of water so that the water level is just below where a bullet would be seated. For bottleneck cases, the level of the water should be just below the shoulder. Next, using a hand-held propane torch, heat the neck of the casing up to a dull red color. Just as this temperature is reached—watch carefully—tip the casing over. This will immediately quench the entire case, soften the brass in the heated area, and increase its ductibility. The end result is a longer case life.

The other way of annealing the case neck, a method said to produce more uniformity, is the immersion method. Basically this method involves immersing the neck of empty and decapped casings in molten lead. Primers should be removed, or trapped air inside the casing will prevent molten lead from rising inside to the proper level. It is also a good idea to clean and dip the neck of the casings in light oil before immersion. The oil will prevent the molten lead from adhering to the case. Hold the casing by the base of the cartridge with a pair of pliers and immerse the neck for about five seconds in the molten lead. Then remove the casing, shake to remove any molten lead, and immediately drop into a pail of cool water to quench. Straight-walled casings should be annealed to a point just below where a bullet would normally be seated. Bottlenecked cartridges should be annealed to just past the shoulder.

## CASE-LENGTH TRIMMING

As you already know, each time a round is fired, the casing becomes slightly longer in the neck area. After repeated firings, the case lengths may have elongated to the point where chambering will be difficult or impossible. To correct this problem and restore the case to the proper length, it should be trimmed. There are a number of special tools for this, but the easiest to use is a case-length trimmer.

Check the casing with a dial caliper or case-length gauge to determine just how much must be removed from the case neck. Then set up the case-length trimmer and remove brass from the neck area until the case is the standard length. Work in stages, removing a bit of lead, checking length, and removing more lead until the proper length is achieved. Once this is accomplished, most case-length trimmers can be locked in position so that case trimming can progress quickly and efficiently. It will be necessary to chamfer (incise a groove: see Glossary) the inside and outside of the neck of the casing. A pocket knife or special deburring tool will make short work of this operation.

Another way of reducing cartridge case length is to purchase a special trim die from one of the reloading equipment manufacturers. A trim die screws into the top of the standard reloading press, and a casing is run up in the usual manner. The part of the casing that sticks up past the top of the die is then filed off with a flat file. If more than one-eighth inch protrudes above the top of the die, use a hacksaw and then a flat file. After this operation, clean the inside and outside of the case neck with a deburring tool or pocket knife to remove the burrs.

Of the two methods described above for case length reduction, the one using the case-length trimmer tool is the most efficient but also the most expensive. The trim-die method is less costly and does almost as good of a job,

**When using a special trim die, the excess cartridge is taken off with a hacksaw. Next remove the burr around the case mouth.**

but one trim die is necessary for each caliber or style of casing. Whichever method is used, check your progress often with a dial caliper or case-length gauge.

### TIME-SAVING SHORTCUTS FOR THE RELOADER

The time you spend reloading is time well spent, for you will realize considerable savings in ammunition costs over the years. Nevertheless, there are a number of things you can do to speed up your output of finished ammunition and in the long run become much more efficient. A number of time-saving ideas based on my over twenty years' reloading experience follow.

- Probably the greatest time-saver for the home reloader is to run all or most of your casings through one operation before going on to the next step, assembly-line style. For example, if you reload .38 special or .357 magnum ammunition, decap and resize all of your casings at one sitting. Do not attempt to accomplish all reloading steps in one reloading session unless, of course, this is necessary. Then, the next time you sit down at the reloading bench, prime all casings. The following time, flare the case mouths. The next time you sit down to reload you can charge the cases with powder and seat and crimp the bullets. This approach will help you to make ammunition move quickly in the long run.
- Another time-saver is to try always to keep a batch of sized, primed casings handy. That way, when you require ammunition, all that is needed is to charge the cases with powder, seat and crimp the bullets in place. Store primed casings in a sealed container to prevent moisture damage to the primer and also to keep the casings free from dirt or insects. I know one fellow who discovered, quite to his surprise, that a spider had spun webs in the bottom of a batch of .45 automatic casings stored in an open container.
- You will save time and do a better job of priming casings if you use a special priming tool. Although empty

casings can be easily primed on a standard reloading press, this work goes much more quickly when a special tool is used for the task. At the very least, an automatic priming tube will speed up priming on a typical C- or O-type reloading press.

- Store empty shotshell casings in a covered cardboard box in a warm, dry area. This is particularly important for the older paper casings, which have a tendency to pick up humidity over time. Plastic hulls, while not affected by moisture unless a fiber base wad is present, will collect dust, which tends to scratch the casings during reloading.
- Always use a reloading block during powder-charging of cases. Not only does a reloading block prevent accidental spills of powder and bullets, its use will also make moving a quantity of bullets possible.
- Consider purchasing a powder-dispensing tool to speed up the charging of cases with powder. There are several types on the market, some bench-mounted, others hand-held. Be sure to check the accuracy of the charge being dispensed on an accurate reloading scale before and during use.
- Use a special funnel when charging cases with powder. The amount of powder you don't spill as a result of using a powder funnel will mean hundreds of extra rounds over the years.
- Keep accurate records of the ammunition you reload. You should include all of the obvious information, such as primer, powder, bullet, date, velocity, chamber pressure, and—later—comments about how this particular ammunition performed. Compare your notes periodically, and you will quickly learn which loads work best in a given firearm.
- One last way to save valuable time and possible pain is always to observe the safety rules given earlier in this book. Ammunition reloading is not particularly complicated, but it is certainly something that should never be approached casually. By always working under controlled conditions, you will greatly reduce the chances of an unfortunate mishap. In the end, you will be able to enjoy more shooting at a lower overall cost.

# Glossary

**Annealing** the heating of a brass cartridge neck, then rapid cooling. This is done to restore ductility to the neck area.

**Anvil** a fixed metallic protrusion inside the primer pocket, against which the priming mix is crushed to initiate ignition. The Berdan primer pocket is a good example.

**Ball** term used to describe projectile in cartridge, especially in U.S. military. Derived from flintlock and percussion firearms (cap and ball).

**Ball powder** the copyrighted tradename for a double base smokeless propellant powder developed by Olin Industries.

**Ballistics** the science of projectiles in motion from ignition to impact.

**Battery cup** a type of primer where anvil and primer mix are all supported in the primer cup. Shotshell primers are of this type.

**Bell** to expand the rim of a cartridge case slightly and make seating of bullet easier.

**Belted cartridge** a cartridge with a band of brass around the base, above the extractor groove. The belt acts to reinforce casing at this point, and headspaces the cartridge. Most commonly magnum calibers.

**Berdan** type of primer developed by an American inventor, Col. Hiram Berdan. Primer cup contains only primer mix; anvil is part of primer pocket.

**Benchrest** a solid shooting platform or table, used for accuracy testing for firearms. Also, a competitive shooting sport.

**Boattail (BT)** bullet design where base is tapered to reduce friction in bore.

**Bore** the inside of a firearm barrel.

**Boxer** type of primer where all parts are present—anvil, primer mix, and primer cup. Commonest primer in United States as yet, invented by an Englishman, Colonel Boxer.

**Brass** common term for metallic casings.

**Bullet** projectile released when a cartridge is fired.

**Caliber** the diameter of a projectile or inside diameter of a bore in a firearm. Expressed in English in decimals of an inch; sometimes in metric. Often used in conjunction with other meanings, such as year of introduction, original designer, and case length, e.g., 30 '06, 380 Kurs, and 8mm/06, respectively.

**Cannelure** grooves around the diameter of a projectile that are packed with lubrication or into which the cartridge case neck is crimped.

**Canister powder** any propellant sold over-the-counter for the purposes of reloading ammunition on a noncommercial level. All of the powders described in this book are "canister" powders.

**Cartridge** a complete piece of ammunition, most commonly for rifle or pistol—primer, powder, casing, and bullet.

**Case** the container that holds all components in a piece of ammunition. Other names include brass, empty, casing, and, in shotshells, hull and shell. Can be made from paper, plastic, brass, steel, or aluminum.

**Case trimming** to shorten a case by removal of material from the mouth.

**Cast bullet** bullets for rifle or pistol that have been cast from molten lead or lead alloy.

**Center fire** the terms used to describe the priming system in ammunition where the primer is centrally located in the base or head. Most center-fire ammunition can be reloaded; RIM FIRE ammunition generally cannot be reloaded at home.

**Chamber** the rearward portion of a bore into which a piece of ammunition is detonated. The chamber centers and supports the cartridge.

**Chamfer** to remove by bevel-cutting the inside of a cartridge casing. This is done to make seating a new bullet easier to accomplish.

**Charge** to add powder to a primed casing; also, the amount of powder used for this purpose.

**Choke** a construction at the muzzle end of a shotgun bore, designed to control or reduce the spreading of shot charge.

**Chronograph** an electronic instrument used to measure bullet velocity.

**Compressed charge** a casing charged with powder to the point that when a bullet is seated, the powder is compressed.

**Components** the ingredients used in bullet- and shotshell-making.

**Core** the inside of a jacketed bullet, most commonly lead or lead alloy though other materials may have been used.

**Cordite** stringy-type smokeless powder little used in the United States but popular in Great Britain.

**Corrosion** the deterioration of a bore due to chemical action of black powder, also rust.

**Corrosive primer** primers whose burnt residue is hygroscopic (attracts moisture) and that cause deterioration of bore and other metal parts unless removed quickly. All primers produced in the United States are noncorrosive.

**Crimp** in shotshells, the closing of the top of the casing after all components have been loaded; in metallic cartridges, the bending in of the case mouth to hold the bullet in place.

**Crimped primer** the crimping of the area around the primer pocket after the primer has been seated. Almost all military ammunition, which is designed for use in automatic weapons, has crimped primer. Crimp must be removed before a new primer can be seated.

**Deburr** removal of burrs around the mouth of a cartridge case to make seating of a bullet easier.

**Decap** or **Deprime** to push out a spent primer from a fired casing (metallic or shotshell).

**Die** a special tool used to reform brass casings, to flare case mouths, seat bullets, and so forth.

**Double-base powder** a smokeless powder made with nitroglycerine and nitrocellulose base.

**Drop** the distance a bullet drops during flight as a result of the force of gravity.

**Erosion** wearing away of a firearms bore, over time, as a result of the friction from bullet, shot, or hot gases.

**Expander (ball)** the rod inside the expander die that flares a brass case mouth slightly to ease bullet insertion and seating.

**FPS** feet per second.

**Fireforming** the shaping of a cartridge case by firing inside a given chamber; has an effect similar to that of sizing a casing but is more precise.

**Firing pin** that part of a firearms mechanism that strikes a primer of a cartridge.

**Flash hole(s)** a hole (or sometimes holes) inside a primer pocket through which the spark from a detonated primer jumps to ignite the powder inside the cartridge.

**Foot pound** a unit of energy equal to the effort required to raise one pound to a height of one foot against the normal force of gravity.

**Freebore** the distance a bullet travels after ignition before it contacts the rifling of the barrel. Sometimes there is no free-bore.

**FMJ** full-metal jacket. See METAL CASE.

**Gas check** a copper or brass cup attached to the base of lead bullet that prevents rearward escape of gases and deterioration of the bullet upon firing.

**Gilding metal** copper alloy metal used for bullet jackets.

**Grain** a weight measure used in reloading. The term may sometimes be used to describe individual particles of powder, but "28 grains of powder" refers to a weight rather than a count of particles or kernels.

**Grand slam** an award in hunting for collecting 4 different North American varieties of wild sheep.

**Grease groove** the lubricating groove or grooves around the circumference of a bullet.

**Group** a pattern of bullet holes in a target as a result of aiming at one point, with one sight-setting. The distance is commonly measured from the centers of the holes farthest apart.

**Handloading** the making of ammunition for small arms on a noncommercial level. While a machine is commonly used, all operations are accomplished by hand.

**Hangfire** the delayed ignition of a cartridge after the primer has been struck by the firing pin.

**Headspace** the distance from the bolt to the chamber or barrel that positions a cartridge and prevents rearward movement of the cartridge. This is probably the most important factor governing safety in shooting.

**Hollow point (HP)** bullet design with a hole present in the forwardmost portion of the bullet.

**Ignition** the spark produced as a result of striking a primer with a firing pin, which results in the firing or lighting of the propellant inside the cartridge case.

**IMR** an abbreviation for "Improved Military Rifle". Also the name given by the Du Pont Company to its line of single-base rifle powders.



**Jacket** the covering of a lead bullet with GILDING METAL; the exterior skin of a bullet.

**Keyhole** the hole design of bullet as it passes through target paper when the bullet is not level in flight.

**Lands** the spiral, raised portion of a bore that remains after the grooves have been cut. RIFLING is composed of lands and grooves.

**Leading** the deposit of lead inside a bore as a result of shooting lead bullets; undesirable lead fouling.

**Loading block** a flat piece of wood or plastic into which half-depth holes have been drilled or formed to hold cartridge cases during the various stages of reloading.

**Lubricant** basically, there are two types: sizing lubricant used to reduce friction of a brass casing during resizing operations, and bullet lubricant, which is used to reduce or eliminate leading of bore when shooting lead bullets.

**Lubricator/sizer** a machine used for lubricating and sizing lead bullets.

**Magnum** generally, a higher-powered cartridge.

**Mercuric primer** a priming mix containing mercury, not generally available or used in handloading.

**Metal case (MC);** also, **full-metal jacket (FMJ)** a bullet that is completely encased in GILDING METAL. Standard military ammunition bullet.

**Midrange trajectory** commonly, the highest point of vertical distance a bullet will travel during straight flight.

**Misfire** a failure of a cartridge to ignite after the primer has been struck by the firing pin.

**Mould** (sometimes mold blocks) generally, two pieces of metal inside which is a cavity the shape of a bullet. Used for casting lead bullets.

**Mushroom** the ability of a bullet to expand upon impact.

**Muzzle** the forwardmost portion of a barrel.

**Muzzle blast** the blast of hot gases at the end of a barrel.

**Muzzle energy** the amount of energy (in FOOT POUNDS or PSI) a bullet possesses as it passes out of the end of a barrel.

**Muzzle velocity (MV)** the speed of a bullet as it exits a barrel.

**Neck** that portion of a cartridge case that is crimped around a bullet in a straight-walled case design. In a bottleneck design casing, that portion in front of the shoulder.

**Neck expansion** the flaring of the neck of a cartridge case—in a special die—to ease bullet insertion.

**Neck reaming** the act of reducing the thickness of a cartridge-case neck with a special tool from inside.

**Neck size** to resize the neck portion of a cartridge case, instead of resizing the entire cartridge case.

**Noncorrosive** the standard type of primers and cartridges used in this country. There is generally no deterioration of bore or other metal parts as a result of using this type of ammunition.

**Nonmercuric** a priming mix that contains no mercury.

**NRA** abbreviation for the National Rifle Association.

**NRMA** abbreviation for the National Reloading Manufacturers Association.

**NSSF** abbreviation for the National Shooting Sports Foundation.

**Ogive** part of a bullet design, the curved portion of the cylinder.

**Oil dent** a dent in a cartridge casing that is the result of using too much resizing lubricant on the casing.

**Partition** a bullet design including a solid portion inside a jacketed bullet that helps to improve bullet weight upon impact.

**Patched** a word used to describe a lead ball or projectile covered with a small piece of cloth during loading of a muzzle loaded firearm. The cloth or "patch" helps to seal gases from powder. It is also used to swab out the bore of a firearm during cleaning.

**Pattern** the design of shot in a paper target inside a 30-inch circle at 40 yards.

**Plinking** nonformal target shooting.

**Point of aim** point at which a firearm's sights are placed prior to shooting.

**Primer pocket** the cavity in a cartridge base into which is seated a primer.

**Primer punch** a rod inside a resizing die that forces a spent primer from the primer pocket.

**Pressure** the amount of energy produced inside a chamber when a cartridge is ignited, expressed in pounds per square inch (PSI) and/or copper units of pressure (CUP), which are roughly the same.

**Primer indent** the depression made in the face of a primer after it has been struck by a firing pin.

**Progressive powders** multibase powders (various mixtures of nitrocellulose, nitroglycerine, and organic nitrates) with a slow burn rate that results in slow chamber pressure buildup.

Progressive powders are commonly used for rifle ammunition. Compare SINGLE-BASE POWDER and DOUBLE-BASE POWDER.

**Projectile** a bullet after it has been started on its way down the barrel.

**Range** a place to shoot; also, the horizontal distance of travel of a projectile.

**Recoil** the rearward movement of a firearm after a primer has been struck by a firing pin, sometimes referred to as "kick."

**Remaining velocity** the speed (in FPS) of a projectile at any given point in its path of travel.

**Rifling** the spiral grooves inside a barrel—composed of LANDS and grooves.

**Rim** the rearward edge or rim of a cartridge case that enables the cartridge to be extracted from the chamber.

**Rimfire cartridge (RF)** cartridge that contains priming mix inside case rim.

**Round** common military term for a bullet or a piece of small arms ammunition.

**Round nose (RN)** bullet design.

**SAAMI** abbreviation for the Small Arms and Ammunition Manufacturers Institute.

**Seating depth** the depth to which a bullet base is seated below the case mouth.

**Sectional density** a bullet's weight, in pounds, divided by the square of its diameter, in inches.

**Shell holder** that part of a reloading press that holds a casing base during the various stages of reloading.

**Shock** the transfer of a bullet's energy upon impact.

**Shot** lead balls in specific diameters used in shotshells.

**Shoulder** the sloped or curved portion of a cartridge case between the body and neck.

**Single-base powder** a powder made from a nitrocellulose base, as compared to a DOUBLE-BASE POWDER. Single-base powders are commonly used for handgun ammunition.

**Sizing;** also **Resizing** the forcing of an empty cartridge casing into its original dimensions. This is done inside a sizing die during first stage of the reloading process.

**Slug** in shotshells, the projectile or bullet.

**Soft point** a bullet with a lead nose.

**Spire point (SPT)** bullet design with the shank narrowing sharply to a pointed nose.

**Spitzer (SP)** bullet design with the shank rounding smoothly to a pointed nose.
**Swage** to form by forcing through or into a special die.

**Throat** the area of the bore immediately in front of the chamber.
**Trajectory** the path of a projectile in flight.
**Twist** angle of rifling in bore, commonly measured in length of one complete revolution.

**Velocity** the speed of a projectile in flight at any given distance, commonly measured in feet per second (FPS).
**Vernier caliper** instrument used for measuring in reloading.

**Wad** a disc or cylinder made from paper, plastic, felt, or other material used most commonly in shotshell reloading. Placed between powder and shot.
**Wad column** the makeup of wads in a shotshell.
**Wadcutter** a sharp-edged lead bullet designed to punch a clean hole in paper targets.
**Web** that part of a cartridge case between the bottom of a primer pocket and interior or cartridge case. One or more holes pass through the web to permit flash from primer to ignite powder.
**Wildcat** a cartridge design not generally available commercially.
**Windage** left or right sight-correction used to adjust for wind forces.
**Work-hardening** the result of repeated stresses on a cartridge case.
**Working-up** the development of a cartridge load by gradually increasing the amount of powder until maximum velocity is attained.

**Zeroing** adjusting the sights of a firearm so that point of aim equals point of impact.

# RELOADING TABLES

# Important Notes on Reloading Data

The information contained in the following reloading tables has been obtained from The Du Pont Company, Hercules Incorporated, Sierra Bullets, Federal Cartridge Corporation, Omark Industries (CCI/RCBS), Nosler Bullets, The Lyman Company, and Remington Arms Company. I am grateful for their cooperation.

For each load—rifle, pistol, and shotshell—you will learn the source of the information in the last column. Sources are indicated with abbreviations listed on page 91.

## A NOTE FROM THE DU PONT COMPANY

> "The information presented is based upon results obtained in the Du Pont ballistics laboratory, and is offered without charge as an aid to handloaders, to be employed at their own discretion and risk. Safe reloading practices should be observed at all times. Since Du Pont has no control over the circumstances of loading, they assume no liability for the results obtained, and they guarantee only that their powder meets their manufacturing standards."

Data for reloading from Du Pont Company came from the "Du Pont Handloader's Guide" (E 53588—5/83) and is current at the time of this writing. New copies of this guide are available from:

Du Pont Company
Petrochemicals Dept.
Explosives Products Division
Wilmington, Delaware 19898

## A NOTE FROM THE FEDERAL CARTRIDGE CORPORATION

The material identified by the abbreviation **FED** in the source column is reprinted from the Federal Cartridge "Shotshell Loading Data" booklet, last updated in April 1983. New copies of this publication are available from:

Federal Cartridge Corporation
2700 Foshay Tower
Minneapolis, Minnesota 55402

## A NOTE ABOUT HERCULES DATA

Data contained in these tables were obtained from Hercules Incorporated and are identified with the abbreviation **HER** in the source column of the tables. The information is current at the time of writing and came from the "Reloaders' Guide for Hercules Smokeless Powders," published in March 1983. Current copies of this publication are available from:

Hercules Incorporated
Hercules Plaza
Wilmington, DE 19899

## OTHER ACKNOWLEDGMENTS

Information identified in the source column with the abbreviation **SR** was obtained from the Sierra Bullet publication *Sierra Bullets, Reloading Manual*, Second Edition, Copyright 1978 and 1985.

Information identified in the source column with the abbreviation **NO** was obtained from the *Nosler Reloading Manual*, Number Two, Copyright 1981.

Information identified in the source column with the abbreviation **SP** was obtained from the *Speer Reloading Manual*, Number Ten, Copyright 1981.

## ABBREVIATIONS USED IN RELOADING DATA TABLES

**HER**—Hercules Incorporated
**DUP**—Du Pont
**SR**—Sierra Bullet
**REM**—Remington Arms Company
**WIN**—Winchester
**CCI**—CCI/RCBS/SPEER
**FED**—Federal Cartridge Corporation
**SP**—Speer Bullet
**NO**—Nosler Bullet

## ABBREVIATIONS USED FOR POWDER TYPE

**HER**—Hercules Powder
**DUP**—Du Pont Powder
**OLN**—Olin
**RDT**—Hercules Red Dot
**UNIQ**—Hercules Unique
**HERC**—Hercules Herco
**GDT**—Hercules Green Dot
**BLDT**—Hercules Blue Dot
**2400**—Hercules 2400
**BEYE**—Hercules Bullseye
**RE7**—Hercules Reloader 7
**PB**—Du Pont PB
**700X**—Du Pont Hi-Skor 700-X
**800X**—Du Pont Hi-Skor 800-X
**4756**—Du Pont IMR 4756
**4227**—Du Pont IMR 4227
**7625**—Du Pont IMR 7625
**4198**—Du Pont IMR 4198
**4064**—Du Pont IMR 4064
**3031**—Du Pont IMR 3031
**4831**—Du Pont IMR 4831
**4759**—Du Pont SR 4759
**4895**—Du Pont IMR 4895
**4320**—Du Pont IMR 4320
**4350**—Du Pont 4350
**680**—Olin 680
**296**—Olin 296
**630**—Olin 630
**7828**—Du Pont IMR 7828

## BULLET ABBREVIATIONS

**SR**—Sierra bullet
**Horn**—Hornet
**HPBT**—Hollow-Point Boattail
**SP**—Spitzer
**BL**—Blitz
**RNC**—Round Nose Carbine
**HPFN**—Hollow-Point Flat-Nose
**JHP**—Jacketed Hollow Point
**FN**—Flat Nose
**HP**—Hollow Point
**JHC**—Jacketed Hollow Cavity
**LWC—Lead Wadcutter**
**LRN**—Lead Round-Nose
**FMJS**—Full Metal Jacket Silhouette
**JSP**—Jacketed Soft Point
**RN**—Round Nose

**SSP**—Semi-Spitzer
**PL**—Plinker Bullet—Speer Bullet
**VAR**—Varminter Bullet—Speer Bullet
**BBWC**—Bevel Base Wadcutter
**SEMP**—Semi-Pointed
**SPZ**—Spitzer
**SBT**—Spitzer Boattail
**SP**—Speer Bullet
**NO**—Nosler Bullet
**SSPP**—Semi-Spitzer Partition—Nosler
**HBWC**—Hollow Base Wadcutter
**SWC**—Semi-Wad Cutter
**RB**—Round Ball
**MHP**—Magnum Hollow Point
**BTSP**—Boattail Solid Base—Nosler Bullet
**RNP**—Round-Nose Partition Bullet—Nosler Bullet
**FP**—Flat-Point Bullet—Nosler Bullet
**PPP**—Partitioned Protected Point Bullet—Nosler Bullet
**GS**—Grand Slam Bullet—Speer Bullet

## USING THE RELOADING DATA TABLES

1. Find the bullet-caliber table for the ammunition you want to reload (such as .38 special).

2. Since the listings begin with the lightest-weight bullets, start at the top of the listing and move down until you find the weight bullet you will be reloading (such as 150 grain hollow-point [HP or JHP]).

3. Each load for a particular bullet weight begins with the lightest charge of powder and increases (down the column) until the maximum load is offered. If more than one powder is used (which is common), each powder is covered on one line.

4. Once you have decided on a bullet weight, you must decide next on the velocity range you would like the bullet to travel at. Look down and across the table to arrive at this velocity range, or the figure closest to it.

5. In the tables you will also find *OAL*, which is maximum overall length of a reloaded or finished piece of ammunition. Your finished round should be this length. A .38 Special, for example, is 1.550 inches.

6. As mentioned in the text, it is advisable to load the lighter ammunition (lower powder charges) and try these in the firearm before loading the maximum-charged ammunition.

# Pistol Data

LOADS FOR .22 REMINGTON JET

| BULLET | | | POWDER | | | | | |
|---|---|---|---|---|---|---|---|---|
| Wt | Mfg | Type | GRN | Mfg | Type | VEL(fps) | OAL | Source |
| 40 | SR | Horn | 6.5 | HER | Uniq | 1600 | 1.659 | SR |
| 40 | SR | Horn | 7.0 | HER | Uniq | 1700 | | SR |
| 40 | SR | Horn | 12.8 | DUP | 4227 | 1800 | 1.659 | SR |
| 40 | SR | Horn | 13.4 | DUP | 4227 | 1900 | | SR |
| 45 | SR | Horn | 6.1 | HER | Uniq | 1500 | 1.659 | SR |
| 45 | SR | Horn | 6.7 | HER | Uniq | 1600 | | SR |
| 45 | SR | Horn | 12.5 | DUP | 4227 | 1600 | | SR |
| 45 | SR | Horn | 13.0 | DUP | 4227 | 1700 | | SR |

LOADS FOR .221 FIRE BALL (XP-100)

| BULLET | | | POWDER | | | | | |
|---|---|---|---|---|---|---|---|---|
| Wt | Mfg | Type | GRN | Mfg | Type | VEL.(fps) | OAL | Source |
| 40 | SR | HP | 14.5 | HER | 2400 | 2700 | 1.830 | SR |
| 40 | SR | HP | 15.2 | HER | 2400 | 2800 | | SR |
| 40 | SR | HP | 15.9 | HER | 2400 | 2900 | | SR |
| 45 | SR | SP | 14.3 | DUP | 4227 | 2300 | 1.830 | SR |
| 45 | SR | SP | 14.8 | DUP | 4227 | 2400 | | SR |
| 45 | SR | SP | 15.3 | DUP | 4227 | 2500 | | SR |
| 45 | SR | SP | 14.3 | HER | 2400 | 2500 | | SR |
| 45 | SR | SP | 14.8 | HER | 2400 | 2600 | | SR |
| 45 | SR | SP | 15.3 | HER | 2400 | 2700 | | SR |
| 45 | SP | SP | 17.3 | HER | R-7 | 2603 | | SP |
| 45 | SP | SP | 18.3 | HER | R-7 | 2775 | | SP |
| 45 | SP | SP | 15.2 | DUP | 4224 | 2571 | | SP |
| 45 | SP | SP | 16.2 | DUP | 4227 | 2727 | | SP |
| 50 | SR | BL | 13.3 | DUP | 4227 | 2200 | | SR |
| 50 | SR | BL | 14.7 | DUP | 4227 | 2400 | | SR |
| 50 | SR | BL | 16.1 | DUP | 4227 | 2600 | | SR |
| 50 | SR | BL | 12.9 | HER | 2400 | 2400 | | SR |
| 50 | SR | BL | 14.3 | HER | 2400 | 2600 | | SR |
| 50 | SP | SP | 16.9 | HER | R-7 | 2497 | | SP |
| 50 | SP | SP | 17.9 | HER | R-7 | 2638 | | SP |
| 50 | SP | SP | 14.9 | DUP | 4227 | 2490 | | SP |
| 50 | SP | SP | 15.9 | DUP | 4227 | 2634 | | SP |
| 52 | SR | HPBT | 13.6 | DUP | 4227 | 2200 | | SR |
| 52 | SR | HPBT | 15.0 | DUP | 4227 | 2400 | | SR |
| 52 | SR | HPBT | 16.2 | DUP | 4227 | 2600 | | SR |
| 52 | SR | HPBT | 12.5 | HER | 2400 | 2300 | | SR |
| 52 | SR | HPBT | 13.2 | HER | 2400 | 2400 | | SR |
| 52 | SR | HPBT | 13.9 | HER | 2400 | 2500 | | SR |
| 52 | SP | HP | 16.7 | HER | R-7 | 2441 | | SP |
| 52 | SP | HP | 17.7 | HER | R-7 | 2606 | | SP |
| 52 | SP | HP | 14.7 | DUP | 4227 | 2404 | | SP |
| 52 | SP | HP | 15.7 | DUP | 4227 | 2585 | | SP |
| 55 | SP | SP | 16.4 | HER | R-7 | 2348 | | SP |
| 55 | SP | SP | 17.4 | HER | R-7 | 2501 | | SP |
| 55 | SP | SP | 14.5 | DUP | 4227 | 2419 | | SP |
| 55 | SP | SP | 15.4 | DUP | 4227 | 2562 | | SP |
| 70 | SP | SSP | 15.5 | HER | R-7 | 2137 | | SP |
| 70 | SP | SSP | 16.5 | HER | R-7 | 2281 | | SP |
| 70 | SP | SSP | 15.0 | DUP | 4198 | 1982 | | SP |
| 70 | SP | SSP | 16.0 | DUP | 4198 | 2255 | | SP |

## LOADS FOR .25 AUTO (.25 ACP)

| BULLET | | | POWDER | | | | | |
|---|---|---|---|---|---|---|---|---|
| Wt | Mfg | Type | GRN | Mfg | Type | VEL(fps) | OAL | Source |
| 50 | SR | FMJ | 1.0 | HER | BEYE | 600 | .900 | SR |
| 50 | SR | FMJ | 1.1 | HER | BEYE | 650 | | SR |
| 50 | SR | FMJ | 1.2 | HER | BEYE | 700 | | SR |
| 50 | SR | FMJ | 1.3 | HER | BEYE | 750 | | SR |
| 50 | SR | FMJ | 1.4 | HER | BEYE | 800 | | SR |
| 50 | SR | FMJ | 1.1 | HER | UNIQ | 600 | | SR |
| 50 | SR | FMJ | 1.3 | HER | UNIQ | 650 | | SR |
| 50 | SR | FMJ | 1.4 | HER | UNIQ | 700 | | SR |
| 50 | SR | FMJ | 1.5 | HER | UNIQ | 750 | | SR |
| 50 | SR | FMJ | 1.6 | HER | UNIQ | 800 | | SR |

LOADS FOR 7MM BR REMINGTON .2840

| BULLET | | | POWDER | | | | | |
|---|---|---|---|---|---|---|---|---|
| Wt | Mfg | Type | GRN | Mfg | Type | VEL(fps) | OAL | Source |
| 120 | SR | SPZ | 28.0 | DUP | 3031 | 2200 | 2.284 | SR |
| 120 | SR | SPZ | 29.0 | DUP | 3031 | 2300 | | SR |
| 120 | SR | SPZ | 29.5 | DUP | 3031 | 2350 | | SR |
| 120 | SR | SPZ | 29.3 | DUP | 4895 | 2200 | | SR |
| 120 | SR | SPZ | 30.3 | DUP | 4895 | 2300 | | SR |
| 120 | SR | SPZ | 30.9 | DUP | 4895 | 2350 | | SR |
| 140 | SR | ALL | 26.2 | DUP | 3031 | 2000 | | SR |
| 140 | SR | ALL | 27.2 | DUP | 3031 | 2100 | | SR |
| 140 | SR | ALL | 28.2 | DUP | 3031 | 2200 | | SR |
| 140 | SR | ALL | 26.8 | DUP | 4895 | 2000 | | SR |
| 140 | SR | ALL | 27.8 | DUP | 4895 | 2100 | | SR |
| 140 | SR | ALL | 28.9 | DUP | 4895 | 2200 | | SR |
| 150 | SR | HP | 24.2 | DUP | 3031 | 1900 | 2.265 | SR |
| 150 | SR | HP | 25.3 | DUP | 3031 | 2000 | | SR |
| 150 | SR | HP | 26.6 | DUP | 3031 | 2100 | | SR |
| 150 | SR | HP | 25.2 | DUP | 4895 | 1900 | | SR |
| 150 | SR | HP | 26.3 | DUP | 4895 | 2000 | | SR |
| 150 | SR | HP | 27.5 | DUP | 4895 | 2100 | | SR |
| 160 | SR | SPBT | 22.9 | DUP | 3031 | 1800 | 2.290 | SR |
| 160 | SR | SPBT | 24.2 | DUP | 3031 | 1900 | | SR |
| 160 | SR | SPBT | 25.5 | DUP | 3031 | 2000 | | SR |
| 160 | SR | SPBT | 25.1 | DUP | 4895 | 1900 | | SR |
| 160 | SR | SPBT | 26.3 | DUP | 4895 | 2000 | | SR |
| 160 | SR | SPBT | 26.8 | DUP | 4895 | 2050 | | SR |

LOADS FOR .30 M1 CARBINE .308 (PISTOL)

| BULLET | | | | POWDER | | | | |
|---|---|---|---|---|---|---|---|---|
| Wgt | Mfg | Type | GRN | Mfg | Type | VEL(fps) | OAL | Source |
| 100 | SP | PL | 4.5 | HER | Uniq | 938 | | SP |
| 100 | SP | PL | 5.4 | HER | Uniq | 1142 | | SP |
| 100 | SP | PL | 13.5 | DUP | 4227 | 1312 | | SP |
| 100 | SP | PL | 14.5 | DUP | 4227 | 1417 | | SP |
| 110 | SR | RNC | 6.8 | HER | Uniq | 1200 | 1.690 | SR |
| 110 | SR | RNC | 7.3 | HER | Uniq | 1300 | | SR |
| 110 | SR | RNC | 7.5 | HER | Uniq | 1350 | | SR |
| 110 | SR | RNC | 7.7 | HER | Herc | 1200 | | SR |
| 110 | SR | RNC | 8.7 | HER | Herc | 1300 | | SR |
| 110 | SR | RNC | 9.2 | HER | Herc | 1350 | | SR |
| 110 | SR | RNC | 10.2 | HER | Herc | 1450 | | SR |
| 110 | SR | RNC | 13.0 | HER | 2400 | 1300 | | SR |
| 110 | SR | RNC | 13.8 | HER | 2400 | 1400 | | SR |
| 110 | SR | RNC | 14.2 | HER | 2400 | 1450 | | SR |
| 110 | SR | RNC | 14.2 | DUP | 4227 | 1300 | | SR |
| 110 | SR | RNC | 14.9 | DUP | 4227 | 1400 | | SR |
| 110 | SR | RNC | 15.2 | DUP | 4227 | 1450 | | SR |
| 110 | SP | VAR | 4.5 | HER | Uniq | 779 | | SP |
| 110 | SP | VAR | 5.5 | HER | Uniq | 945 | | SP |
| 110 | SP | VAR | 13.5 | DUP | 4227 | 1281 | | SP |
| 110 | SP | VAR | 14.5 | DUP | 4227 | 1387 | | SP |

LOADS FOR .32 AUTO (.32 ACP) .312

| BULLET | | | POWDER | | | | | |
|---|---|---|---|---|---|---|---|---|
| Weight | Mfg | Type | GRN | Mfg | Type | VEL(fps) | OAL | Source |
| 71 | SR | FMJ | 1.5 | HER | BEYE | 600 | .978 | SR |
| 71 | SR | FMJ | 1.7 | HER | BEYE | 650 | | SR |
| 71 | SR | FMJ | 1.9 | HER | BEYE | 700 | | SR |
| 71 | SR | FMJ | 2.0 | HER | BEYE | 750 | | SR |
| 71 | SR | FMJ | 2.1 | HER | BEYE | 800 | | SR |
| 71 | SR | FMJ | 2.2 | HER | BEYE | 850 | | SR |
| 71 | SR | FMJ | 2.3 | HER | BEYE | 900 | | SR |
| 71 | SR | FMJ | 2.0 | HER | UNIQ | 600 | | SR |
| 71 | SR | FMJ | 2.2 | HER | UNIQ | 650 | | SR |
| 71 | SR | FMJ | 2.3 | HER | UNIQ | 700 | | SR |
| 71 | SR | FMJ | 2.4 | HER | UNIQ | 750 | | SR |
| 71 | SR | FMJ | 2.5 | HER | UNIQ | 800 | | SR |
| 71 | SR | FMJ | 2.6 | HER | UNIQ | 850 | | SR |
| 71 | SR | FMJ | 2.7 | HER | UNIQ | 900 | | SR |

LOADS FOR 9MM LUGER

| BULLET | | | POWDER | | | | | |
|---|---|---|---|---|---|---|---|---|
| Wgt | Mfg | Type | GRN | Mfg | Type | VEL(fps) | OAL | Source |
| 90 | SR | JHP | 4.7 | DUP | 700x | 1350 | 1.010 | SR |
| 90 | SR | JHP | 5.1 | DUP | 700x | 1400 | | SR |
| 90 | SR | JHP | 5.5 | DUP | 700x | 1450 | - | SR |
| 90 | SR | JHP | 5.5 | HER | Uniq | 1300 | | SR |
| 90 | SR | JHP | 6.0 | HER | Uniq | 1350 | | SR |
| 90 | SR | JHP | 7.0 | HER | Uniq | 1400 | | SR |
| 90 | SR | JHP | 7.5 | HER | Uniq | 1450 | | SR |
| 90 | SR | JHP | 7.4 | HER | Herc | 1400 | | SR |
| 90 | SR | JHP | 7.8 | HER | Herc | 1450 | | SR |
| 90 | SR | JHP | 8.1 | HER | Herc | 1500 | | SR |
| 95 | SR | FMJ | 4.5 | HER | BEYE | 1150 | 1.020 | SR |
| 95 | SR | FMJ | 4.8 | HER | BEYE | 1200 | | SR |
| 95 | SR | FMJ | 5.0 | HER | BEYE | 1250 | | SR |
| 95 | SR | FMJ | 5.4 | HER | UNIQ | 1150 | | SR |
| 95 | SR | FMJ | 5.8 | HER | UNIQ | 1200 | | SR |
| 95 | SR | FMJ | 6.1 | HER | UNIQ | 1250 | | SR |
| 95 | SR | FMJ | 6.4 | HER | UNIQ | 1300 | | SR |
| 95 | SR | FMJ | 6.7 | HER | UNIQ | 1350 | | SR |
| 95 | SR | FMJ | 7.0 | HER | UNIQ | 1400 | | SR |
| 115 | SR | JHP | 4.0 | DUP | 700x | 1150 | 1.100 | SR |
| 115 | SR | JHP | 4.3 | DUP | 700x | 1200 | | SR |
| 115 | SR | JHP | 4.7 | DUP | 700x | 1250 | | SR |
| 115 | SR | JHP | 5.1 | DUP | 700x | 1300 | | SR |
| 115 | SR | JHP | 5.0 | HER | Uniq | 1100 | | SR |
| 115 | SR | JHP | 5.6 | HER | Uniq | 1200 | | SR |
| 115 | SR | JHP | 6.4 | HER | Uniq | 1300 | | SR |
| 115 | SR | JHP | 7.0 | HER | Uniq | 1350 | | SR |
| 115 | SR | JHP | 6.6 | HER | Herc | 1300 | | SR |
| 115 | SR | JHP | 6.9 | HER | Herc | 1350 | | SR |
| 115 | SR | JHP | 7.2 | HER | Herc | 1400 | | SR |
| 125 | SR | FMJ | 3.2 | HER | BEYE | 900 | 1.090 | SR |
| 125 | SR | FMJ | 3.5 | HER | BEYE | 950 | | SR |
| 125 | SR | FMJ | 3.7 | HER | BEYE | 1000 | | SR |
| 125 | SR | FMJ | 3.9 | HER | BEYE | 1050 | | SR |
| 125 | SR | FMJ | 4.1 | HER | BEYE | 1100 | | SR |
| 125 | SR | FMJ | 4.3 | HER | BEYE | 1150 | | SR |
| 125 | SR | FMJ | 4.5 | HER | BEYE | 1200 | | SR |
| 125 | SR | FMJ | 4.1 | HER | UNIQ | 1000 | | SR |
| 125 | SR | FMJ | 4.6 | HER | UNIQ | 1050 | | SR |
| 125 | SR | FMJ | 5.1 | HER | UNIQ | 1100 | | SR |
| 125 | SR | FMJ | 5.5 | HER | UNIQ | 1150 | | SR |
| 125 | SR | FMJ | 5.9 | HER | UNIQ | 1200 | | SR |
| 130 | SR | FMJ | 3.5 | HER | BEYE | 900 | 1.120 | SR |
| 130 | SR | FMJ | 3.8 | HER | BEYE | 950 | | SR |
| 130 | SR | FMJ | 4.1 | HER | BEYE | 1000 | | SR |
| 130 | SR | FMJ | 4.4 | HER | BEYE | 1050 | | SR |
| 130 | SR | FMJ | 4.6 | HER | UNIQ | 1000 | | SR |
| 130 | SR | FMJ | 5.0 | HER | UNIQ | 1050 | | SR |
| 130 | SR | FMJ | 5.3 | HER | UNIQ | 1100 | | SR |
| 130 | SR | FMJ | 5.6 | HER | UNIQ | 1150 | | SR |
| 130 | SR | FMJ | 5.9 | HER | UNIQ | 1200 | | SR |

LOADS FOR .380 AUTO

| BULLET | | | POWDER | | | | | |
|---|---|---|---|---|---|---|---|---|
| Wgt | Mfg | Type | GRN | Mfg | Type | VEL(fps) | OAL | Source |
| 90 | SR | JHP | 2.2 | DUP | 700x | 750 | .930 | SR |
| 90 | SR | JHP | 2.7 | DUP | 700x | 850 | | SR |
| 90 | SR | JHP | 3.2 | DUP | 700x | 950 | | SR |
| 90 | SR | JHP | 3.2 | HER | Uniq | 800 | | SR |
| 90 | SR | JHP | 3.8 | HER | Uniq | 900 | | SR |
| 90 | SR | JHP | 4.6 | HER | Uniq | 1000 | | SR |
| 90 | SR | JHP | 3.5 | DUP | 7625 | 800 | | SR |
| 90 | SR | JHP | 3.8 | DUP | 7625 | 900 | | SR |
| 90 | SR | JHP | 4.2 | DUP | 7625 | 1000 | | SR |
| 90 | SR | JHP | 4.6 | DUP | 7625 | 1100 | | SR |
| 95 | SR | FMJ | 2.6 | HER | BEYE | 750 | .960 | SR |
| 95 | SR | FMJ | 2.8 | HER | BEYE | 800 | | SR |
| 95 | SR | FMJ | 3.0 | HER | BEYE | 850 | | SR |
| 95 | SR | FMJ | 3.2 | HER | BEYE | 900 | | SR |
| 95 | SR | FMJ | 3.4 | HER | BEYE | 950 | | SR |
| 95 | SR | FMJ | 3.3 | HER | UNIQ | 800 | | SR |
| 95 | SR | FMJ | 3.5 | HER | UNIQ | 850 | | SR |
| 95 | SR | FMJ | 3.7 | HER | UNIQ | 900 | | SR |
| 95 | SR | FMJ | 3.9 | HER | UNIQ | 950 | | SR |
| 95 | SR | FMJ | 4.0 | HER | UNIQ | 1000 | | SR |
| 115 | SR | JHP | 1.8 | DUP | 700x | 600 | .930 | SR |
| 115 | SR | JHP | 2.1 | DUP | 700x | 700 | | SR |
| 115 | SR | JHP | 2.4 | DUP | 700x | 800 | | SR |
| 115 | SR | JHP | 2.6 | DUP | 700x | 850 | | SR |
| 115 | SR | JHP | 2.7 | HER | Uniq | 600 | | SR |
| 115 | SR | JHP | 3.0 | HER | Uniq | 700 | | SR |
| 115 | SR | JHP | 3.4 | HER | Uniq | 800 | | SR |
| 115 | SR | JHP | 3.5 | HER | Uniq | 850 | | SR |
| 115 | SR | JHP | 2.6 | DUP | 7625 | 750 | | SR |
| 115 | SR | JHP | 3.2 | DUP | 7625 | 850 | | SR |
| 115 | SR | JHP | 3.4 | DUP | 7625 | 900 | | SR |
| 115 | SR | JHP | 3.7 | DUP | 7625 | 950 | | SR |

LOADS FOR .38 SUPER .355

| BULLET | | | POWDER | | | | | |
|---|---|---|---|---|---|---|---|---|
| Wgt | Mfg | Type | GRN | Mfg | Type | VEL(fps) | OAL | Source |
| 88 | SP | HP | 6.9 | HER | Uniq | 1382 | | SP |
| 88 | SP | HP | 7.5 | HER | Uniq | 1509 | | SP |
| 88 | SP | HP | 5.4 | HER | RDT | 1289 | | SP |
| 88 | SP | HP | 6.0 | HER | RDT | 1431 | | SP |
| 90 | SR | JHP | 5.5 | DUP | PB | 1200 | 1.180 | SR |
| 90 | SR | JHP | 6.3 | DUP | PB | 1300 | | SR |
| 90 | SR | JHP | 7.1 | DUP | PB | 1400 | | SR |
| 90 | SR | JHP | 7.7 | DUP | PB | 1500 | | SR |
| 90 | SR | JHP | 8.0 | DUP | PB | 1550 | | SR |
| 90 | SR | JHP | 7.6 | HER | UNIQ | 1300 | | SR |
| 90 | SR | JHP | 8.1 | HER | UNIQ | 1400 | | SR |
| 90 | SR | JHP | 9.0 | HER | UNIQ | 1500 | | SR |
| 95 | SR | FMJ | 5.5 | HER | BEYE | 1200 | 1.210 | SR |
| 95 | SR | FMJ | 5.9 | HER | BEYE | 1250 | | SR |
| 95 | SR | FMJ | 6.2 | HER | BEYE | 1300 | | SR |
| 95 | SR | FMJ | 6.5 | HER | BEYE | 1350 | | SR |
| 95 | SR | FMJ | 6.8 | HER | BEYE | 1400 | | SR |
| 95 | SR | FMJ | 7.2 | HER | UNIQ | 1300 | | SR |
| 95 | SR | FMJ | 7.7 | HER | UNIQ | 1350 | | SR |
| 95 | SR | FMJ | 8.2 | HER | UNIQ | 1400 | | SR |
| 95 | SR | FMJ | 8.6 | HER | UNIQ | 1450 | | SR |
| 95 | SR | FMJ | 9.0 | HER | UNIQ | 1500 | | SR |
| 100 | SP | HP | 8.0 | HER | HERC | 1237 | | SP |
| 100 | SP | HP | 8.5 | HER | HERC | 1323 | | SP |
| 100 | SP | HP | 5.9 | HER | Uniq | 1217 | | SP |
| 100 | SP | HP | 6.5 | HER | Uniq | 1335 | | SP |
| 115 | SR | JHP | 5.2 | DUP | PB | 1100 | 1.180 | SR |
| 115 | SR | JHP | 6.0 | DUP | PB | 1200 | | SR |
| 115 | SR | JHP | 6.6 | DUP | PB | 1300 | | SR |
| 115 | SR | JHP | 7.3 | DUP | PB | 1400 | | SR |
| 115 | SR | JHP | 6.8 | HER | UNIQ | 1250 | | SR |
| 115 | SR | JHP | 8.0 | HER | UNIQ | 1350 | | SR |
| 115 | SR | JHP | 8.5 | HER | UNIQ | 1400 | | SR |
| 125 | SR | FMJ | 4.6 | HER | BEYE | 1050 | 1.275 | SR |
| 125 | SR | FMJ | 4.9 | HER | BEYE | 1100 | | SR |
| 125 | SR | FMJ | 5.2 | HER | BEYE | 1150 | | SR |
| 125 | SR | FMJ | 5.2 | HER | UNIQ | 1050 | | SR |
| 125 | SR | FMJ | 5.6 | HER | UNIQ | 1100 | | SR |
| 125 | SR | FMJ | 6.0 | HER | UNIQ | 1150 | | SR |
| 125 | SR | FMJ | 6.4 | HER | UNIQ | 1200 | | SR |
| 125 | SR | FMJ | 6.7 | HER | UNIQ | 1250 | | SR |
| 30 | SR | FMJ | 4.6 | HER | BEYE | 1000 | | SR |
| .30 | SR | FMJ | 4.9 | HER | BEYE | 1050 | | SR |
| 130 | SR | FMJ | 5.2 | HER | BEYE | 1100 | | SR |
| 130 | SR | FMJ | 5.0 | HER | UNIQ | 1000 | | SR |
| 130 | SR | FMJ | 5.5 | HER | UNIQ | 1050 | | SR |
| 130 | SR | FMJ | 6.0 | HER | UNIQ | 1100 | | SR |
| 130 | SR | FMJ | 6.5 | HER | UNIQ | 1150 | | SR |
| 130 | SR | FMJ | 6.9 | HER | UNIQ | 1200 | | SR |

LOADS for .38 Special .357"

| BULLET | | | | POWDER | | | | | |
|---|---|---|---|---|---|---|---|---|---|
| Wgt | Mfg | Type | | GRN | Mfg | Type | VEL(fps) | OAL | Source |
| 110 | SR | JHC | | 7.3 | HER | Uniq | 1200 | 1.55 | SR |
| 110 | SR | JHC | | 7.5 | HER | Uniq | 1250 | | SR |
| 110 | SR | JHC | | 6.5 | DUP | PB | 1200 | | SR |
| 110 | SR | JHC | | 8.0 | DUP | PB | 1300 | | SR |
| 110 | SR | JHC | | 8.5 | DUP | PB | 1350 | | SR |
| 110 | SP | HP | +P | 11.6 | DUP | 4227 | 994 | | SP |
| 110 | SP | HP | +P | 13.5 | DUP | 4227 | 1080 | | SP |
| 110 | SP | HP | +P | 5.4 | HER | Uniq | 954 | | SP |
| 110 | SP | HP | +P | 6.2 | HER | Uniq | 1133 | | SP |
| 125 | SR | JHC | | 6.9 | HER | Herc | 1100 | 1.55 | SR |
| 125 | SR | JHC | | 7.2 | HER | Herc | 1150 | | SR |
| 125 | SR | JHC | | 6.0 | DUP | PB | 1000 | | SR |
| 125 | SR | JHC | | 7.1 | DUP | PB | 1100 | | SR |
| 125 | SR | JHC | | 6.8 | HER | Uniq | 1100 | | SR |
| 125 | SR | JHC | | 7.0 | HER | Uniq | 1150 | | SR |
| 125 | SP | all | +P | 11.2 | DUP | 4227 | 905 | | SP |
| 125 | SP | all | +P | 13.0 | DUP | 4227 | 981 | | SP |
| 125 | SP | all | +P | 5.3 | HER | Uniq | 917 | | SP |
| 125 | SP | all | +p | 5.9 | HER | Uniq | 997 | | SP |
| 140 | SP | HP | +P | 10.8 | DUP | 4227 | 880 | | SP |
| 140 | SP | HP | +P | 13.0 | DUP | 4227 | 1003 | | SP |
| 140 | SP | HP | +P | 5.1 | HER | Uniq | 856 | | SP |
| 140 | SP | HP | +P | 5.9 | HER | Uniq | 946 | | SP |
| 140 | SR | HP | | 4.5 | HER | BEYE | 850 | | SR |
| 140 | SR | HP | | 4.8 | HER | BEYE | 900 | | SR |
| 140 | SR | HP | | 5.1 | HER | BEYE | 950 | | SR |
| 140 | SR | HP | | 5.6 | HER | UNIQ | 850 | | SR |
| 140 | SR | HP | | 5.9 | HER | UNIQ | 900 | | SR |
| 140 | SR | HP | | 6.2 | HER | UNIQ | 950 | | SR |
| 140 | SR | HP | | 6.4 | HER | UNIQ | 1000 | | SR |
| 146 | | LWC | | 3.5 | HER | Uniq | 770 | | HR |
| 146 | | LWC | | 4.3 | HER | Uniq | 920 | | HR |
| 146 | | LWC | | 3.5 | HER | Herc | 770 | | HR |
| 146 | | LWC | | 4.2 | HER | Herc | 910 | | HR |
| 146 | SP | HP | +P | 10.8 | DUP | 4227 | 870 | | SP |
| 146 | SP | HP | +P | 12.9 | DUP | 4227 | 1033 | | SP |
| 146 | SP | HP | +P | 5.0 | HER | Uniq | 871 | | SP |
| 146 | SP | HP | +P | 5.9 | HER | Uniq | 1048 | | SP |
| 148 | SP | BBWC | | 4.5 | HER | Uniq | 868 | | SP |
| 148 | SP | BBWC | | 5.1 | HER | Uniq | 981 | | SP |
| 148 | SP | BBWC | | 3.6 | DUP | 700X | 836 | | SP |
| 148 | SP | BBWC | | 4.0 | DUP | 700X | 924 | | SP |
| 148 | SP | HBWC | | 2.7 | HER | RDT | 754 | | SP |
| 148 | SP | HBWC | | 3.0 | HER | RDT | 806 | | SP |
| 148 | SP | HBWC | | 2.6 | DUP | 700X | 739 | | SP |
| 148 | SP | HBWC | | 2.9 | DUP | 700X | 791 | | SP |
| 150 | SR | JHC | | 6.1 | HER | Uniq | 950 | 1.55 | SR |
| 150 | SR | JHC | | 6.4 | HER | Uniq | 1000 | | SR |
| 150 | SR | JHC | | 6.4 | HER | Herc | 950 | 1.55 | SR |
| 150 | SR | JHC | | 6.6 | HER | Herc | 1000 | | SR |

LOADS for .38 Special .357"

| BULLET | | | | POWDER | | | | | |
|---|---|---|---|---|---|---|---|---|---|
| Wgt | Mfg | Type | | GRN | Mfg | Type | VEL(fps) | OAL | Source |
| 158 | | LRN | | 5.5 | HER | Uniq | 980 | | HR |
| 158 | | LRN | | 5.5 | HER | Herc | 1020 | | HR |
| 158 | SR | JSP | | 6.1 | HER | Uniq | 900 | 1.55 | SR |
| 158 | SR | JSP | | 6.4 | HER | Uniq | 950 | | SR |
| 158 | SP | all | +P | 10.2 | DUP | 4227 | 822 | | SP |
| 158 | SP | all | +P | 12.3 | DUP | 4227 | 924 | | SP |
| 158 | SP | all | +P | 4.8 | HER | Uniq | 807 | | SP |
| 158 | SP | all | +P | 5.5 | HER | Uniq | 875 | | SP |
| 160 | SP | SP | +P | 4.9 | HER | Uniq | 858 | | SP |
| 160 | SP | SP | +P | 5.5 | HER | Uniq | 889 | | SP |
| 160 | SP | SP | +P | 4.3 | HER | BEYE | 814 | | SP |
| 160 | SP | SP | +P | 5.0 | HER | BEYE | 924 | | SP |
| 170 | SR | FMJS | | 4.9 | HER | Uniq | 750 | | SR |
| 170 | SR | FMJS | | 5.1 | HER | Uniq | 800 | | SR |
| 170 | SR | FMJS | | 5.5 | HER | Uniq | 900 | | SR |
| 170 | SR | FMJS | | 9.2 | HER | 2400 | 800 | | SR |
| 170 | SR | FMJS | | 9.9 | HER | Uniq | 850 | | SR |
| 170 | SR | FMJS | | 10.6 | HER | Uniq | 900 | | SR |
| 170 | SR | FMJS | | 12.2 | DUP | 4227 | 900 | | SR |
| 170 | SR | FMJS | | 12.7 | DUP | 4227 | 950 | | SR |
| 170 | SR | FMJS | | 13.2 | DUP | 4227 | 1000 | | SR |
| 180 | SR | FPJM | | 4.7 | HER | UNIQ | 700 | | SR |
| 180 | SR | FPJM | | 4.9 | HER | UNIQ | 750 | | SR |
| 180 | SR | FPJM | | 5.1 | HER | UNIQ | 800 | | SR |
| 180 | SR | FPJM | | 5.3 | HER | UNIQ | 850 | | SR |
| 180 | SR | FPJM | | 9.7 | DUP | 4227 | 700 | | SR |
| 180 | SR | FPJM | | 10.1 | DUP | 4227 | 750 | | SR |
| 180 | SR | FPJM | | 10.4 | DUP | 4227 | 800 | | SR |
| 180 | SR | FPJM | | 10.7 | DUP | 4227 | 850 | | SR |

LOADS FOR .357 MAGNUM

| BULLET | | | POWDER | | | | | |
|---|---|---|---|---|---|---|---|---|
| Wgt | Mfg | Type | GRN | Mfg | Type | VEL(fps) | OAL | Source |
| 110 | SR | JHC | 7.4 | HER | PB | 1250 | 1.545 | SR |
| 110 | SR | JHC | 8.1 | HER | PB | 1300 | | SR |
| 110 | SR | JHC | 8.7 | HER | PB | 1350 | | SR |
| 110 | SR | JHC | 9.3 | HER | Uniq | 1350 | | SR |
| 110 | SR | JHC | 10.0 | HER | Uniq | 1400 | | SR |
| 110 | SR | JHC | 10.6 | HER | Uniq | 1450 | | SR |
| 110 | SR | JHC | 18.2 | HER | 2400 | 1450 | | SR |
| 110 | SR | JHC | 18.7 | HER | 2400 | 1500 | | SR |
| 110 | SR | JHC | 19.3 | HER | 2400 | 1550 | | SR |
| 110 | SP | HP | 8.5 | HER | Uniq | 1331 | | SP |
| 110 | SP | HP | 9.5 | HER | Uniq | 1455 | | SP |
| 110 | SP | HP | 7.3 | DUP | 700X | 1333 | | SP |
| 110 | SP | HP | 8.3 | DUP | 700X | 1432 | | SP |
| 125 | SR | JHC | 7.4 | HER | PB | 1200 | 1.570 | SR |
| 125 | SR | JHC | 8.2 | HER | PB | 1250 | | SR |
| 125 | SR | JHC | 8.8 | HER | PB | 1300 | | SR |
| 125 | SR | JHC | 9.0 | HER | Uniq | 1350 | | SR |
| 125 | SR | JHC | 9.3 | HER | Uniq | 1400 | | SR |
| 125 | SR | JHC | 9.7 | HER | Uniq | 1450 | | SR |
| 125 | SR | JHC | 10.0 | HER | Uniq | 1500 | | SR |
| 125 | SR | JHC | 16.2 | HER | 2400 | 1300 | | SR |
| 125 | SR | JHC | 17.8 | HER | 2400 | 1400 | | SR |
| 125 | SR | JHC | 19.0 | HER | 2400 | 1500 | | SR |
| 125 | SP | all | 8.6 | HER | Uniq | 1313 | | SP |
| 125 | SP | all | 9.1 | HER | Uniq | 1390 | | SP |
| 125 | SP | all | 7.3 | DUP | 700X | 1267 | | SP |
| 125 | SP | all | 7.8 | DUP | 700X | 1354 | | SP |
| 140 | SP | HP | 8.3 | HER | Uniq | 1209 | | SP |
| 140 | SP | HP | 8.8 | HER | Uniq | 1301 | | SP |
| 140 | SP | HP | 7.3 | HER | BEYE | 1139 | | SP |
| 140 | SP | HP | 7.8 | HER | BEYE | 1204 | | SP |
| 140 | SR | HP | 7.5 | HER | BEYE | 1150 | | SR |
| 140 | SR | HP | 7.8 | HER | BEYE | 1200 | | SR |
| 140 | SR | HP | 8.2 | HER | UNIQ | 1150 | | SR |
| 140 | SR | HP | 8.6 | HER | UNIQ | 1200 | | SR |
| 140 | SR | HP | 9.0 | HER | UNIQ | 1250 | | SR |
| 146 | | LWC | 4.5 | HER | Uniq | 805 | 1.525 | HR |
| 146 | | LWC | 8.0 | HER | Uniq | 1380 | | HR |
| 146 | SP | HP | 8.0 | HER | Uniq | 1279 | | SP |
| 146 | SP | HP | 8.5 | HER | Uniq | 1363 | | SP |
| 146 | SP | HP | 6.7 | HER | BEYE | 1166 | | SP |
| 146 | SP | HP | 7.2 | HER | BEYE | 1253 | | SP |
| 148 | SP | HBWC | 2.9 | DUP | 700X | 745 | | SP |
| 148 | SP | HBWC | 3.2 | DUP | 700X | 815 | | SP |
| 148 | SP | HBWC | 3.0 | HER | BEYE | 733 | | SP |
| 148 | SP | HBWC | 3.3 | HER | BEYE | 806 | | SP |
| 150 | SR | JHC | 7.3 | HER | Uniq | 1100 | | SR |
| 150 | SR | JHC | 8.0 | HER | Uniq | 1200 | | SR |
| 150 | SR | JHC | 9.0 | HER | Uniq | 1300 | | SR |
| 150 | SR | JHC | 14.7 | HER | 2400 | 1200 | | SR |
| 150 | SR | JHC | 15.3 | HER | 2400 | 1250 | | SR |
| 150 | SR | JHC | 16.0 | HER | 2400 | 1300 | | SR |

LOADS FOR .357 MAGNUM

| BULLET | | | POWDER | | | | | |
|---|---|---|---|---|---|---|---|---|
| Wgt | Mfg | Type | GRN | Mfg | Type | VEL(fps) | OAL | Source |
| 158 | SR | JSP | 7.2 | HER | Uniq | 1000 | 1.565 | SR |
| 158 | SR | JSP | 7.7 | HER | Uniq | 1050 | | SR |
| 158 | SR | JSP | 8.2 | HER | Uniq | 1100 | | SR |
| 158 | SR | JSP | 14.0 | HER | 2400 | 1150 | | SR |
| 158 | SR | JSP | 15.0 | HER | 2400 | 1200 | | SR |
| 158 | SP | all | 7.7 | HER | Uniq | 1116 | | SP |
| 158 | SP | all | 8.2 | HER | Uniq | 1198 | | SP |
| 158 | SP | all | 6.6 | HER | BEYE | 1028 | | SP |
| 158 | SP | all | 7.1 | HER | BEYE | 1123 | | SP |
| 160 | SP | SP | *13.7 | HER | 2400 | 1178 | | SP |
| 160 | SP | SP | *15.7 | HER | 2400 | 1341 | | SP |
| 160 | SP | SP | 11.5 | HER | BLDT | 1221 | | SP |
| 160 | SP | SP | 12.5 | HER | BLDT | 1327 | | SP |
| 170 | SR | FMJS | 6.1 | HER | Uniq | 900 | 1.550 | SR |
| 170 | SR | FMJS | 6.7 | HER | Uniq | 1000 | | SR |
| 170 | SR | FMJS | 7.0 | HER | Uniq | 1050 | | SR |
| 170 | SR | FMJS | 13.1 | HER | 2400 | 1050 | | SR |
| 170 | SR | FMJS | 13.9 | HER | 2400 | 1100 | | SR |
| 170 | SR | FMJS | 14.5 | DUP | 4227 | 1000 | | SR |
| 170 | SR | FMJS | 15.1 | DUP | 4227 | 1050 | | SR |
| 170 | SR | FMJS | 15.6 | DUP | 4227 | 1100 | | SR |
| 180 | SP | FN | *14.0 | DUP | 4227 | 1334 | | SP |
| 180 | SP | FN | *15.0 | DUP | 4227 | 1406 | | SP |
| 180 | SP | FN | *13.0 | HER | 2400 | 1442 | | SP |
| 180 | SP | FN | *13.6 | HER | 2400 | 1469 | | SP |
| 180 | SR | FPJM | 5.2 | HER | UNIQ | 800 | | SR |
| 180 | SR | FPJM | 5.6 | HER | UNIQ | 850 | | SR |
| 180 | SR | FPJM | 6.0 | HER | UNIQ | 900 | | SR |
| 180 | SR | FPJM | 6.3 | HER | UNIQ | 950 | | SR |
| 180 | SR | FPJM | 6.6 | HER | UNIQ | 1000 | | SR |
| 180 | SR | FPJM | 13.3 | DUP | 4227 | 950 | | SR |
| 180 | SR | FPJM | 13.8 | DUP | 4227 | 1000 | | SR |
| 180 | SR | FPJM | 14.2 | DUP | 4227 | 1050 | | SR |

* DENOTES USE OF CCI #550 MAGNUM PRIMER

LOADS FOR .357 MAXIMUM

| BULLET | | | POWDER | | | | | |
|---|---|---|---|---|---|---|---|---|
| Weight | Mfg | Type | GRN | Mfg | Type | VEL(fps) | OAL | Source |
| 180 | SR | FPJM | 14.6 | HER | 2400 | 1300 | 1.990 | SR |
| 180 | SR | FPJM | 15.4 | HER | 2400 | 1350 | | SR |
| 180 | SR | FPJM | 16.2 | HER | 2400 | 1400 | | SR |
| 180 | SR | FPJM | 17.0 | HER | 2400 | 1450 | | SR |
| 180 | SR | FPJM | 17.8 | HER | 2400 | 1500 | | SR |
| 180 | SR | FPJM | 17.2 | DUP | 4227 | 1300 | | SR |
| 180 | SR | FPJM | 18.1 | DUP | 4227 | 1350 | | SR |
| 180 | SR | FPJM | 18.9 | DUP | 4227 | 1400 | | SR |
| 180 | SR | FPJM | 19.7 | DUP | 4227 | 1450 | | SR |
| 180 | SR | FPJM | 20.5 | DUP | 4227 | 1500 | | SR |

LOADS FOR .41 MAGNUM

| BULLET | | | POWDER | | | | | |
|---|---|---|---|---|---|---|---|---|
| Wgt | Mfg | Type | GRN | Mfg | Type | VEL (fps) | OAL | Source |
| 170 | SR | JHC | 10.5 | HER | Uniq | 1250 | 1.590 | SR |
| 70 | SR | JHC | 11.2 | HER | Uniq | 1350 | | SR |
| 170 | SR | JHC | 11.5 | HER | Uniq | 1400 | | SR |
| 170 | SR | JHC | 20.4 | HER | 2400 | 1300 | | SR |
| 170 | SR | JHC | 21.2 | HER | 2400 | 1400 | | SR |
| 170 | SR | JHC | 22.0 | HER | 2400 | 1500 | | SR |
| 170 | SR | JHC | 23.0 | DUP | 4227 | 1350 | | SR |
| 170 | SR | JHC | 23.5 | DUP | 4227 | 1400 | | SR |
| 170 | SR | JHC | 24.0 | DUP | 4227 | 1450 | | SR |
| 200 | SP | HP | *18.0 | DUP | 4227 | 1133 | | SP |
| 200 | SP | HP | *20.0 | DUP | 4227 | 1254 | | SP |
| 200 | SP | HP | 8.5 | HER | Uniq | 1117 | | SP |
| 200 | SP | HP | 9.0 | HER | Uniq | 1181 | | SP |
| 210 | SR | JHC | 9.1 | HER | Uniq | 1100 | 1.590 | SR |
| 210 | SR | JHC | 9.9 | HER | Uniq | 1200 | | SR |
| 210 | SR | JHC | 10.7 | HER | Uniq | 1300 | | SR |
| 210 | SR | JHC | 16.6 | HER | 2400 | 1000 | | SR |
| 210 | SR | JHC | 17.8 | HER | 2400 | 1100 | | SR |
| 210 | SR | JHC | 18.9 | HER | 2400 | 1200 | | SR |
| 210 | SR | JHC | 19.5 | HER | 2400 | 1250 | | SR |
| 210 | SR | JHC | 18.7 | DUP | 4227 | 1000 | | SR |
| 210 | SR | JHC | 19.8 | DUP | 4227 | 1100 | | SR |
| 210 | SR | JHC | 21.0 | DUP | 4227 | 1200 | | SR |
| 220 | SP | SP | *16.0 | HER | 2400 | 1192 | | SP |
| 220 | SP | SP | *17.5 | HER | 2400 | 1297 | | SP |
| 220 | SP | SP | 8.0 | HER | Uniq | 1023 | | SP |
| 220 | SP | SP | 8.5 | HER | Uniq | 1081 | | SP |
| 220 | SR | FPJM | 8.9 | HER | UNIQ | 1000 | | SR |
| 220 | SR | FPJM | 9.3 | HER | UNIQ | 1050 | | SR |
| 220 | SR | FPJM | 9.7 | HER | UNIQ | 1100 | | SR |
| 220 | SR | FPJM | 10.1 | HER | UNIQ | 1150 | | SR |
| 220 | SR | FPJM | 10.5 | HER | UNIQ | 1200 | | SR |
| 220 | SR | FPJM | 9.5 | HER | HERC | 1000 | | SR |
| 220 | SR | FPJM | 9.8 | HER | HERC | 1050 | | SR |
| 220 | SR | FPJM | 10.2 | HER | HERC | 1100 | | SR |
| 220 | SR | FPJM | 10.6 | HER | HERC | 1150 | | SR |
| 220 | SR | FPJM | 11.0 | HER | HERC | 1200 | | SR |

* DENOTES USE OF CCI #350 MAGNUM PRIMER.

LOADS FOR .44 SPECIAL

| BULLET | | | POWDER | | | | | |
|---|---|---|---|---|---|---|---|---|
| Wgt | Mfg | Type | GRN | Mfg | Type | VEL(fps) | OAL | Source |
| 120 | SP | RB | 3.0 | HER | UNIQ | 470 | | SP |
| 120 | SP | RB | 2.6 | DUP | 700X | 509 | | SP |
| 180 | SR | JHC | 16.0 | HER | 2400 | 900 | 1.615 | SR |
| 180 | SR | JHC | 18.0 | HER | 2400 | 1000 | | SR |
| 180 | SR | JHC | 19.7 | HER | 2400 | 1100 | | SR |
| 180 | SR | JHC | 20.7 | HER | 2400 | 1150 | | SR |
| 180 | SR | JHC | 17.6 | DUP | 4227 | 850 | | SR |
| 180 | SR | JHC | 19.4 | DUP | 4227 | 950 | | SR |
| 180 | SR | JHC | 20.4 | DUP | 4227 | 1050 | | SR |
| 200 | SP | MHP | 5.9 | HER | UNIQ | 746 | | SP |
| 200 | SP | MHP | 6.9 | HER | UNIQ | 869 | | SP |
| 200 | SP | MHP | 5.2 | HER | BEYE | 755 | | SP |
| 200 | SP | MHP | 5.7 | HER | BEYE | 826 | | SP |
| 210 | SR | JHC | 5.6 | HER | BEYE | 700 | | SR |
| 210 | SR | JHC | 6.0 | HER | BEYE | 750 | | SR |
| 210 | SR | JHC | 6.3 | HER | BEYE | 800 | | SR |
| 210 | SR | JHC | 6.9 | HER | UNIQ | 700 | | SR |
| 210 | SR | JHC | 7.3 | HER | UNIQ | 750 | | SR |
| 210 | SR | JHC | 7.6 | HER | UNIQ | 800 | | SR |
| 210 | SR | JHC | 7.9 | HER | UNIQ | 850 | | SR |
| 210 | SR | JHC | 8.2 | HER | UNIQ | 900 | | SR |
| 210 | SR | JHC | 8.5 | HER | UNIQ | 950 | | SR |
| 210 | SR | JHC | 8.8 | HER | UNIQ | 1000 | | SR |
| 220 | SR | FPJM | 5.5 | HER | BEYE | 700 | | SR |
| 220 | SR | FPJM | 5.8 | HER | BEYE | 750 | | SR |
| 220 | SR | FPJM | 6.1 | HER | BEYE | 800 | | SR |
| 220 | SR | FPJM | 6.3 | HER | BEYE | 850 | | SR |
| 220 | SR | FPJM | 6.7 | HER | UNIQ | 700 | | SR |
| 220 | SR | FPJM | 7.1 | HER | UNIQ | 750 | | SR |
| 220 | SR | FPJM | 7.5 | HER | UNIQ | 800 | | SR |
| 220 | SR | FPJM | 7.8 | HER | UNIQ | 850 | | SR |
| 220 | SR | FPJM | 8.1 | HER | UNIQ | 900 | | SR |
| 220 | SR | FPJM | 8.4 | HER | UNIQ | 950 | | SR |
| 225 | SP | HP | 6.2 | HER | UNIQ | 736 | | SP |
| 225 | SP | HP | 6.7 | HER | UNIQ | 782 | | SP |
| 225 | SP | HP | 6.5 | HER | HERC | 702 | | SP |
| 225 | SP | HP | 6.9 | HER | HERC | 753 | | SP |
| 240 | SR | JHC | 14.0 | HER | 2400 | 900 | 1.615 | SR |
| 240 | SR | JHC | 15.0 | HER | 2400 | 1000 | | SR |
| 240 | SR | JHC | 16.3 | HER | 2400 | 1100 | | SR |
| 240 | SR | JHC | 15.7 | DUP | 4227 | 750 | | SR |
| 240 | SR | JHC | 16.8 | DUP | 4227 | 850 | | SR |
| 240 | SR | JHC | 17.3 | DUP | 4227 | 900 | | SR |
| 240 | SP | SWC | 5.7 | HER | UNIQ | 739 | | SP |
| 240 | SP | SWC | 6.3 | HER | UNIQ | 809 | | SP |
| 240 | SP | SWC | 4.1 | DUP | 700X | 656 | | SP |
| 240 | SP | SWC | 4.6 | DUP | 700X | 745 | | SP |
| 240 | SP | all | 5.8 | HER | UNIQ | 673 | | SP |
| 240 | SP | all | 6.3 | HER | UNIQ | 726 | | SP |
| 240 | SP | all | 4.5 | HER | RDT | 616 | | SP |
| 240 | SP | all | 5.0 | HER | RDT | 684 | | SP |

LOADS FOR .44 SPECIAL

| BULLET | | | POWDER | | | | | |
|---|---|---|---|---|---|---|---|---|
| Wgt | Mfg | Type | GRN | Mfg | Type | VEL(fps) | OAL | Source |
| 250 | SR | FPJM | 5.6 | HER | BEYE | 700 | | SR |
| 250 | SR | FPJM | 6.0 | HER | BEYE | 750 | | SR |
| 250 | SR | FPJM | 6.5 | HER | UNIQ | 700 | | SR |
| 250 | SR | FPJM | 7.0 | HER | UNIQ | 750 | | SR |
| 250 | SR | FPJM | 7.4 | HER | UNIQ | 800 | | SR |
| 250 | SR | FPJM | 7.8 | HER | UNIQ | 850 | | SR |

# LOADS FOR .44 MAGNUM

| BULLET | | | POWDER | | | | | |
|---|---|---|---|---|---|---|---|---|
| Wgt | Mfg | Type | GRN | Mfg | Type | VEL(fps) | OAL | Source |
| 120 | SP | RB | 3.6 | HER | UNIQ | 566 | | SP |
| 120 | SP | RB | 2.9 | HER | BEYE | 549 | | SP |
| 180 | SR | JHC | 10.4 | HER | UNIQ | 1300 | 1.610 | SR |
| 180 | SR | JHC | 12.6 | HER | UNIQ | 1400 | | SR |
| 180 | SR | JHC | 14.0 | HER | UNIQ | 1500 | | SR |
| 180 | SR | JHC | 14.5 | HER | UNIQ | 1550 | | SR |
| 180 | SR | JHC | 21.8 | HER | 2400 | 1300 | | SR |
| 180 | SR | JHC | 22.8 | HER | 2400 | 1400 | | SR |
| 180 | SR | JHC | 24.5 | HER | 2400 | 1500 | | SR |
| 180 | SR | JHC | 26.3 | HER | 2400 | 1600 | | SR |
| 180 | SR | JHC | 27.0 | HER | 2400 | 1700 | | SR |
| 200 | SP | MHP | *25.0 | DUP | 4227 | 1373 | | SP |
| 200 | SP | MHP | *27.0 | DUP | 4227 | 1492 | | SP |
| 200 | SP | MHP | *12.3 | HER | UNIQ | 1403 | | SP |
| 200 | SP | MHP | *13.3 | HER | UNIQ | 1455 | | SP |
| 210 | SR | JHC | 12.0 | HER | UNIQ | 1200 | | SR |
| 210 | SR | JHC | 12.4 | HER | UNIQ | 1250 | | SR |
| 210 | SR | JHC | 12.8 | HER | UNIQ | 1300 | | SR |
| 210 | SR | JHC | 13.2 | HER | UNIQ | 1350 | | SR |
| 210 | SR | JHC | 24.3 | DUP | 4227 | 1250 | | SR |
| 210 | SR | JHC | 25.2 | DUP | 4227 | 1300 | | SR |
| 210 | SR | JHC | 26.1 | DUP | 4227 | 1350 | | SR |
| 210 | SR | JHC | 27.0 | DUP | 4227 | 1400 | | SR |
| 210 | SR | JHC | 27.9 | DUP | 4227 | 1450 | | SR |
| 220 | SR | FPJM | 11.5 | HER | UNIQ | 1200 | | SR |
| 220 | SR | FPJM | 12.1 | HER | UNIQ | 1250 | | SR |
| 220 | SR | FPJM | 12.6 | HER | UNIQ | 1300 | | SR |
| 220 | SR | FPJM | 20.9 | HER | 2400 | 1250 | | SR |
| 220 | SR | FPJM | 22.2 | HER | 2400 | 1300 | | SR |
| 220 | SR | FPJM | 23.5 | HER | 2400 | 1350 | | SR |
| 220 | SR | FPJM | 24.7 | HER | 2400 | 1400 | | SR |
| 225 | SP | HP | *23.0 | DUP | 4227 | 1348 | | SP |
| 225 | SP | HP | *25.0 | DUP | 4227 | 1471 | | SP |
| 225 | SP | HP | *11.9 | HER | UNIQ | 1350 | | SP |
| 225 | SP | HP | *12.9 | HER | UNIQ | 1432 | | SP |
| 240 | SR | JHC | 20.5 | HER | 2400 | 1300 | 1.610 | SR |
| 240 | SR | JHC | 22.0 | HER | 2400 | 1400 | | SR |
| 240 | SR | JHC | 23.3 | HER | 2400 | 1500 | | SR |
| 240 | SR | JHC | 22.0 | DUP | 4227 | 1200 | | SR |
| 240 | SR | JHC | 22.7 | DUP | 4227 | 1250 | | SR |
| 240 | SR | JHC | 23.3 | DUP | 4227 | 1300 | | SR |
| 240 | SP | SWC | 6.5 | HER | UNIQ | 855 | | SP |
| 240 | SP | SWC | 7.0 | HER | UNIQ | 906 | | SP |
| 240 | SP | SWC | 6.0 | HER | RDT | 914 | | SP |
| 240 | SP | SWC | 6.5 | HER | RDT | 956 | | SP |
| 240 | SP | all | *22.0 | DUP | 4227 | 1286 | | SP |
| 240 | SP | all | *24.0 | DUP | 4227 | 1312 | | SP |
| 240 | SP | all | *11.6 | HER | UNIQ | 1243 | | |
| 240 | SP | all | *12.6 | HER | UNIQ | 1361 | | |

* DENOTES USE OF CCI #350 MAGNUM PRIMER.

## LOADS FOR .44 REMINGTON MAGNUM .4295

| BULLET | | | POWDER | | | | | |
|---|---|---|---|---|---|---|---|---|
| Weight | Mfg | Type | GRN | Mfg | Type | VEL(fps) | OAL | Source |
| 250 | SR | FPJM | 15.6 | HER | BLDT | 1200 | | SR |
| 250 | SR | FPJM | 16.5 | HER | BLDT | 1250 | | SR |
| 250 | SR | FPJM | 17.3 | HER | BLDT | 1300 | | SR |
| 250 | SR | FPJM | 18.1 | HER | BLDT | 1350 | | SR |
| 250 | SR | FPJM | 19.5 | HER | 2400 | 1150 | | SR |
| 250 | SR | FPJM | 20.5 | HER | 2400 | 1200 | | SR |
| 250 | SR | FPJM | 21.4 | HER | 2400 | 1250 | | SR |
| 250 | SR | FPJM | 22.3 | HER | 2400 | 1300 | | SR |
| 250 | SR | FPJM | 23.2 | HER | 2400 | 1350 | | SR |

NOTE: The above loads were tested in a Ruger Super Blackhawk, 7 1/2 inch barrel (1:20" barrel twist) using Federal cases and CCI - 350 primers.

| Weight | Mfg | Type | GRN | Mfg | Type | VEL(fps) | OAL | Source |
|---|---|---|---|---|---|---|---|---|
| 250 | SR | FPJM | 15.2 | HER | BLDT | 1250 | | SR |
| 250 | SR | FPJM | 16.0 | HER | BLDT | 1300 | | SR |
| 250 | SR | FPJM | 16.8 | HER | BLDT | 1350 | | SR |
| 250 | SR | FPJM | 17.6 | HER | BLDT | 1400 | | SR |
| 250 | SR | FPJM | 18.3 | HER | BLDT | 1450 | | SR |
| 250 | SR | FPJM | 21.5 | DUP | 4227 | 1200 | | SR |
| 250 | SR | FPJM | 22.3 | DUP | 4227 | 1250 | | SR |
| 250 | SR | FPJM | 23.1 | DUP | 4227 | 1300 | | SR |
| 250 | SR | FPJM | 23.9 | DUP | 4227 | 1350 | | SR |

NOTE : The above loads were tested in a Dan Wesson revolver with a 10 incn barrel (1:18 3/4 inch barrel twist), using Federal cases and CCI - 350 primers.

## LOADS FOR .44 AUTO MAG

| BULLET | | | POWDER | | | | | |
|---|---|---|---|---|---|---|---|---|
| Wgt | Mfg | Type | GRN | Mfg | Type | VEL(fps) | OAL | Source |
| 180 | SR | JHC | 23.1 | HER | 2400 | 1500 | 1.610 | SR |
| 180 | SR | JHC | 24.9 | HER | 2400 | 1600 | | SR |
| 180 | SR | JHC | 25.8 | HER | 2400 | 1650 | | SR |
| 180 | SR | JHC | 27.8 | OLN | 296 | 1550 | | SR |
| 180 | SR | JHC | 28.6 | OLN | 296 | 1600 | | SR |
| 180 | SR | JHC | 29.3 | OLN | 296 | 1650 | | SR |
| 240 | SR | JHC | 10.2 | HER | Uniq | 1200 | 1.610 | SR |
| 240 | SR | JHC | 11.4 | HER | Uniq | 1300 | | SR |
| 240 | SR | JHC | 12.0 | HER | Uniq | 1350 | | SR |
| 240 | SR | JHC | 17.2 | HER | 2400 | 1200 | | SR |
| 240 | SR | JHC | 18.8 | HER | 2400 | 1300 | | SR |
| 240 | SR | JHC | 20.5 | HER | 2400 | 1400 | | SR |
| 240 | SR | JHC | 22.0 | OLN | 296 | 1200 | | SR |
| 240 | SR | JHC | 23.3 | OLN | 296 | 1300 | | SR |
| 240 | SR | JHC | 24.6 | OLN | 296 | 1400 | | SR |

## LOADS FOR .45 AUTO RIM .4515

| BULLET | | | POWDER | | | | | |
|---|---|---|---|---|---|---|---|---|
| Wgt | Mfg | Type | GRN | Mfg | Type | VEL(fps) | OAL | Source |
| 141 | SP | RB | 3.0 | DUP | 700X | 587 | | SP |
| 141 | SP | RB | 3.0 | HER | RDT | 599 | | SP |
| 185 | SR | JHC | 6.9 | DUP | PB | 950 | 1.133 | SR |
| 185 | SR | JHC | 7.5 | DUP | PB | 1050 | | SR |
| 185 | SR | JHC | 8.0 | DUP | PB | 1100 | | SR |
| 185 | SR | JHC | 7.0 | HER | UNIQ | 900 | | SR |
| 185 | SR | JHC | 7.6 | HER | UNIQ | 1000 | | SR |
| 185 | SR | JHC | 8.3 | HER | UNIQ | 1100 | | SR |
| 200 | SP | SWC | 5.0 | HER | UNIQ | 701 | | SP |
| 200 | SP | SWC | 5.5 | HER | UNIQ | 767 | | SP |
| 200 | SP | SWC | 4.0 | HER | RDT | 643 | | SP |
| 200 | SP | SWC | 4.5 | HER | RDT | 717 | | SP |
| 200 | SP | HP | 6.5 | HER | UNIQ | 858 | | SP |
| 200 | SP | HP | 7.0 | HER | UNIQ | 925 | | SP |
| 200 | SP | HP | 4.7 | HER | RDT | 769 | | SP |
| 200 | SP | HP | 5.2 | HER | RDT | 863 | | SP |
| 200 | SR | FPJM | 4.4 | HER | RDT | 700 | 1.155 | SR |
| 200 | SR | FPJM | 4.8 | HER | RDT | 750 | | SR |
| 200 | SR | FPJM | 5.2 | HER | RDT | 800 | | SR |
| 200 | SR | FPJM | 5.6 | HER | RDT | 850 | | SR |
| 200 | SR | FPJM | 9.1 | HER | BLDT | 750 | | SR |
| 200 | SR | FPJM | 9.6 | HER | BLDT | 800 | | SR |
| 200 | SR | FPJM | 10.1 | HER | BLDT | 850 | | SR |
| 200 | SR | FPJM | 10.6 | HER | BLDT | 900 | | SR |
| 225 | SP | HP | 6.4 | HER | UNIQ | 837 | | SP |
| 225 | SP | HP | 6.9 | HER | UNIQ | 900 | | SP |
| 225 | SP | HP | 4.5 | HER | RDT | 726 | | SP |
| 225 | SP | HP | 5.0 | HER | RDT | 810 | | SP |
| 230 | SP | RN | 6.1 | HER | UNIQ | 723 | | SP |
| 230 | SP | RN | 6.6 | HER | UNIQ | 787 | | SP |
| 230 | SP | RN | 4.5 | HER | RDT | 671 | | SP |
| 230 | SP | RN | 5.0 | HER | RDT | 749 | | SP |
| 230 | SR | FMJ | 4.9 | HER | RDT | 700 | 1.270 | SR |
| 230 | SR | FMJ | 5.3 | HER | RDT | 750 | | SR |
| 230 | SR | FMJ | 5.6 | HER | RDT | 800 | | SR |
| 230 | SR | FMJ | 6.4 | HER | UNIQ | 700 | | SR |
| 230 | SR | FMJ | 6.7 | HER | UNIQ | 750 | | SR |
| 230 | SR | FMJ | 6.9 | HER | UNIQ | 800 | | SR |
| 240 | SR | JHC | 5.7 | HER | UNIQ | 750 | 1.185 | SR |
| 240 | SR | JHC | 6.2 | HER | UNIQ | 800 | | SR |
| 240 | SR | JHC | 6.6 | HER | UNIQ | 850 | | SR |
| 240 | SR | JHC | 7.0 | HER | UNIQ | 900 | | SR |
| 240 | SR | JHC | 9.9 | OLN | 630 | 750 | | SR |
| 240 | SR | JHC | 10.4 | OLN | 630 | 800 | | SR |
| 250 | SP | SWC | 6.2 | HER | HERC | 791 | | SP |
| 250 | SP | SWC | 6.6 | HER | HERC | 842 | | SP |
| 250 | SP | SWC | 5.8 | HER | UNIQ | 773 | | SP |
| 250 | SP | SWC | 6.2 | HER | UNIQ | 824 | | SP |
| 260 | SP | HP | 5.2 | HER | UNIQ | 612 | | SP |
| 260 | SP | HP | 5.7 | HER | UNIQ | 672 | | SP |
| 260 | SP | HP | 4.1 | HER | RDT | 584 | | SP |
| 260 | SP | HP | 4.6 | HER | RDT | 651 | | SP |

LOADS FOR .45 ACP .4515

| BULLET | | | POWDER | | | | | |
|---|---|---|---|---|---|---|---|---|
| Wgt | Mfg | Type | GRN | Mfg | Type | VEL(fps) | OAL | Source |
| 185 | SR | JHC | 5.9 | DUP | PB | 800 | 1.222 | SR |
| 185 | SR | JHC | 6.2 | DUP | PB | 850 | | SR |
| 185 | SR | JHC | 6.6 | DUP | PB | 900 | | SR |
| 185 | SR | JHC | 5.7 | HER | UNIQ | 750 | | SR |
| 185 | SR | JHC | 6.0 | HER | UNIQ | 800 | | SR |
| 185 | SR | JHC | 6.6 | HER | UNIQ | 900 | | SR |
| 185 | SR | JHC | 7.7 | HER | UNIQ | 1000 | | SR |
| 185 | SR | JHC | 8.5 | HER | UNIQ | 1100 | | SR |
| 200 | SP | SWC | 4.9 | HER | UNIQ | 716 | | SP |
| 200 | SP | SWC | 5.4 | HER | UNIQ | 790 | | SP |
| 200 | SP | SWC | 4.1 | HER | RDT | 749 | | SP |
| 200 | SP | SWC | 4.5 | HER | RDT | 831 | | SP |
| 200 | SP | HP | 6.6 | HER | UNIQ | 863 | | SP |
| 200 | SP | HP | 7.2 | HER | UNIQ | 961 | | SP |
| 200 | SP | HP | 4.9 | HER | RDT | 834 | | SP |
| 200 | SP | HP | 5.4 | HER | RDT | 917 | | SP |
| 200 | SR | FPJM | 4.5 | HER | BEYE | 700 | 1.155 | SR |
| 200 | SR | FPJM | 5.0 | HER | BEYE | 750 | | SR |
| 200 | SR | FPJM | 5.4 | HER | BEYE | 800 | | SR |
| 200 | SR | FPJM | 5.8 | HER | BEYE | 850 | | SR |
| 200 | SR | FPJM | 5.2 | HER | UNIQ | 700 | | SR |
| 200 | SR | FPJM | 5.6 | HER | UNIQ | 750 | | SR |
| 200 | SR | FPJM | 6.1 | HER | UNIQ | 800 | | SR |
| 200 | SR | FPJM | 6.6 | HER | UNIQ | 850 | | SR |
| 200 | SR | FPJM | 7.0 | HER | UNIQ | 900 | | SR |
| 200 | SR | FPJM | 7.4 | HER | UNIQ | 950 | | SR |
| 225 | SP | HP | 6.4 | HER | UNIQ | 862 | | SP |
| 225 | SP | HP | 7.2 | HER | UNIQ | 940 | | SP |
| 225 | SP | HP | 4.8 | HER | RDT | 787 | | SP |
| 225 | SP | HP | 5.3 | HER | RDT | 858 | | SP |
| 230 | SP | RN | 5.3 | HER | UNIQ | 764 | | SP |
| 230 | SP | RN | 5.8 | HER | UNIQ | 849 | | SP |
| 230 | SP | RN | 4.7 | HER | RDT | 765 | | SP |
| 230 | SP | RN | 5.1 | HER | RDT | 841 | | SP |
| 230 | SR | FMJ | 4.0 | HER | BEYE | 700 | 1.270 | SR |
| 230 | SR | FMJ | 4.3 | HER | BEYE | 750 | | SR |
| 230 | SR | FMJ | 4.6 | HER | BEYE | 800 | | SR |
| 230 | SR | FMJ | 4.9 | HER | BEYE | 850 | | SR |
| 230 | SR | FMJ | 5.2 | HER | BEYE | 900 | | SR |
| 230 | SR | FMJ | 5.1 | HER | UNIQ | 700 | | SR |
| 230 | SR | FMJ | 5.6 | HER | UNIQ | 750 | | SR |
| 230 | SR | FMJ | 6.0 | HER | UNIQ | 800 | | SR |
| 230 | SR | FMJ | 6.4 | HER | UNIQ | 850 | | SR |
| 230 | SR | FMJ | 6.8 | HER | UNIQ | 900 | | SR |
| 240 | SR | JHC | 5.3 | HER | UNIQ | 700 | 1.185 | SR |
| 240 | SR | JHC | 6.1 | HER | UNIQ | 800 | | SR |
| 240 | SR | JHC | 6.7 | HER | UNIQ | 900 | | SR |
| 240 | SR | JHC | 9.5 | OLN | 630 | 700 | | SR |
| 240 | SR | JHC | 10.1 | OLN | 630 | 750 | | SR |
| 240 | SR | JHC | 10.7 | OLN | 630 | 800 | | SR |
| 260 | SP | HP | 5.8 | HER | UNIQ | 740 | | SP |
| 260 | SP | HP | 6.4 | HER | UNIQ | 822 | | SP |
| 260 | SP | HP | 4.6 | HER | RDT | 722 | | SP |
| 260 | SP | HP | 4.9 | HER | RDT | 794 | | SP |

## LOADS FOR .45 LONG COLT for Colt Revolvers & Replicas

| BULLET | | | POWDER | | | | | |
|---|---|---|---|---|---|---|---|---|
| Wgt | Mfg | Type | GRN | Mfg | Type | VEL(fps) | OAL | Source |
| 185acp | SR | JHC | 7.9 | DUP | PB | 800 | 1.522 | SR |
| 185 | SR | JHC | 8.5 | DUP | PB | 900 | | SR |
| 185 | SR | JHC | 9.1 | DUP | PB | 1000 | | SR |
| 185 | SR | JHC | 9.8 | DUP | PB | 1100 | | SR |
| 185 | SR | JHC | 11.0 | DUP | PB | 1200 | | SR |
| 185 | SR | JHC | 9.8 | HER | Uniq | 1000 | | SR |
| 185 | SR | JHC | 11.3 | HER | Uniq | 1100 | | SR |
| 185 | SR | JHC | 12.0 | HER | Uniq | 1150 | | SR |
| 185 | SR | JHC | 20.3 | DUP | 4227 | 1000 | | SR |
| 185 | SR | JHC | 22.2 | DUP | 4227 | 1100 | | SR |
| 185 | SR | JHC | 23.3 | DUP | 4227 | 1150 | | SR |
| 240LC | SR | JHC | 7.7 | HER | Uniq | 750 | 1.570 | SR |
| 240 | SR | JHC | 8.2 | HER | Uniq | 800 | | SR |
| 240 | SR | JHC | 8.7 | HER | Uniq | 850 | | SR |
| 240 | SR | JHC | 9.2 | HER | Uniq | 900 | | SR |
| 240 | SR | JHC | 8.8 | HER | Herc | 800 | | SR |
| 240 | SR | JHC | 9.3 | HER | Herc | 850 | | SR |
| 240 | SR | JHC | 9.8 | HER | Herc | 900 | | SR |

## LOADS FOR .22 REMINGTON JET (T/C CONTENDER)

| BULLET | | | POWDER | | | | | |
|---|---|---|---|---|---|---|---|---|
| Wt | Mfg | Type | GRN | Mfg | Type | VEL(fps) | OAL | Source |
| 40 | SR | Horn | 7.3 | HER | Uniq | 2200 | 1.723 | SR |
| 40 | SR | Horn | 7.7 | HER | Uniq | 2300 | | SR |
| 40 | SR | Horn | 8.0 | HER | Uniq | 2400 | | SR |
| 40 | SR | Horn | 11.0 | DUP | 4227 | 2000 | 1.723 | SR |
| 40 | SR | Horn | 11.4 | DUP | 4227 | 2100 | | SR |
| 40 | SR | Horn | 11.8 | DUP | 4227 | 2200 | | SR |
| 45 | SR | Horn | 6.8 | HER | Uniq | 2000 | 1.723 | SR |
| 45 | SR | Horn | 7.2 | HER | Uniq | 2100 | | SR |
| 45 | SR | Horn | 7.5 | HER | Uniq | 2200 | | SR |
| 45 | SR | Horn | 10.5 | DUP | 4227 | 1900 | 1.723 | SR |
| 45 | SR | Horn | 11.0 | DUP | 4227 | 2000 | | SR |
| 45 | SR | Horn | 11.5 | DUP | 4227 | 2100 | | SR |

LOADS FOR .22 HORNET (T/C CONTENDER)

| BULLET | | | POWDER | | | | | |
|---|---|---|---|---|---|---|---|---|
| Wt | Mfg | Type | GRN | Mfg | Type | VEL(fps) | OAL | Source |
| 40 | SR | Horn | 8.8 | HER | 2400 | 2250 | 1.723 | SR |
| 40 | SR | Horn | 9.1 | HER | 2400 | 2300 | | SR |
| 40 | SR | Horn | 9.4 | HER | 2400 | 2350 | | SR |
| 40 | SR | Horn | 10.6 | OLN | 680 | 2250 | | SR |
| 40 | SR | Horn | 11.3 | OLN | 680 | 2300 | | SR |
| 40 | SR | Horn | 12.0 | OLN | 680 | 2350 | | SR |
| 40 | SP | SP | 10.5 | DUP | 4227 | 2160 | | SP |
| 40 | SP | SP | 11.5 | DUP | 4227 | 2357 | | SP |
| 40 | SP | SP | 8.5 | HER | 2400 | 2122 | | SP |
| 40 | SP | SP | 9.5 | HER | 2400 | 2366 | | SP |
| 45 | SR | Horn | 8.9 | HER | 2400 | 2200 | | SR |
| 45 | SR | Horn | 9.2 | HER | 2400 | 2250 | | SR |
| 45 | SR | Horn | 11.0 | OLN | 680 | 2200 | | SR |
| 45 | SR | Horn | 11.7 | OLN | 680 | 2250 | | SR |
| 45 | SP | SP | 10.3 | DUP | 4227 | 2086 | | SP |
| 45 | SP | SP | 11.3 | DUP | 4227 | 2294 | | SP |
| 45 | SP | SP | 7.7 | HER | 2400 | 1953 | | SP |
| 45 | SP | SP | 8.7 | HER | 2400 | 2201 | | SP |
| 52 | SR | HPBT | 9.6 | OLN | 296 | 2150 | 1.723 | SR |
| 52 | SR | HPBT | 10.0 | OLN | 296 | 2200 | | SR |
| 52 | SR | HPBT | 10.4 | OLN | 296 | 2250 | | SR |

## LOADS FOR .22 K-HORNET (T/C CONTENDER)

| BULLET | | | POWDER | | | | | |
|---|---|---|---|---|---|---|---|---|
| Wt | Mfg | Type | GRN | Mfg | Type | VEL(fps) | OAL | Source |
| 40 | SR | Horn | 8.5 | HER | 2400 | 2000 | n/a | SR |
| 40 | SR | Horn | 9.5 | HER | 2400 | 2200 | | SR |
| 40 | SR | Horn | 10.3 | HER | 2400 | 2400 | | SR |
| 40 | SR | Horn | 10.0 | DUP | 4227 | 2200 | | SR |
| 40 | SR | Horn | 11.0 | DUP | 4227 | 2400 | | SR |
| 40 | SR | Horn | 12.0 | DUP | 4227 | 2600 | | SR |
| 45 | SR | Horn | 7.5 | HER | 2400 | 1800 | n/a | SR |
| 45 | SR | Horn | 8.7 | HER | 2400 | 2000 | | SR |
| 45 | SR | Horn | 9.7 | HER | 2400 | 2200 | | SR |
| 45 | SR | Horn | 10.0 | DUP | 4227 | 2200 | | SR |
| 45 | SR | Horn | 10.7 | DUP | 4227 | 2300 | | SR |
| 45 | SR | Horn | 11.4 | DUP | 4227 | 2400 | | SR |
| 45 | SR | Horn | 12.0 | DUP | 4227 | 2500 | | SR |

LOADS FOR .221 REMINGTON FIRE BALL (T/C CONTENDER)

| BULLET | | | POWDER | | | | | |
|---|---|---|---|---|---|---|---|---|
| Wt | Mfg | Type | GRN | Mfg | Type | VEL(fps) | OAL | Source |
| 40 | SR | Horn | 14.0 | DUP | 4227 | 2600 | 1.830 | SR |
| 40 | SR | Horn | 15.5 | DUP | 4227 | 2700 | | SR |
| 40 | SR | Horn | 14.2 | HER | 2400 | 2700 | | SR |
| 40 | SR | Horn | 14.9 | HER | 2400 | 2800 | | SR |
| 40 | SR | Horn | 15.5 | HER | 2400 | 2900 | | SR |
| 45 | SR | SP | 13.5 | DUP | 4227 | 2400 | | SR |
| 45 | SR | SP | 14.3 | DUP | 4227 | 2500 | | SR |
| 45 | SR | SP | 15.0 | DUP | 4227 | 2600 | | SR |
| 45 | SR | SP | 13.5 | HER | 2400 | 2550 | | SR |
| 45 | SR | SP | 14.5 | HER | 2400 | 2650 | | SR |
| 45 | SR | SP | 15.0 | HER | 2400 | 2700 | | SR |
| 50 | SR | BL | 13.0 | DUP | 4227 | 2250 | 1.830 | SR |
| 50 | SR | BL | 13.9 | DUP | 4227 | 2400 | | SR |
| 50 | SR | BL | 14.5 | DUP | 4227 | 2500 | | SR |
| 50 | SR | BL | 13.0 | HER | 2400 | 2350 | | SR |
| 50 | SR | BL | 13.8 | HER | 2400 | 2450 | | SR |
| 50 | SR | BL | 14.5 | HER | 2400 | 2550 | | SR |
| 52 | SR | HPBT | 12.9 | DUP | 4227 | 2250 | | SR |
| 52 | SR | HPBT | 13.6 | DUP | 4227 | 2350 | | SR |
| 52 | SR | HPBT | 14.2 | DUP | 4227 | 2450 | | SR |
| 52 | SR | HPBT | 13.3 | HER | 2400 | 2350 | | SR |
| 52 | SR | HPBT | 13.8 | HER | 2400 | 2450 | | SR |
| 52 | SR | HPBT | 14.2 | HER | 2400 | 2550 | | SR |

## LOADS FOR .222 REMINGTON (T/C CONTENDER)

| BULLET | | | POWDER | | | | | |
|---|---|---|---|---|---|---|---|---|
| Wt | Mfg | Type | GRN | Mfg | Type | VEL(fps) | OAL | Source |
| 40 | SR | Horn | 15.5 | HER | 2400 | 2600 | 2.020 | SR |
| 40 | SR | Horn | 16.3 | HER | 2400 | 2700 | | SR |
| 40 | SR | Horn | 16.9 | HER | 2400 | 2800 | | SR |
| 40 | SR | Horn | 17.5 | HER | 2400 | 2900 | | SR |
| 40 | SR | Horn | 16.0 | DUP | 4227 | 2600 | | SR |
| 40 | SR | Horn | 16.6 | DUP | 4227 | 2700 | | SR |
| 40 | SR | Horn | 17.2 | DUP | 4227 | 2800 | | SR |
| 40 | SR | Horn | 17.8 | DUP | 4227 | 2900 | | SR |
| 45 | SR | SP | 15.5 | HER | 2400 | 2500 | 2.110 | SR |
| 45 | SR | SP | 16.3 | HER | 2400 | 2600 | | SR |
| 45 | SR | SP | 16.9 | HER | 2400 | 2700 | | SR |
| 45 | SR | SP | 17.5 | HER | 2400 | 2800 | | SR |
| 45 | SR | SP | 16.3 | DUP | 4227 | 2600 | | SR |
| 45 | SR | SP | 16.9 | DUP | 4227 | 2700 | | SR |
| 45 | SR | SP | 17.5 | DUP | 4227 | 2800 | | SR |
| 50 | SR | BL | 15.3 | HER | 2400 | 2400 | 2.115 | SR |
| 50 | SR | BL | 15.9 | HER | 2400 | 2500 | | SR |
| 50 | SR | BL | 16.5 | HER | 2400 | 2600 | | SR |
| 50 | SR | BL | 17.0 | HER | 2400 | 2700 | | SR |
| 50 | SR | BL | 16.0 | DUP | 4227 | 2400 | | SR |
| 50 | SR | BL | 16.4 | DUP | 4227 | 2500 | | SR |
| 50 | SR | BL | 16.8 | DUP | 4227 | 2600 | | SR |
| 50 | SR | BL | 17.0 | DUP | 4227 | 2650 | | SR |
| 52 | SR | HPBT | 15.3 | HER | 2400 | 2400 | 2.105 | SR |
| 52 | SR | HPBT | 15.9 | HER | 2400 | 2500 | | SR |
| 52 | SR | HPBT | 16.3 | HER | 2400 | 2600 | | SR |
| 52 | SR | HPBT | 15.3 | DUP | 4227 | 2400 | | SR |
| 52 | SR | HPBT | 15.9 | DUP | 4227 | 2500 | | SR |
| 52 | SR | HPBT | 16.5 | DUP | 4227 | 2600 | | SR |

## LOADS FOR 7MM TCU (CONTENDER) .284

| BULLET | | | POWDER | | | | | |
|---|---|---|---|---|---|---|---|---|
| Weight | Mfg | Type | GRN | Mfg | Type | VEL (fps) | OAL | Source |
| 120 | SR | SPZ | 21.9 | HER | RE-7 | 2000 | 2.600 | SR |
| 120 | SR | SPZ | 23.1 | HER | RE-7 | 2050 | | SR |
| 120 | SR | SPZ | 26.0 | DUP | 3031 | 2000 | | SR |
| 120 | SR | SPZ | 26.6 | DUP | 3031 | 2050 | | SR |
| 120 | SR | SPZ | 27.2 | DUP | 3031 | 2100 | | SR |
| 120 | SR | SPZ | 27.8 | DUP | 3031 | 2150 | | SR |
| 120 | SR | SPZ | 28.3 | DUP | 3031 | 2200 | | SR |
| 120 | SR | SPZ | 27.9 | DUP | 4895 | 2000 | | SR |
| 120 | SR | SPZ | 28.5 | DUP | 4895 | 2050 | | SR |
| 120 | SR | SPZ | 29.1 | DUP | 4895 | 2100 | | SR |
| 120 | SR | SPZ | 29.7 | DUP | 4895 | 2150 | | SR |
| 120 | SR | SPZ | 30.2 | DUP | 4895 | 2200 | | SR |
| 140 | SR | SPZ | 24.6 | DUP | 3031 | 1900 | | SR |
| 140 | SR | SPZ | 25.4 | DUP | 3031 | 1950 | | SR |
| 140 | SR | SPZ | 26.2 | DUP | 3031 | 2000 | | SR |
| 140 | SR | SPZ | 26.9 | DUP | 3031 | 2050 | | SR |
| 140 | SR | SPZ | 26.1 | DUP | 4895 | 1900 | | SR |
| 140 | SR | SPZ | 26.8 | DUP | 4895 | 1950 | | SR |
| 140 | SR | SPZ | 27.5 | DUP | 4895 | 2000 | | SR |
| 140 | SR | SPZ | 28.1 | DUP | 4895 | 2050 | | SR |
| 140 | SR | SPZ | 28.7 | DUP | 4895 | 2100 | | SR |
| 140 | SR | SPZ | 29.3 | DUP | 4895 | 2150 | | SR |
| 140 | SR | SBT | 24.9 | DUP | 3031 | 1900 | 2.650 | SR |
| 140 | SR | SBT | 25.6 | DUP | 3031 | 1950 | | SR |
| 140 | SR | SBT | 26.3 | DUP | 3031 | 2000 | | SR |
| 140 | SR | SBT | 27.0 | DUP | 3031 | 2050 | | SR |
| 140 | SR | SBT | 26.7 | DUP | 4895 | 1900 | | SR |
| 140 | SR | SBT | 27.2 | DUP | 4895 | 1950 | | SR |
| 140 | SR | SBT | 27.7 | DUP | 4895 | 2000 | | SR |
| 140 | SR | SBT | 28.2 | DUP | 4895 | 2050 | | SR |
| 140 | SR | SBT | 28.7 | DUP | 4895 | 2100 | | SR |
| 140 | SR | SBT | 29.1 | DUP | 4895 | 2150 | | SR |
| 150 | SR | HP | 23.6 | DUP | 3031 | 1800 | 2.700 | SR |
| 150 | SR | HP | 24.2 | DUP | 3031 | 1850 | | SR |
| 150 | SR | HP | 24.7 | DUP | 3031 | 1900 | | SR |
| 150 | SR | HP | 25.2 | DUP | 3031 | 1950 | | SR |
| 150 | SR | HP | 25.7 | DUP | 3031 | 2000 | | SR |
| 150 | SR | HP | 26.2 | DUP | 3031 | 2050 | | SR |
| 150 | SR | HP | 26.7 | DUP | 3031 | 2100 | | SR |
| 150 | SR | HP | 26.1 | DUP | 4895 | 1900 | | SR |
| 150 | SR | HP | 26.8 | DUP | 4895 | 1950 | | SR |
| 150 | SR | HP | 27.4 | DUP | 4895 | 2000 | | SR |
| 150 | SR | HP | 28.0 | DUP | 4895 | 2050 | | SR |
| 150 | SR | HP | 28.6 | DUP | 4895 | 2100 | | SR |
| 150 | SR | HP | 29.2 | DUP | 4895 | 2150 | | SR |
| 160 | SR | SBT | 23.6 | DUP | 3031 | 1800 | 2.700 | SR |
| 160 | SR | SBT | 24.2 | DUP | 3031 | 1850 | | SR |
| 160 | SR | SBT | 24.8 | DUP | 3031 | 1900 | | SR |
| 160 | SR | SBT | 25.4 | DUP | 3031 | 1950 | | SR |
| 160 | SR | SBT | 26.0 | DUP | 3031 | 2000 | | SR |
| 160 | SR | SBT | 26.6 | DUP | 3031 | 2050 | | SR |

LOADS FOR 7MM TCU (CONTENDER) .284

| BULLET | | | POWDER | | | | | |
|---|---|---|---|---|---|---|---|---|
| Weight | Mfg | Type | GRN | Mfg | Type | VEL(fps) | OAL | Source |
| 160 | SR | SBT | 27.2 | DUP | 3031 | 2100 | | SR |
| 160 | SR | SBT | 25.7 | DUP | 4895 | 1900 | | SR |
| 160 | SR | SBT | 26.5 | DUP | 4895 | 1950 | | SR |
| 160 | SR | SBT | 27.3 | DUP | 4895 | 2000 | | SR |
| 160 | SR | SBT | 28.0 | DUP | 4895 | 2050 | | SR |
| 160 | SR | SBT | 28.7 | DUP | 4895 | 2100 | | SR |

LOADS FOR .30-30 WINCHESTER (T/C CONTENDER)

| BULLET | | | POWDER | | | | | |
|---|---|---|---|---|---|---|---|---|
| Wgt | Mfg | Type | GRN | Mfg | Type | VEL(fps) | OAL | Source |
| 110 | SR | RN | 20.5 | DUP | 2400 | 1800 | 2.505 | SR |
| 110 | SR | RN | 22.0 | DUP | 2400 | 1900 | | SR |
| 110 | SR | RN | 23.5 | DUP | 2400 | 2000 | | SR |
| 110 | SR | RN | 28.0 | HER | RE7 | 2000 | | SR |
| 110 | SR | RN | 29.5 | HER | RE7 | 2100 | | SR |
| 110 | SR | RN | 31.0 | HER | RE7 | 2200 | | SR |
| 110 | SP | all | 24.0 | HER | RE7 | 1746 | | SP |
| 110 | SP | all | 26.0 | HER | RE7 | 1912 | | SP |
| 110 | SP | all | 16.0 | DUP | 4227 | 1512 | | SP |
| 110 | SP | all | 18.0 | DUP | 4227 | 1708 | | SP |
| 125 | SR | HPFN | 23.0 | DUP | 4227 | 1900 | 2.490 | SR |
| 125 | SR | HPFN | 23.8 | DUP | 4227 | 2000 | | SR |
| 125 | SR | HPFN | 24.5 | DUP | 4227 | 2100 | | SR |
| 125 | SR | HPFN | 25.0 | HER | RE7 | 1600 | | SR |
| 125 | SR | HPFN | 25.6 | HER | RE7 | 1700 | | SR |
| 125 | SR | HPFN | 26.2 | HER | RE7 | 1800 | | SR |
| 125 | SR | HPFN | 26.8 | HER | RE7 | 1900 | | SR |
| 125 | SR | HPFN | 27.4 | HER | RE7 | 2000 | | SR |
| 125 | SR | HPFN | 28.0 | HER | RE7 | 2100 | | SR |
| 130 | SP | all | 23.0 | HER | RE7 | 1706 | | SP |
| 130 | SP | all | 25.0 | HER | RE7 | 1848 | | SP |
| 130 | SP | all | 16.0 | DUP | 4227 | 1447 | | SP |
| 130 | SP | all | 18.0 | DUP | 4227 | 1632 | | SP |
| 150 | SR | FN | 21.5 | DUP | 4227 | 1700 | 2.745 | SR |
| 150 | SR | FN | 23.0 | DUP | 4227 | 1800 | | SR |
| 150 | SR | FN | 22.0 | HER | RE7 | 1500 | | SR |
| 150 | SR | FN | 23.0 | HER | RE7 | 1600 | | SR |
| 150 | SR | FN | 24.0 | HER | RE7 | 1700 | | SR |
| 150 | SR | FN | 25.0 | HER | RE7 | 1800 | | SR |
| 150 | SP | all | 22.0 | HER | RE7 | 1586 | | SP |
| 150 | SP | all | 24.0 | HER | RE7 | 1740 | | SP |
| 150 | SP | all | 16.0 | DUP | 4227 | 1344 | | SP |
| 150 | SP | all | 18.0 | DUP | 4227 | 1520 | | SP |
| 165 | SP | all | 21.0 | HER | RE7 | 1568 | | SP |
| 165 | SP | all | 23.0 | HER | RE7 | 1726 | | SP |
| 165 | SP | all | 15.0 | DUP | 4227 | 1222 | | SP |
| 165 | SP | all | 17.0 | DUP | 4227 | 1383 | | SP |

LOADS FOR .30 HERRETT (T/C CONTENDER).308

| BULLET | | | POWDER | | | | | |
|---|---|---|---|---|---|---|---|---|
| Wgt | Mfg | Type | GRN | Mfg | Type | VEL(fps) | OAL | Source |
| 110 | SR | HP | 22.0 | HER | 2400 | 2250 | 2.180 | SR |
| 110 | SR | HP | 22.5 | HER | 2400 | 2300 | | SR |
| 110 | SR | HP | 23.0 | HER | 2400 | 2350 | | SR |
| 110 | SR | HP | 23.0 | DUP | 4227 | 2250 | | SR |
| 110 | SR | HP | 24.0 | DUP | 4227 | 2300 | | SR |
| 110 | SR | HP | 25.0 | DUP | 4227 | 2350 | | SR |
| 110 | SR | HPFN | 20.0 | HER | 2400 | 2000 | 2.090 | SR |
| 110 | SR | HPFN | 21.0 | HER | 2400 | 2100 | | SR |
| 110 | SR | HPFN | 22.0 | HER | 2400 | 2200 | | SR |
| 110 | SR | HPFN | 21.5 | DUP | 4227 | 2100 | | SR |
| 110 | SR | HPFN | 22.9 | DUP | 4227 | 2200 | | SR |
| 110 | SR | HPFN | 23.5 | DUP | 4227 | 2250 | | SR |
| 110 | SP | all | 26.0 | HER | RE7 | 1911 | | SP |
| 110 | SP | all | 28.0 | HER | RE7 | 2045 | | SP |
| 110 | SP | all | 21.0 | DUP | 4227 | 2176 | | SP |
| 110 | SP | all | 23.0 | DUP | 4227 | 2340 | | SP |
| 130 | SP | all | 24.0 | HER | RE7 | 1896 | | SP |
| 130 | SP | all | 26.0 | HER | RE7 | 2027 | | SP |
| 130 | SP | all | 19.0 | DUP | 4227 | 1823 | | SP |
| 130 | SP | all | 21.0 | DUP | 4227 | 1999 | | SP |
| 150 | SR | FN | 18.5 | HER | 2400 | 1850 | 2.100 | SR |
| 150 | SR | FN | 19.2 | HER | 2400 | 1900 | | SR |
| 150 | SR | FN | 20.5 | HER | 2400 | 2000 | | SR |
| 150 | SR | FN | 20.2 | DUP | 4227 | 1900 | | SR |
| 150 | SR | FN | 20.9 | DUP | 4227 | 1950 | | SR |
| 150 | SR | FN | 21.5 | DUP | 4227 | 2000 | | SR |
| 150 | SP | all | 21.0 | HER | RE7 | 1670 | | SP |
| 150 | SP | all | 23.0 | HER | RE7 | 1812 | | SP |
| 150 | SP | all | 17.0 | DUP | 4227 | 1626 | | SP |
| 150 | SP | all | 19.0 | DUP | 4227 | 1803 | | SP |
| 165 | SP | all | 18.0 | HER | RE7 | 1526 | | SP |
| 165 | SP | all | 20.0 | HER | RE7 | 1685 | | SP |
| 165 | SP | all | 15.0 | DUP | 4227 | 1514 | | SP |
| 165 | SP | all | 17.0 | DUP | 4227 | 1665 | | SP |

LOADS FOR .357 MAXIMUM (CONTENDER) .357

| BULLET | | | POWDER | | | | | |
|---|---|---|---|---|---|---|---|---|
| Weight | Mfg | Type | GRN | Mfg | Type | VEL(fps) | OAL | Source |
| 180 | SR | FPJM | 17.9 | HER | 2400 | 1500 | 1.990 | SR |
| 180 | SR | FPJM | 18.7 | HER | 2400 | 1550 | | SR |
| 180 | SR | FPJM | 19.5 | HER | 2400 | 1600 | | SR |
| 180 | SR | FPJM | 20.2 | HER | 2400 | 1650 | | SR |
| 180 | SR | FPJM | 20.9 | HER | 2400 | 1700 | | SR |
| 180 | SR | FPJM | 20.1 | DUP | 4227 | 1500 | | SR |
| 180 | SR | FPJM | 20.8 | DUP | 4227 | 1550 | | SR |
| 180 | SR | FPJM | 21.4 | DUP | 4227 | 1600 | | SR |
| 180 | SR | FPJM | 22.0 | DUP | 4227 | 1650 | | SR |

## LOADS FOR .35 REMINGTON (T/C CONTENDER)

| BULLET | | | POWDER | | | | | |
|---|---|---|---|---|---|---|---|---|
| Wgt | Mfg | Type | GRN | Mfg | Type | VEL(fps) | OAL | Source |
| 158 | SR | JSP | 28.8 | DUP | 4198 | 2000 | 2.295 | SR |
| 158 | SR | JSP | 30.4 | DUP | 4198 | 2100 | | SR |
| 158 | SR | JSP | 32.0 | DUP | 4198 | 2200 | | SR |
| 170 | SR | FMJS | 28.8 | DUP | 4198 | 2000 | 2.340 | SR |
| 170 | SR | FMJS | 30.4 | DUP | 4198 | 2100 | | SR |
| 170 | SR | FMJS | 32.0 | DUP | 4198 | 2200 | | SR |
| 200 | SR | RN | 27.8 | DUP | 4198 | 1800 | 2.490 | SR |
| 200 | SR | RN | 29.4 | DUP | 4198 | 1900 | | SR |
| 200 | SR | RN | 31.0 | DUP | 4198 | 2000 | | SR |

LOADS FOR .357 HERRETT (T/C CONTENDER)

| BULLET | | | POWDER | | | | | |
|---|---|---|---|---|---|---|---|---|
| Wgt | Mfg | Type | GRN | Mfg | Type | VEL(fps) | OAL | Source |
| 110 | SR | JHC | 25.0 | HER | 2400 | 2400 | 2.100 | SR |
| 110 | SR | JHC | 26.6 | HER | 2400 | 2500 | | SR |
| 110 | SR | JHC | 28.2 | HER | 2400 | 2600 | | SR |
| 110 | SR | JHC | 29.0 | HER | 2400 | 2650 | | SR |
| 110 | SR | JHC | 29.5 | OLN | 296 | 2600 | | SR |
| 110 | SR | JHC | 31.5 | OLN | 296 | 2700 | | SR |
| 110 | SR | JHC | 32.5 | OLN | 296 | 2750 | | SR |
| 110 | SP | HP | 33.0 | DUP | 4227 | 2389 | | SP |
| 110 | SP | HP | 35.0 | DUP | 4227 | 2523 | | SP |
| 110 | SP | HP | 31.0 | HER | 2400 | 2404 | | SP |
| 110 | SP | HP | 33.0 | HER | 2400 | 2552 | | SP |
| 125 | SR | JHC | 24.7 | HER | 2400 | 2300 | 2.090 | SR |
| 125 | SR | JHC | 26.1 | HER | 2400 | 2400 | | SR |
| 125 | SR | JHC | 27.4 | HER | 2400 | 2500 | | SR |
| 125 | SR | JHC | 28.0 | HER | 2400 | 2550 | | SR |
| 125 | SR | JHC | 28.5 | OLN | 296 | 2500 | | SR |
| 125 | SR | JHC | 30.5 | OLN | 296 | 2600 | | SR |
| 125 | SR | JHC | 31.5 | OLN | 296 | 2650 | | SR |
| 125 | SP | all | 32.0 | DUP | 4227 | 2255 | | SP |
| 125 | SP | all | 34.0 | DUP | 4227 | 2376 | | SP |
| 125 | SP | all | 29.0 | HER | 2400 | 2242 | | SP |
| 125 | SP | all | 31.0 | HER | 2400 | 2389 | | SP |
| 140 | SP | HP | 30.0 | DUP | 4227 | 2131 | | SP |
| 140 | SP | HP | 32.0 | DUP | 4227 | 2260 | | SP |
| 140 | SP | HP | 27.0 | HER | 2400 | 2081 | | SP |
| 140 | SP | HP | 29.0 | HER | 2400 | 2218 | | SP |
| 150 | SR | JHC | 23.2 | HER | 2400 | 2100 | 2.115 | SR |
| 150 | SR | JHC | 24.6 | HER | 2400 | 2200 | | SR |
| 150 | SR | JHC | 25.9 | HER | 2400 | 2300 | | SR |
| 150 | SR | JHC | 26.5 | HER | 2400 | 2350 | | SR |
| 150 | SR | JHC | 27.9 | OLN | 296 | 2300 | | SR |
| 150 | SR | JHC | 29.2 | OLN | 296 | 2350 | | SR |
| 150 | SR | JHC | 30.5 | OLN | 296 | 2400 | | SR |
| 158 | SR | JSP | 23.0 | HER | 2400 | 2100 | 2.100 | SR |
| 158 | SR | JSP | 25.0 | HER | 2400 | 2200 | | SR |
| 158 | SR | JSP | 26.0 | HER | 2400 | 2250 | | SR |
| 158 | SR | JSP | 26.0 | OLN | 296 | 2200 | | SR |
| 158 | SR | JSP | 28.0 | OLN | 296 | 2300 | | SR |
| 158 | SR | JSP | 30.0 | OLN | 296 | 2400 | | SR |
| 158 | SP | all | 27.0 | DUP | 4227 | 1921 | | SP |
| 158 | SP | all | 29.0 | DUP | 4227 | 2075 | | SP |
| 158 | SP | all | 24.0 | HER | 2400 | 1968 | | SP |
| 158 | SP | all | 26.0 | HER | 2400 | 2144 | | SP |
| 170 | SR | FMJS | 20.8 | HER | 2400 | 1900 | 2.145 | SR |
| 170 | SR | FMJS | 22.4 | HER | 2400 | 2000 | | SR |
| 170 | SR | FMJS | 24.0 | HER | 2400 | 2100 | | SR |
| 170 | SR | FMJS | 24.8 | OLN | 296 | 2100 | | SR |
| 170 | SR | FMJS | 26.4 | OLN | 296 | 2200 | | SR |
| 170 | SR | FMJS | 28.0 | OLN | 296 | 2300 | | SR |
| 180 | SP | FN | 25.0 | DUP | 4227 | 1687 | | SP |
| 180 | SP | FN | 27.0 | DUP | 4227 | 1817 | | SP |
| 180 | SP | FN | 22.0 | HER | 2400 | 1735 | | SP |
| 180 | SP | FN | 24.0 | HER | 2400 | 1887 | | SP |

LOADS FOR .357 HERRETT (T/C CONTENDER)

| BULLET | | | POWDER | | | | | |
|---|---|---|---|---|---|---|---|---|
| Wgt | Mfg | Type | GRN | Mfg | Type | VEL(fps) | OAL | Source |
| 200 | SR | RN | 24.0 | OLN | 680 | 1800 | 2.345 | SR |
| 200 | SR | RN | 26.0 | OLN | 680 | 1900 | | SR |
| 200 | SR | RN | 28.0 | OLN | 680 | 2000 | | SR |

WARNING: THESE LOADS ARE HOT ! Use this information only as a guide. These loads were developed with correctly headspaced cases. Starting loads in any other gun should be reduced by 10 %.

LOADS FOR .45 LONG COLT (RUGER AND T/C CONTENDER)

| BULLET | | | POWDER | | | | | |
|---|---|---|---|---|---|---|---|---|
| Wgt | Mfg | Type | GRN | Mfg | Type | VEL(fps) | OAL | Source |
| 200 | SP | HP | 11.3 | HER | Uniq | 1259 | | SP |
| 200 | SP | HP | 11.8 | HER | Uniq | 1301 | | SP |
| 200 | SP | HP | 9.5 | HER | BEYE | 1125 | | SP |
| 200 | SP | HP | 10.0 | HER | BEYE | 1187 | | SP |
| 225 | SP | HP | 10.0 | HER | Uniq | 1082 | | SP |
| 225 | SP | HP | 11.0 | HER | Uniq | 1185 | | SP |
| 225 | SP | HP | 9.2 | HER | BEYE | 1096 | | SP |
| 225 | SP | HP | 9.7 | HER | BEYE | 1151 | | SP |
| 240 | SR | JHC | 8.7 | HER | Uniq | 850 | 1.570 | SR |
| 240 | SR | JHC | 9.7 | HER | Uniq | 950 | | SR |
| 240 | SR | JHC | 10.9 | HER | Uniq | 1050 | | SR |
| 240 | SR | JHC | 11.5 | HER | Uniq | 1100 | | SR |
| 240* | SR | JHC | 14.2 | HER | BLDT | 1000 | | SR |
| 240* | SR | JHC | 15.3 | HER | BLDT | 1100 | | SR |
| 240* | SR | JHC | 16.4 | HER | BLDT | 1200 | | SR |
| 240* | SR | JHC | 16.9 | HER | BLDT | 1250 | | SR |
| 240* | SR | JHC | 17.1 | HER | 2400 | 900 | | SR |
| 240* | SR | JHC | 18.7 | HER | 2400 | 1000 | | SR |
| 240* | SR | JHC | 20.3 | HER | 2400 | 1100 | | SR |
| 240 | SR | JHC | 19.7 | DUP | 4227 | 1000 | | SR |
| 240 | SR | JHC | 22.1 | DUP | 4227 | 1100 | | SR |
| 240 | SR | JHC | 24.3 | DUP | 4227 | 1200 | | SR |
| 250 | SP | SWC | 9.8 | HER | Uniq | 1053 | | SP |
| 250 | SP | SWC | 10.5 | HER | Uniq | 1136 | | SP |
| 250 | SP | SWC | 8.8 | HER | BEYE | 1011 | | SP |
| 250 | SP | SWC | 9.3 | HER | BEYE | 1076 | | SP |
| 260 | SP | HP | 9.5 | HER | Uniq | 987 | | SP |
| 260 | SP | HP | 10.5 | HER | Uniq | 1079 | | SP |
| 260 | SP | HP | 8.5 | HER | BEYE | 950 | | SP |
| 260 | SP | HP | 9.2 | HER | BEYE | 1031 | | SP |

*- CCI 350 PRIMERS

LOADS FOR .22 HORNET .224

| BULLET | | | POWDER | | | | | |
|---|---|---|---|---|---|---|---|---|
| Wgt | Mfg | Type | GRN | Mfg | Type | VEL(fps) | OAL | Source |
| 40 | SR | Horn. | 9.5 | HER | 2400 | 2500 | 1.723 | SR |
| 40 | SR | Horn. | 10.0 | HER | 2400 | 2600 | | SR |
| 40 | SR | Horn. | 10.5 | HER | 2400 | 2700 | | SR |
| 40 | SR | Horn. | 11.0 | HER | 2400 | 2800 | | SR |
| 40 | SR | Horn. | 9.8 | DUP | 4227 | 2500 | | SR |
| 40 | SR | Horn. | 10.4 | DUP | 4227 | 2600 | | SR |
| 40 | SR | Horn. | 11.0 | DUP | 4227 | 2700 | | SR |
| 40 | SR | Horn. | 11.6 | DUP | 4227 | 2800 | | SR |
| 40 | SP | SP | 11.0 | DUP | 4227 | 2655 | | SP |
| 40 | SP | SP | 11.5 | DUP | 4227 | 2803 | | SP |
| 40 | SP | SP | 12.0 | DUP | 4227 | 2884 | | SP |
| 40 | SP | SP | 10.5 | DUP | 4198 | 2391 | | SP |
| 40 | SP | SP | 11.0 | DUP | 4198 | 2517 | | SP |
| 40 | SP | SP | 11.5 | DUP | 4198 | 2617 | | SP |
| 45 | SR | Horn. | 8.0 | HER | 2400 | 2300 | | SR |
| 45 | SR | Horn. | 8.6 | HER | 2400 | 2400 | | SR |
| 45 | SR | Horn. | 9.2 | HER | 2400 | 2500 | | SR |
| 45 | SR | Horn. | 10.0 | HER | 2400 | 2600 | | SR |
| 45 | SR | Horn. | 9.5 | DUP | 4227 | 2300 | | SR |
| 45 | SR | Horn. | 10.0 | DUP | 4227 | 2400 | | SR |
| 45 | SR | Horn. | 10.5 | DUP | 4227 | 2500 | | SR |
| 45 | SR | Horn. | 11.0 | DUP | 4227 | 2600 | | SR |
| 45 | SR | Horn. | 11.7 | DUP | 4227 | 2700 | | SR |
| 45 | SP | SPT | 10.5 | DUP | 4198 | 2276 | | SP |
| 45 | SP | SPT | 11.0 | DUP | 4198 | 2388 | | SP |
| 45 | SP | SPT | 11.5 | DUP | 4198 | 2484 | | SP |
| 45 | SP | SPT | 10.5 | DUP | 4227 | 2503 | | SP |
| 45 | SP | SPT | 11.0 | DUP | 4227 | 2616 | | SP |
| 45 | SP | SPT | 11.5 | DUP | 4227 | 2729 | | SP |
| 45 | NO | BTSB | 10.5 | DUP | 4227 | 2522 | | NO |
| 45 | NO | BTSB | 11.0 | DUP | 4227 | 2587 | | NO |
| 45 | NO | BTSB | 11.5 | DUP | 4227 | 2652 | | NO |
| 45 | NO | BTSB | 10.5 | DUP | 4198 | 2248 | | NO |
| 45 | NO | BTSB | 11.0 | DUP | 4198 | 2331 | | NO |
| 45 | NO | BTSB | 11.5 | DUP | 4198 | 2413 | | NO |
| 50 | SR | ALL | 9.1 | HER | 2400 | 2400 | | SR |
| 50 | SR | ALL | 10.0 | HER | 2400 | 2500 | | SR |
| 50 | SR | ALL | 10.4 | DUP | 4227 | 2400 | | SR |
| 50 | SR | ALL | 10.8 | DUP | 4227 | 2500 | | SR |
| 50 | SR | ALL | 11.5 | DUP | 4227 | 2600 | | SR |
| 50 | SP | SPT | 10.5 | DUP | 4198 | 2254 | | SP |
| 50 | SP | SPT | 11.0 | DUP | 4198 | 2362 | | SP |
| 50 | SP | SPT | 11.5 | DUP | 4198 | 2458 | | SP |
| 50 | SP | SPT | 10.0 | DUP | 4227 | 2376 | | SP |
| 50 | SP | SPT | 10.5 | DUP | 4227 | 2490 | | SP |
| 50 | SP | SPT | 11.0 | DUP | 4227 | 2603 | | SP |

LOADS FOR .22 HORNET .224

| BULLET | | | POWDER | | | | | |
|---|---|---|---|---|---|---|---|---|
| Wgt | Mfg | Type | GRN | Mfg | Type | VEL(fps) | OAL | Source |
| 50 | NO | ALL | 10.5 | DUP | 4227 | 2491 | | NO |
| 50 | NO | ALL | 11.0 | DUP | 4227 | 2593 | | NO |
| 50 | NO | ALL | 11.5 | DUP | 4227 | 2695 | | NO |
| 50 | NO | ALL | 10.0 | DUP | 4198 | 2161 | | NO |
| 50 | NO | ALL | 10.5 | DUP | 4198 | 2285 | | NO |
| 50 | NO | ALL | 11.0 | DUP | 4198 | 2410 | | NO |
| 52 | SP | ALL | 10.5 | DUP | 4198 | 2222 | | SP |
| 52 | SP | ALL | 11.0 | DUP | 4198 | 2335 | | SP |
| 52 | SP | ALL | 11.5 | DUP | 4198 | 2434 | | SP |
| 52 | SP | ALL | 10.0 | DUP | 4227 | 2392 | | SP |
| 52 | SP | ALL | 10.0 | DUP | 4227 | 2521 | | SP |
| 52 | SP | ALL | 11.0 | DUP | 4227 | 2623 | | SP |
| 52 | NO | HPBT | 10.5 | DUP | 4227 | 2458 | | NO |
| 52 | NO | HPBT | 11.0 | DUP | 4227 | 2547 | | NO |
| 52 | NO | HPBT | 11.5 | DUP | 4227 | 2652 | | NO |
| 52 | NO | HPBT | 10.0 | DUP | 4198 | 2122 | | NO |
| 52 | NO | HPBT | 10.5 | DUP | 4198 | 2242 | | NO |
| 52 | NO | HPBT | 11.0 | DUP | 4198 | 2370 | | NO |
| 53 | SR | HP | 9.4 | HER | 2400 | 2400 | | SR |
| 53 | SR | HP | 10.1 | HER | 2400 | 2500 | | SR |
| 53 | SR | HP | 9.5 | HER | 4227 | 2200 | | SR |
| 53 | SR | HP | 10.0 | HER | 4227 | 2300 | | SR |
| 53 | SR | HP | 10.5 | HER | 4227 | 2400 | | SR |
| 53 | SR | HP | 11.0 | HER | 4227 | 2500 | | SR |
| 53 | SR | HP | 11.6 | HER | 4227 | 2600 | | SR |
| 55 | SR | ALL | 7.8 | HER | 2400 | 2100 | | SR |
| 55 | SR | ALL | 8.5 | HER | 2400 | 2200 | | SR |
| 55 | SR | ALL | 9.3 | HER | 2400 | 2300 | | SR |
| 55 | SR | ALL | 9.6 | DUP | 4227 | 2200 | | SR |
| 55 | SR | ALL | 10.3 | DUP | 4227 | 2300 | | SR |
| 55 | SR | ALL | 11.2 | DUP | 4227 | 2400 | | SR |

LOADS FOR .218 BEE .224

| BULLET | | | POWDER | | | | | |
|---|---|---|---|---|---|---|---|---|
| Wgt | Mfg | Type | GRN | Mfg | Type | VEL(fps) | OAL | Source |
| 40 | SR | horn | 10.9 | HER | 2400 | 2600 | 1.680 | SR |
| 40 | SR | horn | 11.4 | HER | 2400 | 2700 | | SR |
| 40 | SR | horn | 12.0 | HER | 2400 | 2800 | | SR |
| 40 | SR | horn | 12.5 | HER | 2400 | 2900 | | SR |
| 40 | SR | horn | 10.5 | DUP | 4227 | 2400 | | SR |
| 40 | SR | horn | 11.0 | DUP | 4227 | 2500 | | SR |
| 40 | SR | horn | 11.6 | DUP | 4227 | 2600 | | SR |
| 40 | SR | horn | 12.2 | DUP | 4227 | 2700 | | SR |
| 40 | SR | horn | 12.8 | DUP | 4227 | 2800 | | SR |
| 40 | SR | horn | 13.4 | DUP | 4227 | 2900 | | SR |
| 45 | SR | all | 10.0 | HER | 2400 | 2400 | | SR |
| 45 | SR | all | 10.4 | HER | 2400 | 2500 | | SR |
| 45 | SR | all | 11.0 | HER | 2400 | 2600 | | SR |
| 45 | SR | all | 11.6 | HER | 2400 | 2700 | | SR |
| 45 | SR | all | 12.0 | HER | 2400 | 2800 | | SR |
| 45 | SR | all | 11.0 | DUP | 4227 | 2400 | | SR |
| 45 | SR | all | 11.5 | DUP | 4227 | 2500 | | SR |
| 45 | SR | all | 12.1 | DUP | 4227 | 2600 | | SR |
| 45 | SR | all | 12.7 | DUP | 4227 | 2700 | | SR |
| 45 | SR | all | 13.0 | DUP | 4227 | 2800 | | SR |
| 45 | NO | BTSB | 11.0 | DUP | 4227 | 2432 | | NO |
| 45 | NO | BTSB | 12.0 | DUP | 4227 | 2608 | | NO |
| 45 | NO | BTSB | 13.0 | DUP | 4227 | 2783 | | NO |
| 50 | SR | SEMP | 9.0 | HER | 2400 | 2100 | | SR |
| 50 | SR | SEMP | 9.5 | HER | 2400 | 2200 | | SR |
| 50 | SR | SEMP | 10.0 | HER | 2400 | 2300 | | SR |
| 50 | SR | SEMP | 10.5 | HER | 2400 | 2400 | | SR |
| 50 | SR | SEMP | 11.0 | HER | 2400 | 2500 | | SR |
| 50 | SR | SEMP | 11.5 | HER | 2400 | 2600 | | SR |
| 50 | SR | SEMP | 10.8 | DUP | 4227 | 2300 | | SR |
| 50 | SR | SEMP | 11.3 | DUP | 4227 | 2400 | | SR |
| 50 | SR | SEMP | 11.9 | DUP | 4227 | 2500 | | SR |
| 50 | SR | SEMP | 12.5 | DUP | 4227 | 2600 | | SR |
| 50 | NO | all | 10.5 | DUP | 4227 | 2340 | | NO |
| 50 | NO | all | 11.5 | DUP | 4227 | 2483 | | NO |
| 50 | NO | all | 12.5 | DUP | 4227 | 2625 | | NO |
| 52 | NO | HPBT | 10.5 | DUP | 4227 | 2297 | | NO |
| 52 | NO | HPBT | 11.5 | DUP | 4227 | 2433 | | NO |
| 52 | NO | HPBT | 12.5 | DUP | 4227 | 2579 | | NO |
| 53 | SR | HP | 9.5 | HER | 2400 | 2200 | | SR |
| 53 | SR | HP | 9.8 | HER | 2400 | 2300 | | SR |
| 53 | SR | HP | 10.1 | HER | 2400 | 2400 | | SR |
| 53 | SR | HP | 10.5 | HER | 2400 | 2500 | | SR |
| 53 | SR | HP | 10.2 | DUP | 4227 | 2100 | | SR |
| 53 | SR | HP | 10.6 | DUP | 4227 | 2200 | | SR |
| 53 | SR | HP | 11.0 | DUP | 4227 | 2300 | | SR |
| 53 | SR | HP | 11.5 | DUP | 4227 | 2400 | | SR |
| 53 | SR | HP | 12.0 | DUP | 4227 | 2500 | | SR |
| 53 | SR | HP | 12.5 | DUP | 4227 | 2600 | | SR |
| 55 | SR | SEMP | 9.5 | HER | 2400 | 2200 | | SR |
| 55 | SR | SEMP | 10.0 | HER | 2400 | 2300 | | SR |

LOADS FOR .218 BEE .224

| BULLET | | | POWDER | | | | | |
|---|---|---|---|---|---|---|---|---|
| Wgt | Mfg | Type | GRN | Mfg | Type | VEL(fps) | OAL | Source |
| 55 | SR | SEMP | 10.5 | HER | 2400 | 2400 | | SR |
| 55 | SR | SEMP | 10.2 | DUP | 4227 | 2100 | | SR |
| 55 | SR | SEMP | 10.7 | DUP | 4227 | 2200 | | SR |
| 55 | SR | SEMP | 11.3 | DUP | 4227 | 2300 | | SR |
| 55 | SR | SEMP | 11.9 | DUP | 4227 | 2400 | | SR |
| 55 | SR | SEMP | 12.5 | DUP | 4227 | 2500 | | SR |

LOADS FOR .222 REMINGTON .224

| BULLET | | | POWDER | | | | | |
|---|---|---|---|---|---|---|---|---|
| Wgt | Mfg | Type | GRN | Mfg | Type | VEL(fps) | OAL | Source |
| 40 | SR | horn | 19.3 | DUP | 4198 | 2900 | 2.130 | SR |
| 40 | SR | horn | 19.7 | DUP | 4198 | 3000 | | SR |
| 40 | SR | horn | 20.1 | DUP | 4198 | 3100 | | SR |
| 40 | SR | horn | 20.5 | DUP | 4198 | 3200 | | SR |
| 40 | SR | horn | 21.0 | DUP | 4198 | 3300 | | SR |
| 40 | SR | horn | 21.5 | DUP | 4198 | 3400 | | SR |
| 40 | SR | horn | 19.5 | HER | RE-7 | 3000 | | SR |
| 40 | SR | horn | 20.2 | HER | RE-7 | 3100 | | SR |
| 40 | SR | horn | 20.9 | HER | RE-7 | 3200 | | SR |
| 40 | SR | horn | 21.6 | HER | RE-7 | 3300 | | SR |
| 40 | SR | horn | 22.3 | HER | RE-7 | 3400 | | SR |
| 40 | SP | SP | 20.0 | HER | RE-7 | 3331 | | SP |
| 40 | SP | SP | 21.0 | HER | RE-7 | 3502 | | SP |
| 40 | SP | SP | 22.0 | HER | RE-7 | 3652 | | SP |
| 40 | SP | SP | 19.0 | DUP | 4198 | 3201 | | SP |
| 40 | SP | SP | 20.0 | DUP | 4198 | 3360 | | SP |
| 40 | SP | SP | 21.0 | DUP | 4198 | 3513 | | SP |
| 45 | SR | all | 18.9 | DUP | 4198 | 2800 | | SR |
| 45 | SR | all | 19.3 | DUP | 4198 | 2900 | | SR |
| 45 | SR | all | 19.7 | DUP | 4198 | 3000 | | SR |
| 45 | SR | all | 20.1 | DUP | 4198 | 3100 | | SR |
| 45 | SR | all | 20.5 | DUP | 4198 | 3200 | | SR |
| 45 | SR | all | 21.0 | DUP | 4198 | 3300 | | SR |
| 45 | SR | all | 19.4 | HER | RE-7 | 2800 | | SR |
| 45 | SR | all | 19.9 | HER | RE-7 | 2900 | | SR |
| 45 | SR | all | 20.4 | HER | RE-7 | 3000 | | SR |
| 45 | SR | all | 20.9 | HER | RE-7 | 3100 | | SR |
| 45 | SR | all | 21.4 | HER | RE-7 | 3200 | | SR |
| 45 | SR | all | 21.9 | HER | RE-7 | 3300 | | SR |
| 45 | SP | SPT | 20.0 | HER | RE-7 | 3222 | | SP |
| 45 | SP | SPT | 21.0 | HER | RE-7 | 3366 | | SP |
| 45 | SP | SPT | 22.0 | HER | RE-7 | 3522 | | SP |
| 45 | SP | SPT | 19.0 | DUP | 4198 | 3142 | | SP |
| 45 | SP | SPT | 20.0 | DUP | 4198 | 3318 | | SP |
| 45 | SP | SPT | 21.0 | DUP | 4198 | 3472 | | SP |
| 45 | NO | BTSB | 19.0 | DUP | 4198 | 2990 | | NO |
| 45 | NO | BTSB | 20.0 | DUP | 4198 | 3172 | | NO |
| 45 | NO | BTSB | 21.0 | DUP | 4198 | 3358 | | NO |
| 50 | SR | all | 18.5 | DUP | 4198 | 2800 | | SR |
| 50 | SR | all | 19.0 | DUP | 4198 | 2900 | | SR |
| 50 | SR | all | 19.5 | DUP | 4198 | 3000 | | SR |
| 50 | SR | all | 20.0 | DUP | 4198 | 3100 | | SR |
| 50 | SR | all | 20.5 | DUP | 4198 | 3200 | | SR |
| 50 | SR | all | 18.7 | HER | RE-7 | 2800 | | SR |
| 50 | SR | all | 19.2 | HER | RE-7 | 2900 | | SR |
| 50 | SR | all | 19.7 | HER | RE-7 | 3000 | | SR |
| 50 | SR | all | 20.3 | HER | RE-7 | 3100 | | SR |
| 50 | SR | all | 20.9 | HER | RE-7 | 3200 | | SR |
| 50 | SR | all | 21.5 | HER | RE-7 | 3300 | | SR |
| 50 | SP | SPT | 22.0 | DUP | 4064 | 2942 | | SP |
| 50 | SP | SPT | 23.0 | DUP | 4064 | 3074 | | SP |

LOADS FOR .222 REMINGTON .224

| BULLET | | | POWDER | | | | | |
|---|---|---|---|---|---|---|---|---|
| Wgt | Mfg | Type | GRN | Mfg | Type | VEL(fps) | OAL | Source |
| 50 | SP | SPT | 24.0 | DUP | 4064 | 3196 | | SP |
| 50 | SP | SPT | 18.0 | DUP | 4198 | 2931 | | SP |
| 50 | SP | SPT | 19.0 | DUP | 4198 | 3095 | | SP |
| 50 | SP | SPT | 20.0 | DUP | 4198 | 3251 | | SP |
| 50 | NO | SP | 18.0 | DUP | 4198 | 2863 | | NO |
| 50 | NO | SP | 19.0 | DUP | 4198 | 3021 | | NO |
| 50 | NO | SP | 20.0 | DUP | 4198 | 3165 | | NO |
| 50 | NO | HP | 18.0 | DUP | 4198 | 2859 | | NO |
| 50 | NO | HP | 19.0 | DUP | 4198 | 3024 | | NO |
| 50 | NO | HP | 20.0 | DUP | 4198 | 3172 | | NO |
| 52 | SP | all | 21.5 | DUP | 4064 | 2890 | | SP |
| 52 | SP | all | 22.5 | DUP | 4064 | 3019 | | SP |
| 52 | SP | all | 23.5 | DUP | 4064 | 3139 | | SP |
| 52 | SP | all | 17.5 | DUP | 4198 | 2849 | | SP |
| 52 | SP | all | 18.5 | DUP | 4198 | 3001 | | SP |
| 52 | SP | all | 19.5 | DUP | 4198 | 3161 | | SP |
| 52 | NO | HPBT | 18.0 | DUP | 4198 | 2854 | | NO |
| 52 | NO | HPBT | 19.0 | DUP | 4198 | 3003 | | NO |
| 52 | NO | HPBT | 20.0 | DUP | 4198 | 3131 | | NO |
| 53 | SR | HPBT | 18.3 | DUP | 4198 | 2700 | | SR |
| 53 | SR | HPBT | 18.8 | DUP | 4198 | 2800 | | SR |
| 53 | SR | HPBT | 19.3 | DUP | 4198 | 2900 | | SR |
| 53 | SR | HPBT | 19.8 | DUP | 4198 | 3000 | | SR |
| 53 | SR | HPBT | 20.4 | DUP | 4198 | 3100 | | SR |
| 53 | SR | HPBT | 18.9 | HER | RE-7 | 2700 | | SR |
| 53 | SR | HPBT | 19.4 | HER | RE-7 | 2800 | | SR |
| 53 | SR | HPBT | 19.9 | HER | RE-7 | 2900 | | SR |
| 53 | SR | HPBT | 20.4 | HER | RE-7 | 3000 | | SR |
| 53 | SR | HPBT | 20.9 | HER | RE-7 | 3100 | | SR |
| 53 | SR | HPBT | 21.4 | HER | RE-7 | 3200 | | SR |
| 55 | SR | all | 17.5 | DUP | 4198 | 2600 | | SR |
| 55 | SR | all | 18.0 | DUP | 4198 | 2700 | | SR |
| 55 | SR | all | 18.5 | DUP | 4198 | 2800 | | SR |
| 55 | SR | all | 19.0 | DUP | 4198 | 2900 | | SR |
| 55 | SR | all | 19.5 | DUP | 4198 | 3000 | | SR |
| 55 | SR | all | 18.0 | HER | RE-7 | 2700 | | SR |
| 55 | SR | all | 18.6 | HER | RE-7 | 2800 | | SR |
| 55 | SR | all | 19.2 | HER | RE-7 | 2900 | | SR |
| 55 | SR | all | 19.9 | HER | RE-7 | 3000 | | SR |
| 55 | SR | all | 20.6 | HER | RE-7 | 3100 | | SR |
| 55 | SP | all | 21.5 | DUP | 4064 | 2845 | | SP |
| 55 | SP | all | 22.5 | DUP | 4064 | 2990 | | SP |
| 55 | SP | all | 23.5 | DUP | 4064 | 3109 | | SP |
| 55 | SP | all | 17.5 | DUP | 4198 | 2748 | | SP |
| 55 | SP | all | 18.5 | DUP | 4198 | 2896 | | SP |
| 55 | SP | all | 19.5 | DUP | 4198 | 3053 | | SP |
| 55 | NO | SP | 17.5 | DUP | 4198 | 2774 | | NO |
| 55 | NO | SP | 18.5 | DUP | 4198 | 2893 | | NO |
| 55 | NO | SP | 19.5 | DUP | 4198 | 3011 | | NO |
| 60 | NO | SP | 17.0 | DUP | 4198 | 2562 | | NO |
| 60 | NO | SP | 18.0 | DUP | 4198 | 2704 | | NO |

LOADS FOR .222 REMINGTON .224

| BULLET | | | POWDER | | | | | |
|---|---|---|---|---|---|---|---|---|
| Wgt | Mfg | Type | GRN | Mfg | Type | VEL(fps) | OAL | Source |
| 60 | NO | SP | 19.0 | DUP | 4198 | 2854 | | NO |
| 63 | SR | SEMP | 16.6 | DUP | 4198 | 2400 | | SR |
| 63 | SR | SEMP | 17.2 | DUP | 4198 | 2500 | | SR |
| 63 | SR | SEMP | 17.8 | DUP | 4198 | 2600 | | SR |
| 63 | SR | SEMP | 18.4 | DUP | 4198 | 2700 | | SR |
| 63 | SR | SEMP | 19.0 | DUP | 4198 | 2800 | | SR |
| 63 | 6R | SEMP | 16.5 | HER | RE-7 | 2400 | | SR |
| 63 | SR | SEMP | 17.2 | HER | RE-7 | 2500 | | SR |
| 63 | SR | SEMP | 17.9 | HER | RE-7 | 2600 | | SR |
| 63 | SR | SEMP | 18.6 | HER | RE-7 | 2700 | | SR |
| 63 | SR | SEMP | 19.3 | HER | RE-7 | 2800 | | SR |
| 63 | SR | SEMP | 20.0 | HER | RE-7 | 2900 | | SR |
| 70 | SP | SSP | 19.5 | DUP | 4064 | 2519 | | SP |
| 70 | SP | SSP | 20.5 | DUP | 4064 | 2641 | | SP |
| 70 | SP | SSP | 21.5 | DUP | 4064 | 2775 | | SP |

LOADS FOR .223 REMINGTON .224

| BULLET | | | POWDER | | | | | |
|---|---|---|---|---|---|---|---|---|
| Weight | Mfg | Type | GRN | Mfg | Type | VEL(fps) | OAL | Source |
| 40 | SR | horn | 22.0 | DUP | 4198 | 3300 | 2.260 | SR |
| 40 | SR | horn | 22.6 | DUP | 4198 | 3400 | | SR |
| 40 | SR | horn | 23.3 | DUP | 4198 | 3500 | | SR |
| 40 | SR | horn | 21.0 | HER | RE-7 | 3300 | | SR |
| 40 | SR | horn | 21.8 | HER | RE-7 | 3400 | | SR |
| 40 | SR | horn | 22.6 | HER | RE-7 | 3500 | | SR |
| 40 | SP | SP | 25.0 | DUP | 4895 | 3011 | | SP |
| 40 | SP | SP | 26.0 | DUP | 4895 | 3148 | | SP |
| 40 | SP | SP | 27.0 | DUP | 4895 | 3274 | | SP |
| 40 | SP | SP | 21.5 | DUP | 4198 | 3056 | | SP |
| 40 | SP | SP | 22.5 | DUP | 4198 | 3193 | | SP |
| 40 | SP | SP | 23.5 | DUP | 4198 | 3352 | | SP |
| 45 | SR | all | 20.3 | DUP | 4198 | 3000 | | SR |
| 45 | SR | all | 20.9 | DUP | 4198 | 3100 | | SR |
| 45 | SR | all | 21.5 | DUP | 4198 | 3200 | | SR |
| 45 | SR | all | 22.1 | DUP | 4198 | 3300 | | SR |
| 45 | SR | all | 22.8 | DUP | 4198 | 3400 | | SR |
| 45 | SR | all | 21.1 | HER | RE-7 | 3100 | | SR |
| 45 | SR | all | 21.9 | HER | RE-7 | 3200 | | SR |
| 45 | SR | all | 22.7 | HER | RE-7 | 3300 | | SR |
| 45 | SP | SP | 25.0 | DUP | 4895 | 2956 | | SP |
| 45 | SP | SP | 26.0 | DUP | 4895 | 3088 | | SP |
| 45 | SP | SP | 27.0 | DUP | 4895 | 3213 | | SP |
| 45 | SP | SP | 21.0 | DUP | 4198 | 2982 | | SP |
| 45 | SP | SP | 22.0 | DUP | 4198 | 3111 | | SP |
| 45 | SP | SP | 23.0 | DUP | 4198 | 3256 | | SP |
| 50 | NO | SP | 23.0 | DUP | 4064 | 2802 | | NO |
| 50 | NO | SP | 25.0 | DUP | 4064 | 3009 | | NO |
| 50 | NO | SP | 27.0 | DUP | 4064 | 3244 | | NO |
| 50 | NO | HP | 23.0 | DUP | 4064 | 2800 | | NO |
| 50 | NO | HP | 25.0 | DUP | 4064 | 3010 | | NO |
| 50 | NO | HP | 27.0 | DUP | 4064 | 3246 | | NO |
| 50 | SR | all | 19.8 | DUP | 4198 | 2800 | | SR |
| 50 | SR | all | 20.3 | DUP | 4198 | 2900 | | SR |
| 50 | SR | all | 20.8 | DUP | 4198 | 3000 | | SR |
| 50 | SR | all | 21.4 | DUP | 4198 | 3100 | | SR |
| 50 | SR | all | 22.0 | DUP | 4198 | 3200 | | SR |
| 50 | SR | all | 20.5 | HER | RE-7 | 3000 | | SR |
| 50 | SR | all | 21.3 | HER | RE-7 | 3100 | | SR |
| 50 | SR | all | 22.1 | HER | RE-7 | 3200 | | SR |
| 50 | SP | SP | 25.0 | DUP | 4895 | 2933 | | SP |
| 50 | SP | SP | 26.0 | DUP | 4895 | 3036 | | SP |
| 50 | SP | SP | 27.0 | DUP | 4895 | 3149 | | SP |
| 50 | SP | SP | 20.5 | DUP | 4198 | 2917 | | SP |
| 50 | SP | SP | 21.5 | DUP | 4198 | 3053 | | SP |
| 50 | SP | SP | 22.5 | DUP | 4198 | 3185 | | SP |
| 52 | NO | HPBT | 23.0 | DUP | 4064 | 2807 | | NO |

LOADS FOR .22 PPC .224

| BULLET | | | POWDER | | | | | |
|---|---|---|---|---|---|---|---|---|
| Wgt | Mfg | Type | GRN | Mfg | Type | VEL(fps) | OAL | Source |
| 52 | SR | HPBT | 22.1 | DUP | 4198 | 3050 | 1.520 | SR |
| 52 | SR | HPBT | 22.5 | DUP | 4198 | 3100 | | SR |
| 52 | SR | HPBT | 22.9 | DUP | 4198 | 3150 | | SR |
| 52 | SR | HPBT | 23.3 | DUP | 4198 | 3200 | | SR |
| 52 | SR | HPBT | 23.7 | DUP | 4198 | 3250 | | SR |
| 52 | SR | HPBT | 24.1 | DUP | 4198 | 3300 | | SR |
| 52 | SR | HPBT | 24.5 | DUP | 4198 | 3350 | | SR |
| 52 | SR | HPBT | 24.8 | DUP | 4198 | 3400 | | SR |
| 52 | SR | HPBT | 22.2 | HER | RE-7 | 3050 | | SR |
| 52 | SR | HPBT | 22.6 | HER | RE-7 | 3100 | | SR |
| 52 | SR | HPBT | 23.0 | HER | RE-7 | 3150 | | SR |
| 52 | SR | HPBT | 23.4 | HER | RE-7 | 3200 | | SR |
| 52 | SR | HPBT | 23.8 | HER | RE-7 | 3250 | | SR |
| 52 | SR | HPBT | 24.2 | HER | RE-7 | 3300 | | SR |
| 52 | SR | HPBT | 24.6 | HER | RE-7 | 3350 | | SR |
| 52 | SR | HPBT | 25.0 | HER | RE-7 | 3400 | | SR |
| 52 | SP | all | 25.0 | DUP | 4895 | 3010 | | SP |
| 52 | SP | all | 26.0 | DUP | 4895 | 3126 | | SP |
| 52 | SP | all | 27.0 | DUP | 4895 | 3237 | | SP |
| 52 | SP | all | 21.5 | DUP | 4198 | 3053 | | SP |
| 52 | SP | all | 22.5 | DUP | 4198 | 3198 | | SP |
| 52 | SP | all | 23.5 | DUP | 4198 | 3323 | | SP |
| 53 | SR | HP | 22.1 | DUP | 4198 | 3050 | 1.520 | SR |
| 53 | SR | HP | 22.5 | DUP | 4198 | 3100 | | SR |
| 53 | SR | HP | 22.9 | DUP | 4198 | 3150 | | SR |
| 53 | SR | HP | 23.3 | DUP | 4198 | 3200 | | SR |
| 53 | SR | HP | 23.7 | DUP | 4198 | 3250 | | SR |
| 53 | SR | HP | 24.1 | DUP | 4198 | 3300 | | SR |
| 53 | SR | HP | 24.5 | DUP | 4198 | 3350 | | SR |
| 53 | SR | HP | 24.8 | DUP | 4198 | 3400 | | SR |
| 53 | SR | HP | 22.2 | HER | RE-7 | 3050 | | SR |
| 53 | SR | HP | 22.6 | HER | RE-7 | 3100 | | SR |
| 53 | SR | HP | 23.0 | HER | RE-7 | 3150 | | SR |
| 53 | SR | HP | 23.4 | HER | RE-7 | 3200 | | SR |
| 53 | SR | HP | 23.8 | HER | RE-7 | 3250 | | SR |
| 53 | SR | HP | 24.2 | HER | RE-7 | 3300 | | SR |
| 53 | SR | HP | 24.6 | HER | RE-7 | 3350 | | SR |
| 53 | SR | HP | 25.0 | HER | RE-7 | 3400 | | SR |
| 55 | SP | SP | 25.0 | DUP | 4895 | 2966 | | SP |
| 55 | SP | SP | 26.0 | DUP | 4895 | 3080 | | SP |
| 55 | SP | SP | 27.0 | DUP | 4895 | 3186 | | SP |
| 55 | SP | SP | 21.0 | DUP | 4198 | 2914 | | SP |
| 55 | SP | SP | 22.0 | DUP | 4198 | 3049 | | SP |
| 55 | SP | SP | 23.0 | DUP | 4198 | 3174 | | SP |
| 52 | NO | HPBT | 25.0 | DUP | 4064 | 3015 | | NO |
| 52 | NO | HPBT | 27.0 | DUP | 4064 | 3284 | | NO |
| 52 | SR | HPBT | 19.4 | DUP | 4198 | 2800 | | SR |
| 52 | SR | HPBT | 20.0 | DUP | 4198 | 2900 | | SR |
| 52 | SR | HPBT | 20.6 | DUP | 4198 | 3000 | | SR |
| 52 | SR | HPBT | 21.2 | DUP | 4198 | 3100 | | SR |
| 52 | SR | HPBT | 19.3 | HER | RE-7 | 3000 | | SR |
| 52 | SR | HPBT | 20.9 | HER | RE-7 | 3100 | | SR |
| 52 | SR | HPBT | 22.5 | HER | RE-7 | 3200 | | SR |

LOADS FOR .22 PPC .224

| BULLET | | | POWDER | | | | | |
|---|---|---|---|---|---|---|---|---|
| Wgt | Mfg | Type | GRN | Mfg | Type | VEL(fps) | OAL | Source |
| 52 | SP | all | 24.5 | DUP | 4895 | 2850 | | SP |
| 52 | SP | all | 25.5 | DUP | 4895 | 2960 | | SP |
| 52 | SP | all | 26.5 | DUP | 4895 | 3064 | | SP |
| 52 | SP | all | 20.0 | DUP | 4198 | 2851 | | SP |
| 52 | SP | all | 21.0 | DUP | 4198 | 2983 | | SP |
| 52 | SP | all | 22.0 | DUP | 4198 | 3127 | | SP |
| 53 | SR | HP | 19.4 | DUP | 4198 | 2800 | | SR |
| 53 | SR | HP | 20.0 | DUP | 4198 | 2900 | | SR |
| 53 | SR | HP | 20.6 | DUP | 4198 | 3000 | | SR |
| 53 | SR | HP | 21.2 | DUP | 4198 | 3100 | | SR |
| 53 | SR | HP | 19.3 | HER | RE-7 | 3000 | | SR |
| 53 | SR | HP | 20.9 | HER | RE-7 | 3100 | | SR |
| 53 | SR | HP | 22.5 | HER | RE-7 | 3200 | | SR |
| 55 | NO | SP | 22.5 | DUP | 4064 | 2897 | | NO |
| 55 | NO | SP | 24.5 | DUP | 4064 | 2963 | | NO |
| 55 | NO | SP | 26.5 | DUP | 4064 | 3178 | | NO |
| 55 | SP | all | 24.5 | DUP | 4064 | 2785 | | SP |
| 55 | SP | all | 25.5 | DUP | 4064 | 2888 | | SP |
| 55 | SP | all | 26.5 | DUP | 4064 | 3002 | | SP |
| 55 | SP | all | 20.0 | DUP | 4198 | 2826 | | SP |
| 55 | SP | all | 21.0 | DUP | 4198 | 2970 | | SP |
| 55 | SP | all | 22.0 | DUP | 4198 | 3101 | | SP |
| 55 | SR | all | 19.3 | DUP | 4198 | 2800 | | SR |
| 55 | SR | all | 19.9 | DUP | 4198 | 2900 | | SR |
| 55 | SR | all | 20.5 | DUP | 4198 | 3000 | | SR |
| 55 | SR | all | 21.2 | DUP | 4198 | 3100 | | SR |
| 55 | SR | all | 21.9 | DUP | 4198 | 3200 | | SR |
| 55 | SR | all | 20.2 | HER | RE-7 | 2900 | | SR |
| 55 | SR | all | 21.0 | HER | RE-7 | 3000 | | SR |
| 55 | SR | all | 21.8 | HER | RE-7 | 3100 | | SR |
| 60 | NO | SP | 17.0 | DUP | 4198 | 2516 | | NO |
| 60 | NO | SP | 19.0 | DUP | 4198 | 2791 | | NO |
| 60 | NO | SP | 21.0 | DUP | 4198 | 3056 | | NO |
| 63 | SR | SEMP | 18.5 | DUP | 4198 | 2600 | | SR |
| 63 | SR | SEMP | 20.2 | DUP | 4198 | 2700 | | SR |
| 63 | SR | SEMP | 18.5 | HER | RE-7 | 2700 | | SR |
| 63 | SR | SEMP | 19.5 | HER | RE-7 | 2800 | | SR |
| 63 | SR | SEMP | 20.5 | HER | RE-7 | 2900 | | SR |
| 63 | SR | SEMP | 21.5 | HER | RE-7 | 3000 | | SR |
| 70 | SP | SSP | 22.0 | DUP | 4064 | 2398 | | SP |
| 70 | SP | SSP | 23.0 | DUP | 4064 | 2510 | | SP |
| 70 | SP | SSP | 24.0 | DUP | 4064 | 2623 | | SP |
| 70 | SP | SSP | 18.5 | DUP | 4198 | 2413 | | SP |
| 70 | SP | SSP | 19.5 | DUP | 4198 | 2548 | | SP |
| 70 | SP | SSP | 20.5 | DUP | 4198 | 2665 | | SP |

LOADS FOR .222 REMINGTON MAGNUM .224

| BULLET | | | POWDER | | | | | |
|---|---|---|---|---|---|---|---|---|
| Wgt | Mfg | Type | GRN | Mfg | Type | VEL(fps) | OAL | Source |
| 40 | SR | horn | 21.9 | DUP | 4198 | 3300 | 2.280 | SR |
| 40 | SR | horn | 22.4 | DUP | 4198 | 3400 | | SR |
| 40 | SR | horn | 22.9 | DUP | 4198 | 3500 | | SR |
| 40 | SR | horn | 21.8 | HER | RE-7 | 3300 | | SR |
| 40 | SR | horn | 22.5 | HER | RE-7 | 3400 | | SR |
| 40 | SR | horn | 23.2 | HER | RE-7 | 3500 | | SR |
| 40 | SP | SP | 26.0 | DUP | 4064 | 3255 | | SP |
| 40 | SP | SP | 27.0 | DUP | 4064 | 3380 | | SP |
| 40 | SP | SP | 28.0 | DUP | 4064 | 3489 | | SP |
| 40 | SP | SP | 21.5 | DUP | 4198 | 3251 | | SP |
| 40 | SP | SP | 22.5 | DUP | 4198 | 3404 | | SP |
| 40 | SP | SP | 23.5 | DUP | 4198 | 3540 | | SP |
| 45 | SR | all | 22.1 | DUP | 4198 | 3200 | | SR |
| 45 | SR | all | 22.5 | DUP | 4198 | 3300 | | SR |
| 45 | SR | all | 22.9 | DUP | 4198 | 3400 | | SR |
| 45 | SR | all | 23.3 | DUP | 4198 | 3500 | | SR |
| 45 | SR | all | 23.7 | DUP | 4198 | 3600 | | SR |
| 45 | SR | all | 21.7 | HER | RE-7 | 3200 | | SR |
| 45 | SR | all | 22.5 | HER | RE-7 | 3300 | | SR |
| 45 | SR | all | 23.3 | HER | RE-7 | 3400 | | SR |
| 45 | SP | SP | 25.0 | DUP | 4895 | 3161 | | SP |
| 45 | SP | SP | 26.0 | DUP | 4895 | 3290 | | SP |
| 45 | SP | SP | 27.0 | DUP | 4895 | 3396 | | SP |
| 45 | SP | SP | 21.5 | DUP | 4198 | 3197 | | SP |
| 45 | SP | SP | 22.5 | DUP | 4198 | 3350 | | SP |
| 45 | SP | SP | 23.5 | DUP | 4198 | 3488 | | SP |
| 50 | SR | all | 21.0 | DUP | 4198 | 3000 | | SR |
| 50 | SR | all | 21.5 | DUP | 4198 | 3100 | | SR |
| 50 | SR | all | 22.0 | DUP | 4198 | 3200 | | SR |
| 50 | SR | all | 22.5 | DUP | 4198 | 3300 | | SR |
| 50 | SR | all | 23.1 | DUP | 4198 | 3400 | | SR |
| 50 | SR | all | 20.8 | HER | RE-7 | 3000 | | SR |
| 50 | SR | all | 21.5 | HER | RE-7 | 3100 | | SR |
| 50 | SR | all | 22.2 | HER | RE-7 | 3200 | | SR |
| 50 | SR | all | 23.0 | HER | RE-7 | 3300 | | SR |
| 50 | SP | SP | 26.0 | DUP | 4064 | 3190 | | SP |
| 50 | SP | SP | 27.0 | DUP | 4064 | 3299 | | SP |
| 50 | SP | SP | 28.0 | DUP | 4064 | 3414 | | SP |
| 50 | SP | SP | 21.0 | DUP | 4198 | 3132 | | SP |
| 50 | SP | SP | 22.0 | DUP | 4198 | 3285 | | SP |
| 50 | SP | SP | 23.0 | DUP | 4198 | 3415 | | SP |
| 50 | NO | SP | 25.0 | DUP | 4064 | 3103 | | NO |
| 50 | NO | SP | 26.0 | DUP | 4064 | 3284 | | NO |
| 50 | NO | SP | 27.0 | DUP | 4064 | 3415 | | NO |
| 50 | NO | HP | 25.0 | DUP | 4064 | 3110 | | NO |
| 50 | NO | HP | 26.0 | DUP | 4064 | 3282 | | NO |
| 50 | NO | HP | 27.0 | DUP | 4064 | 3417 | | NO |
| 52 | SP | all | 26.0 | DUP | 4064 | 3189 | | SP |
| 52 | SP | all | 27.0 | DUP | 4064 | 3304 | | SP |
| 52 | SP | all | 28.0 | DUP | 4064 | 3415 | | SP |
| 52 | SP | all | 25.0 | DUP | 4895 | 3098 | | SP |
| 52 | SP | all | 26.0 | DUP | 4895 | 3216 | | SP |
| 52 | SP | all | 27.0 | DUP | 4895 | 3330 | | SP |

LOADS FOR .222 REMINGTON MAGNUM .224

| BULLET | | | POWDER | | | | | |
|---|---|---|---|---|---|---|---|---|
| Wgt | Mfg | Type | GRN | Mfg | Type | VEL(fps) | OAL | Source |
| 52 | NO | HPBT | 24.5 | DUP | 4064 | 3029 | | NO |
| 52 | NO | HPBT | 25.5 | DUP | 4064 | 3171 | | NO |
| 52 | NO | HPBT | 26.5 | DUP | 4064 | 3297 | | NO |
| 52&53 | SR | all | 21.0 | DUP | 4198 | 3000 | | SR |
| 52&53 | SR | all | 21.5 | DUP | 4198 | 3100 | | SR |
| 52&53 | SR | all | 22.0 | DUP | 4198 | 3200 | | SR |
| 52&53 | SR | all | 22.5 | DUP | 4198 | 3300 | | SR |
| 52&53 | SR | all | 23.1 | DUP | 4198 | 3400 | | SR |
| 52&53 | SR | all | 20.1 | HER | RE-7 | 2900 | | SR |
| 52&53 | SR | all | 20.8 | HER | RE-7 | 3000 | | SR |
| 52&53 | SR | all | 21.5 | HER | RE-7 | 3100 | | SR |
| 52&53 | SR | all | 22.3 | HER | RE-7 | 3200 | | SR |
| 55 | SR | all | 20.8 | DUP | 4198 | 3000 | | SR |
| 55 | SR | all | 21.3 | DUP | 4198 | 3100 | | SR |
| 55 | SR | all | 21.9 | DUP | 4198 | 3200 | | SR |
| 55 | SR | all | 22.5 | DUP | 4198 | 3300 | | SR |
| 55 | SR | all | 20.1 | HER | RE-7 | 2900 | | SR |
| 55 | SR | all | 20.9 | HER | RE-7 | 3000 | | SR |
| 55 | SR | all | 21.7 | HER | RE-7 | 3100 | | SR |
| 55 | SR | all | 22.5 | HER | RE-7 | 3200 | | SR |
| 55 | SP | all | 25.5 | DUP | 4064 | 3077 | | SP |
| 55 | SP | all | 26.5 | DUP | 4064 | 3191 | | SP |
| 55 | SP | all | 27.5 | DUP | 4064 | 3294 | | SP |
| 55 | SP | all | 24.5 | DUP | 4895 | 3040 | | SP |
| 55 | SP | all | 25.5 | DUP | 4895 | 3166 | | SP |
| 55 | SP | all | 26.5 | DUP | 4895 | 3275 | | SP |
| 55 | NO | SP | 24.0 | DUP | 4064 | 2970 | | NO |
| 55 | NO | SP | 25.0 | DUP | 4064 | 3089 | | NO |
| 55 | NO | SP | 3192 | DUP | 4064 | 3192 | | NO |
| 63 | SR | SEMP | 20.4 | DUP | 4198 | 2900 | | SR |
| 63 | SR | SEMP | 21.1 | DUP | 4198 | 3000 | | SR |
| 63 | SR | SEMP | 21.8 | DUP | 4198 | 3100 | | SR |
| 63 | SR | SEMP | 22.5 | DUP | 4198 | 3200 | | SR |
| 63 | SR | SEMP | 19.9 | HER | RE-7 | 2800 | | SR |
| 63 | SR | SEMP | 20.8 | HER | RE-7 | 2900 | | SR |
| 63 | SR | SEMP | 21.7 | HER | RE-7 | 3000 | | SR |
| 70 | SP | SSP | 24.5 | DUP | 4064 | 2820 | | SP |
| 70 | SP | SSP | 25.5 | DUP | 4064 | 2939 | | SP |
| 70 | SP | SSP | 26.5 | DUP | 4064 | 3043 | | SP |
| 70 | SP | SSP | 23.0 | DUP | 4895 | 2723 | | SP |
| 70 | SP | SSP | 24.0 | DUP | 4895 | 2849 | | SP |
| 70 | SP | SSP | 25.0 | DUP | 4895 | 2938 | | SP |

LOADS FOR .224 WEATHERBY MAGNUM .224

| BULLET | | | | POWDER | | | | |
|---|---|---|---|---|---|---|---|---|
| Wgt | Mfg | Type | GRN | Mfg | Type | VEL(fps) | OAL | Source |
| 45 | SR | all | 25.4 | DUP | 4198 | 3500 | 2.312 | SR |
| 45 | SR | all | 26.0 | DUP | 4198 | 3600 | | SR |
| 45 | SR | all | 26.6 | DUP | 4198 | 3700 | | SR |
| 45 | SR | all | 27.3 | DUP | 4198 | 3800 | | SR |
| 45 | SR | all | 28.0 | DUP | 4198 | 3900 | | SR |
| 45 | SR | all | 30.3 | DUP | 4064 | 3500 | | SR |
| 45 | SR | all | 31.1 | DUP | 4064 | 3600 | | SR |
| 45 | SR | all | 32.0 | DUP | 4064 | 3700 | | SR |
| 45 | SR | all | 32.9 | DUP | 4064 | 3800 | | SR |
| 50 | SR | all | 24.4 | DUP | 4198 | 3300 | | SR |
| 50 | SR | all | 25.1 | DUP | 4198 | 3400 | | SR |
| 50 | SR | all | 25.9 | DUP | 4198 | 3500 | | SR |
| 50 | SR | all | 26.7 | DUP | 4198 | 3600 | | SR |
| 50 | SR | all | 27.5 | DUP | 4198 | 3700 | | SR |
| 50 | SR | all | 30.1 | DUP | 4064 | 3500 | | SR |
| 50 | SR | all | 31.0 | DUP | 4064 | 3600 | | SR |
| 50 | SR | all | 32.0 | DUP | 4064 | 3700 | | SR |
| 50 | SR | all | 33.0 | DUP | 4064 | 3800 | | SR |
| 52&53 | SR | all | 23.9 | DUP | 4198 | 3200 | | SR |
| 52&53 | SR | all | 24.6 | DUP | 4198 | 3300 | | SR |
| 52&53 | SR | all | 25.3 | DUP | 4198 | 3400 | | SR |
| 52&53 | SR | all | 26.0 | DUP | 4198 | 3500 | | SR |
| 52&53 | SR | all | 26.8 | DUP | 4198 | 3600 | | SR |
| 52&53 | SR | all | 29.9 | DUP | 4064 | 3400 | | SR |
| 52&53 | SR | all | 30.7 | DUP | 4064 | 3500 | | SR |
| 52&53 | SR | all | 31.6 | DUP | 4064 | 3600 | | SR |
| 52&53 | SR | all | 32.5 | DUP | 4064 | 3700 | | SR |
| 55 | SR | all | 23.8 | DUP | 4198 | 3200 | | SR |
| 55 | SR | all | 24.7 | DUP | 4198 | 3300 | | SR |
| 55 | SR | all | 25.6 | DUP | 4198 | 3400 | | SR |
| 55 | SR | all | 26.5 | DUP | 4198 | 3500 | | SR |
| 55 | SR | all | 29.2 | DUP | 4064 | 3300 | | SR |
| 55 | SR | all | 30.0 | DUP | 4064 | 3400 | | SR |
| 55 | SR | all | 30.9 | DUP | 4064 | 3500 | | SR |
| 55 | SR | all | 31.8 | DUP | 4064 | 3600 | | SR |
| 55 | SR | all | 32.7 | DUP | 4064 | 3700 | | SR |
| 63 | SR | SEMP | 23.3 | DUP | 4198 | 3000 | | SR |
| 63 | SR | SEMP | 24.0 | DUP | 4198 | 3100 | | SR |
| 63 | SR | SEMP | 24.8 | DUP | 4198 | 3200 | | SR |
| 63 | SR | SEMP | 25.6 | DUP | 4198 | 3300 | | SR |
| 63 | SR | SEMP | 29.5 | DUP | 4064 | 3300 | | SR |
| 63 | SR | SEMP | 30.4 | DUP | 4064 | 3400 | | SR |
| 63 | SR | SEMP | 31.4 | DUP | 4064 | 3500 | | SR |

LOADS FOR .225 WINCHESTER .224

| BULLET | | | POWDER | | | | | |
|---|---|---|---|---|---|---|---|---|
| Wgt | Mfg | Type | GRN | Mfg | Type | VEL(fps) | OAL | Source |
| 40 | SR | horn | 22.4 | DUP | 4198 | 3000 | 2.500 | SR |
| 40 | SR | horn | 23.1 | DUP | 4198 | 3100 | | SR |
| 40 | SR | horn | 23.8 | DUP | 4198 | 3200 | | SR |
| 40 | SR | horn | 24.6 | DUP | 4198 | 3300 | | SR |
| 40 | SR | horn | 25.4 | DUP | 4198 | 3400 | | SR |
| 40 | SR | horn | 30.8 | DUP | 4064 | 3300 | | SR |
| 40 | SR | horn | 31.3 | DUP | 4064 | 3400 | | SR |
| 40 | SR | horn | 31.9 | DUP | 4064 | 3500 | | SR |
| 40 | SP | SP | 30.5 | DUP | 4064 | 3344 | | SP |
| 40 | SP | SP | 32.5 | DUP | 4064 | 3592 | | SP |
| 40 | SP | SP | 34.5 | DUP | 4064 | 3791 | | SP |
| 45 | SR | all | 25.0 | DUP | 4198 | 3200 | | SR |
| 45 | SR | all | 25.9 | DUP | 4198 | 3300 | | SR |
| 45 | SR | all | 30.2 | DUP | 4064 | 3200 | | SR |
| 45 | SR | all | 30.8 | DUP | 4064 | 3300 | | SR |
| 45 | SR | all | 31.4 | DUP | 4064 | 3400 | | SR |
| 45 | SR | all | 32.0 | DUP | 4064 | 3500 | | SR |
| 45 | SR | all | 32.6 | DUP | 4064 | 3600 | | SR |
| 45 | SR | all | 33.2 | DUP | 4064 | 3700 | | SR |
| 45 | SP | SP | 30.0 | DUP | 4064 | 3284 | | SP |
| 45 | SP | SP | 32.0 | DUP | 4064 | 3496 | | SP |
| 45 | SP | SP | 34.0 | DUP | 4064 | 3737 | | SP |
| 50 | SR | all | 24.2 | DUP | 4198 | 3100 | | SR |
| 50 | SR | all | 25.0 | DUP | 4198 | 3200 | | SR |
| 50 | SR | all | 30.2 | DUP | 4064 | 3200 | | SR |
| 50 | SR | all | 30.8 | DUP | 4064 | 3300 | | SR |
| 50 | SR | all | 31.4 | DUP | 4064 | 3400 | | SR |
| 50 | SR | all | 32.1 | DUP | 4064 | 3500 | | SR |
| 50 | SR | all | 32.8 | DUP | 4064 | 3600 | | SR |
| 50 | SP | SP | 29.5 | DUP | 4064 | 3202 | | SP |
| 50 | SP | SP | 31.5 | DUP | 4064 | 3419 | | SP |
| 50 | SP | SP | 33.5 | DUP | 4064 | 3641 | | SP |
| 50 | NO | SP | 30.5 | DUP | 4064 | 3173 | | NO |
| 50 | NO | SP | 32.5 | DUP | 4064 | 3445 | | NO |
| 50 | NO | SP | 34.5 | DUP | 4064 | 3707 | | NO |
| 50 | NO | HP | 30.5 | DUP | 4064 | 3186 | | NO |
| 50 | NO | HP | 32.5 | DUP | 4064 | 3450 | | NO |
| 50 | NO | HP | 34.5 | DUP | 4064 | 3713 | | NO |
| 52 | SP | all | 29.0 | DUP | 4064 | 3134 | | NO |
| 52 | SP | all | 31.0 | DUP | 4064 | 3350 | | SP |
| 52 | SP | all | 33.0 | DUP | 4064 | 3551 | | SP |
| 52 | NO | HPBT | 30.0 | DUP | 4064 | 3229 | | NO |
| 52 | NO | HPBT | 32.0 | DUP | 4064 | 3447 | | NO |
| 52 | NO | HPBT | 34.0 | DUP | 4064 | 3647 | | NO |
| 52&53 | SR | all | 29.7 | DUP | 4064 | 3100 | | SR |
| 52&53 | SR | all | 30.2 | DUP | 4064 | 3200 | | SR |
| 52&53 | SR | all | 30.7 | DUP | 4064 | 3300 | | SR |
| 52&53 | SR | all | 31.2 | DUP | 4064 | 3400 | | SR |
| 52&53 | SR | all | 31.8 | DUP | 4064 | 3500 | | SR |
| 52&53 | SR | all | 32.4 | DUP | 4064 | 3600 | | SR |
| 55 | SR | all | 29.3 | DUP | 4064 | 3100 | | SR |
| 55 | SR | all | 29.9 | DUP | 4064 | 3200 | | SR |

LOADS FOR .225 WINCHESTER .224

| BULLET | | | POWDER | | | | | |
|---|---|---|---|---|---|---|---|---|
| Wgt | Mfg | Type | GRN | Mfg | Type | VEL(fps) | OAL | Source |
| 55 | SR | all | 30.5 | DUP | 4064 | 33oo | | SR |
| 55 | SR | all | 31.1 | DUP | 4064 | 3400 | | SR |
| 55 | SR | all | 31.8 | DUP | 4064 | 3500 | | SR |
| 55 | SP | all | 28.5 | DUP | 4064 | 3095 | | SP |
| 55 | SP | all | 30.5 | DUP | 4064 | 3331 | | SP |
| 55 | SP | all | 32.5 | DUP | 4064 | 3503 | | SP |
| 55 | NO | SP | 29.5 | DUP | 4064 | 3041 | | NO |
| 55 | NO | SP | 31.5 | DUP | 4064 | 3246 | | NO |
| 55 | NO | SP | 33.5 | DUP | 4064 | 3517 | | NO |
| 63 | SR | SEMP | 27.8 | DUP | 4064 | 2900 | | SR |
| 63 | SR | SEMP | 28.5 | DUP | 4064 | 3000 | | SR |
| 63 | SR | SEMP | 29.3 | DUP | 4064 | 3100 | | SR |
| 63 | SR | SEMP | 30.1 | DUP | 4064 | 3200 | | SR |
| 63 | SR | SEMP | 30.9 | DUP | 4064 | 3300 | | SR |
| 70 | SP | SSP | 26.0 | DUP | 4064 | 2585 | | SP |
| 70 | SP | SSP | 28.0 | DUP | 4064 | 2784 | | SP |
| 70 | SP | SSP | 30.0 | DUP | 4064 | 2980 | | SP |

LOADS FOR .22-250 REMINGTON .224

| BULLET | | | POWDER | | | | | |
|---|---|---|---|---|---|---|---|---|
| Wgt | Mfg | Type | GRN | Mfg | Type | VEL(fps) | OAL | Source |
| 40 | SP | SP | 34.0 | DUP | 4064 | 3656 | | SP |
| 40 | SP | SP | 36.0 | DUP | 4064 | 3865 | | SP |
| 40 | SP | SP | 38.0 | DUP | 4064 | 4053 | | SP |
| 45 | SR | all | 26.9 | DUP | 4198 | 3500 | 2.350 | SR |
| 45 | SR | all | 27.9 | DUP | 4198 | 3600 | | SR |
| 45 | SR | all | 28.9 | DUP | 4198 | 3700 | | SR |
| 45 | SR | all | 29.9 | DUP | 4198 | 3800 | | SR |
| 45 | SR | all | 30.9 | DUP | 4198 | 3900 | | SR |
| 45 | SR | all | 28.9 | HER | RE-7 | 3600 | | SR |
| 45 | SR | all | 30.0 | HER | RE-7 | 3700 | | SR |
| 45 | SR | all | 31.2 | HER | RE-7 | 3800 | | SR |
| 45 | SP | SP | 33.5 | DUP | 4064 | 3535 | | SP |
| 45 | SP | SP | 35.5 | DUP | 4064 | 3740 | | SP |
| 45 | SP | SP | 37.5 | DUP | 4064 | 3932 | | SP |
| 50 | SR | all | 26.0 | DUP | 4198 | 3400 | | SR |
| 50 | SR | all | 27.0 | DUP | 4198 | 3500 | | SR |
| 50 | SR | all | 28.0 | DUP | 4198 | 3600 | | SR |
| 50 | SR | all | 29.0 | DUP | 4198 | 3700 | | SR |
| 50 | SR | all | 28.3 | HER | RE-7 | 3500 | | SR |
| 50 | SR | all | 29.4 | HER | RE-7 | 3600 | | SR |
| 50 | SR | all | 30.6 | HER | RE-7 | 3700 | | SR |
| 50 | SP | SP | 33.0 | DUP | 4064 | 3421 | | SP |
| 50 | SP | SP | 35.0 | DUP | 4064 | 3639 | | SP |
| 50 | SP | SP | 37.0 | DUP | 4064 | 3810 | | SP |
| 50 | NO | SP | 31.0 | DUP | 4064 | 3320 | | NO |
| 50 | NO | SP | 33.0 | DUP | 4064 | 3542 | | NO |
| 50 | NO | SP | 35.0 | DUP | 4064 | 3769 | | NO |
| 50 | NO | HP | 31.0 | DUP | 4064 | 3323 | | NO |
| 50 | NO | HP | 33.0 | DUP | 4064 | 3545 | | NO |
| 50 | NO | HP | 35.0 | DUP | 4064 | 3771 | | NO |
| 52 | SP | all | 32.5 | DUP | 4064 | 3393 | | SP |
| 52 | SP | all | 34.5 | DUP | 4064 | 3599 | | SP |
| 52 | SP | all | 36.5 | DUP | 4064 | 3788 | | SP |
| 52 | NO | HPBT | 30.5 | DUP | 4064 | 3248 | | NO |
| 52 | NO | HPBT | 32.5 | DUP | 4064 | 3451 | | NO |
| 52 | NO | HPBT | 34.5 | DUP | 4064 | 3692 | | NO |
| 52&53 | SR | all | 26.6 | HER | RE-7 | 3300 | | SR |
| 52&53 | SR | all | 27.8 | HER | RE-7 | 3400 | | SR |
| 52&53 | SR | all | 29.0 | HER | RE-7 | 3500 | | SR |
| 52&53 | SR | all | 30.3 | HER | RE-7 | 3600 | | SR |
| 52&53 | SR | all | 31.0 | DUP | 3031 | 3300 | | SR |
| 52&53 | SR | all | 31.8 | DUP | 3031 | 3400 | | SR |
| 52&53 | SR | all | 32.6 | DUP | 3031 | 3500 | | SR |
| 52&53 | SR | all | 33.4 | DUP | 3031 | 3600 | | SR |
| 52&53 | SR | all | 34.2 | DUP | 3031 | 3700 | | SR |
| 55 | SR | all | 26.5 | HER | RE-7 | 3300 | | SR |
| 55 | SR | all | 27.7 | HER | RE-7 | 3400 | | SR |
| 55 | SR | all | 29.0 | HER | RE-7 | 3500 | | SR |
| 55 | SR | all | 30.3 | HER | RE-7 | 3600 | | SR |
| 55 | SR | all | 30.8 | DUP | 3031 | 3300 | | SR |
| 55 | SR | all | 31.6 | DUP | 3031 | 3400 | | SR |
| 55 | SR | all | 32.5 | DUP | 3031 | 3500 | | SR |

LOADS FOR .22-250 REMINGTON .224

| BULLET | | | POWDER | | | | | |
|---|---|---|---|---|---|---|---|---|
| Wgt | Mfg | Type | GRN | Mfg | Type | VEL(fps) | OAL | Source |
| 55 | SR | all | 33.4 | DUP | 3031 | 3600 | | SR |
| 55 | SR | all | 34.3 | DUP | 3031 | 3700 | | SR |
| 55 | SP | all | 32.0 | DUP | 4064 | 3339 | | SP |
| 55 | SP | all | 34.0 | DUP | 4064 | 3528 | | SP |
| 55 | SP | all | 36.0 | DUP | 4064 | 3726 | | SP |
| 55 | NO | SP | 30.5 | DUP | 4064 | 3197 | | NO |
| 55 | NO | SP | 32.5 | DUP | 4064 | 3409 | | NO |
| 55 | NO | SP | 34.5 | DUP | 4064 | 3612 | | NO |
| 60 | NO | SP | 29.0 | DUP | 4064 | 3208 | | NO |
| 60 | NO | SP | 31.0 | DUP | 4064 | 3341 | | NO |
| 60 | NO | SP | 33.0 | DUP | 4064 | 3486 | | NO |
| 63 | SR | SEMP | 25.2 | HER | RE-7 | 3000 | | SR |
| 63 | SR | SEMP | 26.2 | HER | RE-7 | 3100 | | SR |
| 63 | SR | SEMP | 27.2 | HER | RE-7 | 3200 | | SR |
| 63 | SR | SEMP | 28.3 | HER | RE-7 | 3300 | | SR |
| 63 | SR | SEMP | 30.1 | DUP | 4064 | 3000 | | SR |
| 63 | SR | SEMP | 31.0 | DUP | 4064 | 3100 | | SR |
| 63 | SR | SEMP | 32.0 | DUP | 4064 | 3200 | | SR |
| 63 | SR | SEMP | 33.0 | DUP | 4064 | 3300 | | SR |
| 63 | SR | SEMP | 34.0 | DUP | 4064 | 3400 | | SR |
| 63 | SR | SEMP | 35.0 | DUP | 4064 | 3500 | | SR |
| 70 | SP | SSP | 29.5 | DUP | 4064 | 2970 | | SP |
| 70 | SP | SSP | 31.5 | DUP | 4064 | 3168 | | SP |
| 70 | SP | SSP | 33.5 | DUP | 4064 | 3341 | | SP |

## LOADS FOR .220 SWIFT .224

| BULLET | | | POWDER | | | | | |
|---|---|---|---|---|---|---|---|---|
| Wgt | Mfg | Type | GRN | Mfg | Type | VEL(fps) | OAL | Source |
| 40 | SP | SP | 35.0 | DUP | 4064 | 3449 | | SP |
| 40 | SP | SP | 37.0 | DUP | 4064 | 3670 | | SP |
| 40 | SP | SP | 39.0 | DUP | 4064 | 3888 | | SP |
| 45 | SR | all | 33.9 | DUP | 3031 | 3700 | 2.680 | SR |
| 45 | SR | all | 34.7 | DUP | 3031 | 3800 | | SR |
| 45 | SR | all | 35.6 | DUP | 3031 | 3900 | | SR |
| 45 | SR | all | 36.5 | DUP | 3031 | 4000 | | SR |
| 45 | SR | all | 37.4 | DUP | 3031 | 4100 | | SR |
| 45 | SR | all | 38.3 | DUP | 3031 | 4200 | | SR |
| 45 | SR | all | 36.6 | DUP | 4064 | 3700 | | SR |
| 45 | SR | all | 37.4 | DUP | 4064 | 3800 | | SR |
| 45 | SR | all | 38.2 | DUP | 4064 | 3900 | | SR |
| 45 | SR | all | 39.0 | DUP | 4064 | 4000 | | SR |
| 45 | SR | all | 39.8 | DUP | 4064 | 4100 | | SR |
| 45 | SP | SP | 35.0 | DUP | 4064 | 3502 | | SP |
| 45 | SP | SP | 37.0 | DUP | 4064 | 3680 | | SP |
| 45 | SP | SP | 39.0 | DUP | 4064 | 3867 | | SP |
| 50 | SR | all | 33.5 | DUP | 3031 | 3600 | | SR |
| 50 | SR | all | 34.4 | DUP | 3031 | 3700 | | SR |
| 50 | SR | all | 35.4 | DUP | 3031 | 3800 | | SR |
| 50 | SR | all | 36.4 | DUP | 3031 | 3900 | | SR |
| 50 | SR | all | 37.4 | DUP | 3031 | 4000 | | SR |
| 50 | SR | all | 35.5 | DUP | 4064 | 3600 | | SR |
| 50 | SR | all | 36.4 | DUP | 4064 | 3700 | | SR |
| 50 | SR | all | 37.3 | DUP | 4064 | 3800 | | SR |
| 50 | SR | all | 38.2 | DUP | 4064 | 3900 | | SR |
| 50 | SR | all | 39.2 | DUP | 4064 | 4000 | | SR |
| 50 | SR | all | 40.2 | DUP | 4064 | 4100 | | SR |
| 50 | SP | SP | 40.0 | DUP | 4831 | 3339 | | SP |
| 50 | SP | SP | 42.0 | DUP | 4831 | 3495 | | SP |
| 50 | SP | SP | 44.0 | DUP | 4831 | 3661 | | SP |
| 50 | NO | SP | 40.0 | DUP | 4831 | 3413 | | NO |
| 50 | NO | SP | 42.0 | DUP | 4831 | 3717 | | NO |
| 50 | NO | SP | 44.0 | DUP | 4831 | 3870 | | NO |
| 50 | NO | HP | 40.0 | DUP | 4831 | 3416 | | NO |
| 50 | NO | HP | 42.0 | DUP | 4831 | 3720 | | NO |
| 50 | NO | HP | 44.0 | DUP | 4831 | 3871 | | NO |
| 52 | SP | all | 40.0 | DUP | 4831 | 3429 | | SP |
| 52 | SP | all | 42.0 | DUP | 4831 | 3551 | | SP |
| 52 | SP | all | 44.0 | DUP | 4831 | 3653 | | SP |
| 52 | NO | HPBT | 40.0 | DUP | 4831 | 3433 | | NO |
| 52 | NO | HPBT | 42.0 | DUP | 4831 | 3653 | | NO |
| 52 | NO | HPBT | 44.0 | DUP | 4831 | 3821 | | NO |
| 52&53 | SR | all | 33.6 | DUP | 3031 | 3600 | | SR |
| 52&53 | SR | all | 34.5 | DUP | 3031 | 3700 | | SR |
| 52&53 | SR | all | 35.5 | DUP | 3031 | 3800 | | SR |
| 52&53 | SR | all | 36.5 | DUP | 3031 | 3900 | | SR |
| 52&53 | SR | all | 34.5 | DUP | 4064 | 3500 | | SR |
| 52&53 | SR | all | 35.4 | DUP | 4064 | 3600 | | SR |
| 52&53 | SR | all | 36.3 | DUP | 4064 | 3700 | | SR |
| 52&53 | SR | all | 37.2 | DUP | 4064 | 3800 | | SR |
| 52&53 | SR | all | 38.1 | DUP | 4064 | 3900 | | SR |
| 52&53 | SR | all | 39.0 | DUP | 4064 | 4000 | | SR |

LOADS FOR .220 SWIFT .224

| BULLET | | | POWDER | | | | | |
|---|---|---|---|---|---|---|---|---|
| Wgt | Mfg | Type | GRN | Mfg | Type | VEL(fps) | OAL | Source |
| 55 | SR | all | 32.9 | DUP | 3031 | 3500 | | SR |
| 55 | SR | all | 33.8 | DUP | 3031 | 3600 | | SR |
| 55 | SR | all | 34.8 | DUP | 3031 | 3700 | | SR |
| 55 | SR | all | 35.8 | DUP | 3031 | 3800 | | SR |
| 55 | SR | all | 36.8 | DUP | 3031 | 3900 | | SR |
| 55 | SR | all | 35.3 | DUP | 4064 | 3500 | | SR |
| 55 | SR | all | 36.2 | DUP | 4064 | 3600 | | SR |
| 55 | SR | all | 37.1 | DUP | 4064 | 3700 | | SR |
| 55 | SR | all | 38.0 | DUP | 4064 | 3800 | | SR |
| 55 | SR | all | 39.0 | DUP | 4064 | 3900 | | SR |
| 55 | SP | all | 39.5 | DUP | 4831 | 3302 | | SP |
| 55 | SP | all | 41.5 | DUP | 4831 | 3463 | | SP |
| 55 | SP | all | 43.5 | DUP | 4831 | 3629 | | SP |
| 55 | NO | SP | 40.0 | DUP | 4831 | 3480 | | NO |
| 55 | NO | SP | 42.0 | DUP | 4831 | 3632 | | NO |
| 55 | NO | SP | 44.0 | DUP | 4831 | 3873 | | NO |
| 60 | NO | SP | 39.0 | DUP | 4831 | 3338 | | NO |
| 60 | NO | SP | 41.0 | DUP | 4831 | 3559 | | NO |
| 60 | NO | SP | 43.0 | DUP | 4831 | 3763 | | NO |
| 63 | SR | SEMP | 29.8 | DUP | 3031 | 3200 | | SR |
| 63 | SR | SEMP | 31.1 | DUP | 3031 | 3300 | | SR |
| 63 | SR | SEMP | 32.5 | DUP | 3031 | 3400 | | SR |
| 63 | SR | SEMP | 33.9 | DUP | 3031 | 3500 | | SR |
| 63 | SR | SEMP | 30.5 | DUP | 4064 | 3100 | | SR |
| 63 | SR | SEMP | 31.7 | DUP | 4064 | 3200 | | SR |
| 63 | SR | SEMP | 32.9 | DUP | 4064 | 3300 | | SR |
| 63 | SR | SEMP | 34.1 | DUP | 4064 | 3400 | | SR |
| 63 | SR | SEMP | 35.3 | DUP | 4064 | 3500 | | SR |
| 70 | SP | SSP | 36.0 | DUP | 4831 | 2885 | | SP |
| 70 | SP | SSP | 38.0 | DUP | 4831 | 3046 | | SP |
| 70 | SP | SSP | 40.0 | DUP | 4831 | 3223 | | SP |

LOADS FOR 6 X 47 .243

| BULLET | | | POWDER | | | | | |
|---|---|---|---|---|---|---|---|---|
| Wgt | Mfg | Type | GRN | Mfg | Type | VEL(fps) | OAL | Source |
| 60 | SR | HP | 20.1 | DUP | 4198 | 2800 | n/a | SR |
| 60 | SR | HP | 20.9 | DUP | 4198 | 2900 | | SR |
| 60 | SR | HP | 21.7 | DUP | 4198 | 3000 | | SR |
| 60 | SR | HP | 22.5 | DUP | 4198 | 3100 | | SR |
| 60 | SR | HP | 23.2 | DUP | 4198 | 3200 | | SR |
| 60 | SR | HP | 23.9 | DUP | 4198 | 3300 | | SR |
| 60 | SR | HP | 24.6 | DUP | 4198 | 3400 | | SR |
| 60 | SR | HP | 22.9 | DUP | 3031 | 2800 | | SR |
| 60 | SR | HP | 23.6 | DUP | 3031 | 2900 | | SR |
| 60 | SR | HP | 24.3 | DUP | 3031 | 3000 | | SR |
| 60 | SR | HP | 25.1 | DUP | 3031 | 3100 | | SR |
| 60 | SR | HP | 25.9 | DUP | 3031 | 3200 | | SR |
| 60 | SR | HP | 26.7 | DUP | 3031 | 3300 | | SR |
| 70 | SR | HPBT | 20.3 | DUP | 4198 | 2700 | | SR |
| 70 | SR | HPBT | 21.1 | DUP | 4198 | 2800 | | SR |
| 70 | SR | HPBT | 21.9 | DUP | 4198 | 2900 | | SR |
| 70 | SR | HPBT | 22.7 | DUP | 4198 | 3000 | | SR |
| 70 | SR | HPBT | 23.5 | DUP | 4198 | 3100 | | SR |
| 70 | SR | HPBT | 24.3 | DUP | 4198 | 3200 | | SR |
| 70 | SR | HPBT | 22.8 | DUP | 3031 | 2700 | | SR |
| 70 | SR | HPBT | 23.5 | DUP | 3031 | 2800 | | SR |
| 70 | SR | HPBT | 24.2 | DUP | 3031 | 2900 | | SR |
| 70 | SR | HPBT | 24.9 | DUP | 3031 | 3000 | | SR |
| 70 | SR | HPBT | 25.6 | DUP | 3031 | 3100 | | SR |
| 70 | NO | HPBT | 20.0 | DUP | 4198 | 2648 | | NO |
| 70 | NO | HPBT | 22.0 | DUP | 4198 | 2953 | | NO |
| 70 | NO | HPBT | 24.0 | DUP | 4198 | 3173 | | NO |
| 75 | SR | HP | 20.7 | DUP | 4198 | 2700 | | SR |
| 75 | SR | HP | 21.5 | DUP | 4198 | 2800 | | SR |
| 75 | SR | HP | 22.3 | DUP | 4198 | 2900 | | SR |
| 75 | SR | HP | 23.0 | DUP | 4198 | 3000 | | SR |
| 75 | SR | HP | 23.5 | DUP | 3031 | 2700 | | SR |
| 75 | SR | HP | 23.8 | DUP | 3031 | 2800 | | SR |
| 75 | SR | HP | 24.1 | DUP | 3031 | 2900 | | SR |
| 75 | SR | HP | 24.9 | DUP | 3031 | 3000 | | SR |
| 75 | SR | HP | 25.7 | DUP | 3031 | 3100 | | SR |
| 75 | SP | HP | 23.0 | DUP | 4064 | 2421 | | SP |
| 75 | SP | HP | 25.0 | DUP | 4064 | 2647 | | SP |
| 75 | SP | HP | 27.0 | DUP | 4064 | 2836 | | SP |
| 80 | SP | SP | 23.0 | DUP | 4064 | 2404 | | SP |
| 80 | SP | SP | 25.0 | DUP | 4064 | 2595 | | SP |
| 80 | SP | SP | 27.0 | DUP | 4064 | 2786 | | SP |
| 85 | SP | BT | 22.5 | DUP | 4064 | 2414 | | SP |
| 85 | SP | BT | 24.5 | DUP | 4064 | 2634 | | SP |
| 85 | SP | BT | 26.5 | DUP | 4064 | 2809 | | SP |
| 90 | SP | all | 22.0 | DUP | 4064 | 2288 | | SP |
| 90 | SP | all | 24.0 | DUP | 4064 | 2453 | | SP |
| 90 | SP | all | 26.0 | DUP | 4064 | 2682 | | SP |

LOADS FOR 6mm PPC .243

| BULLET | | | POWDER | | | | | |
|---|---|---|---|---|---|---|---|---|
| Wgt | Mfg | Type | GRN | Mfg | Type | VEL(fps) | OAL | Source |
| 70 | SR | HPBT | 22.5 | DUP | 4198 | 2850 | n/a | SR |
| 70 | SR | HPBT | 23.0 | DUP | 4198 | 2900 | | SR |
| 70 | SR | HPBT | 23.5 | DUP | 4198 | 2950 | | SR |
| 70 | SR | HPBT | 24.0 | DUP | 4198 | 3000 | | SR |
| 70 | SR | HPBT | 24.5 | DUP | 4198 | 3050 | | SR |
| 70 | SR | HPBT | 25.0 | DUP | 4198 | 3100 | | SR |
| 70 | SR | HPBT | 25.5 | DUP | 4198 | 3150 | | SR |
| 70 | SR | HPBT | 26.0 | DUP | 4198 | 3200 | | SR |
| 70 | SR | HPBT | 23.7 | HER | RE-7 | 2900 | | SR |
| 70 | SR | HPBT | 24.2 | HER | RE-7 | 2950 | | SR |
| 70 | SR | HPBT | 24.6 | HER | RE-7 | 3000 | | SR |
| 70 | SR | HPBT | 25.0 | HER | RE-7 | 3050 | | SR |
| 70 | SR | HPBT | 25.4 | HER | RE-7 | 3100 | | SR |
| 70 | SR | HPBT | 25.8 | HER | RE-7 | 3150 | | SR |
| 75 | SP | HP | 24.0 | DUP | 4198 | 2620 | | SP |
| 75 | SP | HP | 25.0 | DUP | 4198 | 2738 | | SP |
| 75 | SP | HP | 26.0 | DUP | 4198 | 2910 | | SP |
| 80 | SP | SP | 24.0 | DUP | 4198 | 2562 | | SP |
| 80 | SP | SP | 25.0 | DUP | 4198 | 2666 | | SP |
| 80 | SP | SP | 26.0 | DUP | 4198 | 2766 | | SP |
| 85 | SP | BT | 23.5 | DUP | 4198 | 2838 | | SP |
| 85 | SP | BT | 24.5 | DUP | 4198 | 2961 | | SP |
| 85 | SP | BT | 25.5 | DUP | 4198 | 3068 | | SP |
| 90 | SP | all | 23.5 | DUP | 4064 | 2308 | | SP |
| 90 | SP | all | 24.5 | DUP | 4064 | 2423 | | SP |
| 90 | SP | all | 25.5 | DUP | 4064 | 2510 | | SP |

LOADS FOR 6mm INTERNATIONAL .243

| BULLET | | | POWDER | | | | | |
|---|---|---|---|---|---|---|---|---|
| Wgt | Mfg | Type | GRN | Mfg | Type | VEL(fps) | OAL | Source |
| 60 | SR | HP | 27.4 | DUP | 4198 | 3100 | 2.450 | SR |
| 60 | SR | HP | 28.5 | DUP | 4198 | 3200 | | SR |
| 60 | SR | HP | 29.6 | DUP | 4198 | 3300 | | SR |
| 60 | SR | HP | 30.6 | DUP | 4198 | 3400 | | SR |
| 60 | SR | HP | 31.6 | DUP | 4198 | 3500 | | SR |
| 60 | SR | HP | 30.7 | DUP | 3031 | 3100 | | SR |
| 60 | SR | HP | 31.6 | DUP | 3031 | 3200 | | SR |
| 60 | SR | HP | 32.5 | DUP | 3031 | 3300 | | SR |
| 60 | SR | HP | 33.4 | DUP | 3031 | 3400 | | SR |
| 60 | SR | HP | 34.3 | DUP | 3031 | 3500 | | SR |
| 60 | SR | HP | 35.2 | DUP | 3031 | 3600 | | SR |
| 70 | NO | HPBT | 32.0 | DUP | 4064 | 3100 | | NO |
| 70 | NO | HPBT | 34.0 | DUP | 4064 | 3215 | | NO |
| 70 | NO | HPBT | 36.0 | DUP | 4064 | 3387 | | NO |
| 70&75 | SR | all | 25.6 | DUP | 4198 | 2800 | 2.450 | SR |
| 70&75 | SR | all | 26.9 | DUP | 4198 | 2900 | | SR |
| 70&75 | SR | all | 28.1 | DUP | 4198 | 3000 | | SR |
| 70&75 | SR | all | 28.8 | DUP | 3031 | 2800 | | SR |
| 70&75 | SR | all | 29.9 | DUP | 3031 | 2900 | | SR |
| 70&75 | SR | all | 31.0 | DUP | 3031 | 3000 | | SR |
| 70&75 | SR | all | 32.0 | DUP | 3031 | 3100 | | SR |
| 70&75 | SR | all | 33.0 | DUP | 3031 | 3200 | | SR |
| 70&75 | SR | all | 34.0 | DUP | 3031 | 3300 | | SR |
| 85 | NO | SP | 30.5 | DUP | 4064 | 2788 | | NO |
| 85 | NO | SP | 32.5 | DUP | 4064 | 2943 | | NO |
| 85 | NO | SP | 34.5 | DUP | 4064 | 3113 | | NO |

LOADS FOR .243 WINCHESTER .243

| BULLET | | | POWDER | | | | | |
|---|---|---|---|---|---|---|---|---|
| Wgt | Mfg | Type | GRN | Mfg | Type | VEL(fps) | OAL | Source |
| 60 | SR | HP | 37.9 | DUP | 4064 | 3400 | 2.710 | SR |
| 60 | SR | HP | 39.1 | DUP | 4064 | 3500 | | SR |
| 60 | SR | HP | 40.3 | DUP | 4064 | 3600 | | SR |
| 60 | SR | HP | 41.6 | DUP | 4064 | 3700 | | SR |
| 60 | SR | HP | 43.8 | DUP | 4831 | 3300 | | SR |
| 60 | SR | HP | 44.9 | DUP | 4831 | 3400 | | SR |
| 60 | SR | HP | 46.1 | DUP | 4831 | 3500 | | SR |
| 60 | SR | HP | 47.3 | DUP | 4831 | 3600 | | SR |
| 70 | NO | HPBT | 37.5 | DUP | 4064 | 3171 | | NO |
| 70 | NO | HPBT | 39.5 | DUP | 4064 | 3326 | | NO |
| 70 | NO | HPBT | 41.5 | DUP | 4064 | 3489 | | NO |
| 70&75 | SR | all | 34.0 | DUP | 3031 | 3100 | | SR |
| 70&75 | SR | all | 35.3 | DUP | 3031 | 3200 | | SR |
| 70&75 | SR | all | 36.6 | DUP | 3031 | 3300 | | SR |
| 70&75 | SR | all | 38.0 | DUP | 3031 | 3400 | | SR |
| 70&75 | SR | all | 35.5 | DUP | 4064 | 3100 | | SR |
| 70&75 | SR | all | 36.8 | DUP | 4064 | 3200 | | SR |
| 70&75 | SR | all | 38.1 | DUP | 4064 | 3300 | | SR |
| 70&75 | SR | all | 39.4 | DUP | 4064 | 3400 | | SR |
| 75 | SP | HP | 36.0 | DUP | 4064 | 3005 | | SP |
| 75 | SP | HP | 38.0 | DUP | 4064 | 3171 | | SP |
| 75 | SP | HP | 40.0 | DUP | 4064 | 3312 | | SP |
| 80 | SP | SP | 35.0 | DUP | 4064 | 2931 | | SP |
| 80 | SP | SP | 37.0 | DUP | 4064 | 3097 | | SP |
| 80 | SP | SP | 39.0 | DUP | 4064 | 3246 | | SP |
| 85 | NO | SP | 34.0 | DUP | 4064 | 2908 | | NO |
| 85 | NO | SP | 36.0 | DUP | 4064 | 3053 | | NO |
| 85 | NO | SP | 38.0 | DUP | 4064 | 3220 | | NO |
| 85 | SP | BT | 36.0 | DUP | 4064 | 2930 | | SP |
| 85 | SP | BT | 38.0 | DUP | 4064 | 3091 | | SP |
| 85 | SP | BT | 40.0 | DUP | 4064 | 3217 | | SP |
| 85 | SR | all | 31.4 | DUP | 3031 | 2800 | | SR |
| 85 | SR | all | 32.9 | DUP | 3031 | 2900 | | SR |
| 85 | SR | all | 34.4 | DUP | 3031 | 3000 | | SR |
| 85 | SR | all | 35.9 | DUP | 3031 | 3100 | | SR |
| 85 | SR | all | 37.5 | DUP | 3031 | 3200 | | SR |
| 85 | SR | all | 33.7 | DUP | 4064 | 2800 | | SR |
| 85 | SR | all | 34.9 | DUP | 4064 | 2900 | | SR |
| 85 | SR | all | 36.1 | DUP | 4064 | 3000 | | SR |
| 85 | SR | all | 37.3 | DUP | 4064 | 3100 | | SR |
| 85 | SR | all | 38.6 | DUP | 4064 | 3200 | | SR |
| 90 | SP | all | 33.5 | DUP | 4064 | 2688 | | SP |
| 90 | SP | all | 35.5 | DUP | 4064 | 2847 | | SP |
| 90 | SP | BT | 37.5 | DUP | 4064 | 2982 | | SP |
| 95 | NO | SP | 33.0 | DUP | 4064 | 2679 | | NO |
| 95 | NO | SP | 35.0 | DUP | 4064 | 2790 | | NO |
| 95 | NO | SP | 37.0 | DUP | 4064 | 2905 | | NO |
| 95 | NO | SP | 47.0C | DUP | 7828 | 3110 | | DUP |
| 100 | SR | all | 30.6 | DUP | 3031 | 2600 | | SR |
| 100 | SR | all | 32.2 | DUP | 3031 | 2700 | | SR |
| 100 | SR | all | 33.8 | DUP | 3031 | 2800 | | SR |
| 100 | SR | all | 35.5 | DUP | 3031 | 2900 | | SR |

## LOADS FOR .243 WINCHESTER .243

| BULLET | | | POWDER | | | | | |
|---|---|---|---|---|---|---|---|---|
| Wgt | Mfg | Type | GRN | Mfg | Type | VEL(fps) | OAL | Source |
| 100 | SR | all | 31.0 | DUP | 4064 | 2600 | | SR |
| 100 | SR | all | 32.7 | DUP | 4064 | 2700 | | SR |
| 100 | SR | all | 34.5 | DUP | 4064 | 2800 | | SR |
| 100 | SR | all | 36.3 | DUP | 4064 | 2900 | | SR |
| 100 | SR | all | 39.2 | DUP | 4831 | 2700 | | SR |
| 100 | SR | all | 40.9 | DUP | 4831 | 2800 | | SR |
| 100 | SR | all | 42.6 | DUP | 4831 | 2900 | | SR |
| 100 | SR | all | 44.2 | DUP | 4831 | 3000 | | SR |
| 100 | NO | SSP | 33.0 | DUP | 4064 | 2676 | | NO |
| 100 | NO | SSP | 35.0 | DUP | 4064 | 2793 | | NO |
| 100 | NO | SSP | 37.0 | DUP | 4064 | 2904 | | NO |
| 100 | NO | SP | 33.0 | DUP | 4064 | 2672 | | NO |
| 100 | NO | SP | 35.0 | DUP | 4064 | 2788 | | NO |
| 100 | NO | SP | 37.0 | DUP | 4064 | 2900 | | NO |
| 100 | NO | SP | 47.0C | DUP | 7828 | 3050 | | DUP |
| 105 | SP | all | 40.0 | DUP | 4831 | 2717 | | SP |
| 105 | SP | all | 42.0 | DUP | 4831 | 2844 | | SP |
| 105 | SP | all | 2844 | DUP | 4831 | 2970 | | SP |
| 105 | SP | SP | 46.0C | DUP | 7828 | 2980 | | DUP |

Letter C indicates compressed powder charge.

LOADS FOR .244 REMINGTON/6 mm REMINGTON .243

| BULLET | | | POWDER | | | | | |
|---|---|---|---|---|---|---|---|---|
| Wgt | Mfg | Type | GRN | Mfg | Type | VEL(fps) | OAL | Source |
| 60 | SR | HP | 37.1 | DUP | 3031 | 3300 | 2.825 | SR |
| 60 | SR | HP | 38.3 | DUP | 3031 | 3400 | | SR |
| 60 | SR | HP | 39.5 | DUP | 3031 | 3500 | | SR |
| 60 | SR | HP | 40.7 | DUP | 3031 | 3600 | | SR |
| 60 | SR | HP | 38.4 | DUP | 4064 | 3200 | | SR |
| 60 | SR | HP | 39.4 | DUP | 4064 | 3300 | | SR |
| 60 | SR | HP | 40.4 | DUP | 4064 | 3400 | | SR |
| 60 | SR | HP | 41.4 | DUP | 4064 | 3500 | | SR |
| 60 | SR. | HP | 42.4 | DUP | 4064 | 3600 | | SR |
| 60 | SR | HP | 43.4 | DUP | 4064 | 3700 | | SR |
| 70 | NO | HPBT | 37.5 | DUP | 4064 | 3300 | | NO |
| 70 | NO | HPBT | 39.5 | DUP | 4064 | 3411 | | NO |
| 70 | NO | HPBT | 41.5 | DUP | 4064 | 3505 | | NO |
| 75 | SP | HP | 37.5 | DUP | 4064 | 3121 | | SP |
| 75 | SP | HP | 39.5 | DUP | 4064 | 3298 | | SP |
| 75 | SP | HP | 41.5 | DUP | 4064 | 3461 | | SP |
| 70&75 | SR | all | 36.0 | DUP | 3031 | 3100 | | SR |
| 70&75 | SR | all | 37.3 | DUP | 3031 | 3200 | | SR |
| 70&75 | SR | all | 38.6 | DUP | 3031 | 3300 | | SR |
| 70&75 | SR | all | 39.9 | DUP | 3031 | 3400 | | SR |
| 70&75 | SR | all | 37.7 | DUP | 4064 | 3100 | | SR |
| 70&75 | SR | all | 39.1 | DUP | 4064 | 3200 | | SR |
| 70&75 | SR | all | 40.5 | DUP | 4064 | 3300 | | SR |
| 70&75 | SR | all | 41.9 | DUP | 4064 | 3400 | | SR |
| 80 | SP | SP | 37.0 | DUP | 4064 | 3033 | | SP |
| 80 | SP | SP | 39.0 | DUP | 4064 | 3214 | | SP |
| 80 | SP | SP | 41.0 | DUP | 4064 | 3367 | | SP |
| 85 | SR | all | 34.0 | DUP | 3031 | 2900 | | SR |
| 85 | SR | all | 35.3 | DUP | 3031 | 3000 | | SR |
| 85 | SR | all | 36.7 | DUP | 3031 | 3100 | | SR |
| 85 | SR | all | 38.1 | DUP | 3031 | 3200 | | SR |
| 85 | SR | all | 36.4 | DUP | 4064 | 2900 | | SR |
| 85 | SR | all | 37.7 | DUP | 4064 | 3000 | | SR |
| 85 | SR | all | 38.9 | DUP | 4064 | 3100 | | SR |
| 85 | SR | all | 40.1 | DUP | 4064 | 3200 | | SR |
| 85 | SP | BT | 36.0 | DUP | 4064 | 3017 | | SP |
| 85 | SP | BT | 38.0 | DUP | 4064 | 3142 | | SP |
| 85 | SP | BT | 40.0 | DUP | 4064 | 3307 | | SP |
| 85 | NO | SP | 35.5 | DUP | 4064 | 2914 | | NO |
| 85 | NO | SP | 37.5 | DUP | 4064 | 3057 | | NO |
| 85 | NO | SP | 39.5 | DUP | 4064 | 3207 | | NO |
| 90 | SP | all | 35.0 | DUP | 4064 | 2819 | | SP |
| 90 | SP | all | 37.0 | DUP | 4064 | 2978 | | SP |
| 90 | SP | all | 39.0 | DUP | 4064 | 3132 | | SP |
| 95 | NO | SP | 34.0 | DUP | 4064 | 2719 | | NO |
| 95 | NO | SP | 36.0 | DUP | 4064 | 2860 | | NO |
| 95 | NO | SP | 38.0 | DUP | 4064 | 3005 | | NO |
| 100 | SR | all | 31.2 | DUP | 3031 | 2600 | | SR |
| 100 | SR | all | 32.8 | DUP | 3031 | 2700 | | SR |
| 100 | SR | all | 34.4 | DUP | 3031 | 2800 | | SR |
| 100 | SR | all | 36.0 | DUP | 3031 | 2900 | | SR |
| 100 | SR | all | 34.7 | DUP | 4064 | 2700 | | SR |
| 100 | SR | all | 36.2 | DUP | 4064 | 2800 | | SR |

LOADS FOR .244 REMINGTON/6 mm REMINGTON .243

| BULLET | | | POWDER | | | | | |
|---|---|---|---|---|---|---|---|---|
| Wgt | Mfg | Type | GRN | Mfg | Type | VEL.(fps) | OAL | Source |
| 100 | SR | all | 37.7 | DUP | 4064 | 2900 | | SR |
| 100 | SR | all | 39.3 | DUP | 4064 | 3000 | | SR |
| 100 | NO | SSP | 34.0 | DUP | 4064 | 2708 | | NO |
| 100 | NO | SSP | 36.0 | DUP | 4064 | 2854 | | NO |
| 100 | NO | SSP | 38.0 | DUP | 4064 | 3002 | | NO |
| 100 | NO | SP | 34.0 | DUP | 4064 | 2708 | | NO |
| 100 | NO | SP | 36.0 | DUP | 4064 | 2855 | | NO |
| 100 | NO | SP | 38.0 | DUP | 4064 | 2991 | | NO |
| 100 | SR | SP | 48.0C | DUP | 7828 | 3040 | | DUP |
| 105 | SP | all | 33.0 | DUP | 4064 | 2583 | | SP |
| 105 | SP | all | 35.0 | DUP | 4064 | 2756 | | SP |
| 105 | SP | all | 37.0 | DUP | 4064 | 2907 | | SP |
| 105 | SP | SP | 47.4C | DUP | 7828 | 2975 | | DUP |

Letter C indicates compressed powder charge.

## LOADS FOR .240 WEATHERBY MAGNUM .243

| BULLET | | | POWDER | | | | | |
|---|---|---|---|---|---|---|---|---|
| Wgt | Mfg | Type | GRN | Mfg | Type | VEL(fps) | OAL | Source |
| 70 | NO | HPBT | 51.0 | DUP | 4831 | 3344 | | NO |
| 70 | NO | HPBT | 53.0 | DUP | 4831 | 3519 | | NO |
| 70 | NO | HPBT | 55.0 | DUP | 4831 | 3694 | | NO |
| 70&75 | SR | all | 40.1 | DUP | 4064 | 3100 | 3.062 | SR |
| 70&75 | SR | all | 41.8 | DUP | 4064 | 3200 | | SR |
| 70&75 | SR | all | 43.5 | DUP | 4064 | 3300 | | SR |
| 70&75 | SR | all | 45.2 | DUP | 4064 | 3400 | | SR |
| 70&75 | SR | all | 49.9 | DUP | 4831 | 3200 | | SR |
| 70&75 | SR | all | 51.0 | DUP | 4831 | 3300 | | SR |
| 70&75 | SR | all | 52.1 | DUP | 4831 | 3400 | | SR |
| 70&75 | SR | all | 53.2 | DUP | 4831 | 3500 | | SR |
| 70&75 | SR | all | 54.2 | DUP | 4831 | 3600 | | SR |
| 75 | SP | HP | 42.0 | DUP | 4064 | 3126 | | SP |
| 75 | SP | HP | 44.0 | DUP | 4064 | 3289 | | SP |
| 75 | SP | HP | 46.0 | DUP | 4064 | 3481 | | SP |
| 80 | SP | SP | 41.0 | DUP | 4064 | 3132 | | SP |
| 80 | SP | SP | 43.0 | DUP | 4064 | 3271 | | SP |
| 80 | SP | SP | 45.0 | DUP | 4064 | 3408 | | SP |
| 85 | SR | all | 37.5 | DUP | 4064 | 2900 | | SR |
| 85 | SR | all | 39.1 | DUP | 4064 | 3000 | | SP |
| 85 | SR | all | 40.8 | DUP | 4064 | 3100 | | SR |
| 85 | SR | all | 42.5 | DUP | 4064 | 3200 | | SR |
| 85 | SR | all | 47.4 | DUP | 4831 | 2900 | | SR |
| 85 | SR | all | 48.7 | DUP | 4831 | 3100 | | SR |
| 85 | SR | all | 50.0 | DUP | 4831 | 3200 | | SR |
| 85 | SR | all | 51.3 | DUP | 4831 | 3300 | | SR |
| 85 | SR | all | 52.5 | DUP | 4831 | 3400 | | SR |
| 85 | SP | BT | 49.0 | DUP | 4831 | 3304 | | SP |
| 85 | SP | BT | 51.0 | DUP | 4831 | 3358 | | SP |
| 85 | SP | BT | 53.0 | DUP | 4831 | 3473 | | SP |
| 85 | NO | SP | 49.0 | DUP | 4831 | 3195 | | NO |
| 85 | NO | SP | 51.0 | DUP | 4831 | 3344 | | NO |
| 85 | NO | SP | 53.0 | DUP | 4831 | 3493 | | NO |
| 90 | SP | all | 48.0 | DUP | 4831 | 3176 | | SP |
| 90 | SP | all | 50.0 | DUP | 4831 | 3292 | | SP |
| 90 | SP | all | 52.0 | DUP | 4831 | 3420 | | SP |
| 95 | NO | SP | 47.0 | DUP | 4831 | 3004 | | NO |
| 95 | NO | SP | 49.0 | DUP | 4831 | 3127 | | NO |
| 95 | NO | SP | 51.0 | DUP | 4831 | 3268 | | NO |
| 100 | SR | all | 37.0 | DUP | 4064 | 2700 | | SR |
| 100 | SR | all | 38.4 | DUP | 4064 | 2800 | | SR |
| 100 | SR | all | 39.8 | DUP | 4064 | 2900 | | SR |
| 100 | SR | all | 41.2 | DUP | 4064 | 3000 | | SR |
| 100 | SR | all | 43.3 | DUP | 4831 | 2700 | | SR |
| 100 | SR | all | 44.8 | DUP | 4831 | 2800 | | SR |
| 100 | SR | all | 46.3 | DUP | 4831 | 2900 | | SR |
| 100 | SR | all | 47.8 | DUP | 4831 | 3000 | | SR |
| 100 | SR | all | 49.2 | DUP | 4831 | 3100 | | SR |
| 100 | SR | all | 50.6 | DUP | 4831 | 3200 | | SR |
| 100 | NO | SSP | 47.0 | DUP | 4831 | 2990 | | NO |
| 100 | NO | SSP | 49.0 | DUP | 4831 | 3114 | | NO |

LOADS FOR .240 WEATHERBY MAGNUM .243

| BULLET | | | POWDER | | | | | |
|---|---|---|---|---|---|---|---|---|
| Wgt | Mfg | Type | GRN | Mfg | Type | VEL(fps) | OAL | Source |
| 100 | NO | SSP | 51.0 | DUP | 4831 | 3259 | | NO |
| 100 | NO | SP | 47.0 | DUP | 4831 | 2988 | | NO |
| 100 | NO | SP | 49.0 | DUP | 4831 | 3115 | | NO |
| 100 | NO | SP | 51.0 | DUP | 4831 | 3254 | | NO |
| 105 | SP | all | 45.5 | DUP | 4831 | 2915 | | SP |
| 105 | SP | all | 47.5 | DUP | 4831 | 3044 | | SP |
| 105 | SP | all | 49.5 | DUP | 4831 | 3154 | | SP |

LOADS FOR .256 WINCHESTER .257

| BULLET | | | POWDER | | | | | |
|---|---|---|---|---|---|---|---|---|
| Wgt | Mfg | Type | GRN | Mfg | Type | VEL(fps) | OAL | Source |
| 75 | SR | HP | 11.6 | HER | 2400 | 2100 | 1.590 | SR |
| 75 | SR | HP | 12.2 | HER | 2400 | 2200 | | SR |
| 75 | SR | HP | 12.9 | HER | 2400 | 2300 | | SR |
| 75 | SR | HP | 13.6 | HER | 2400 | 2400 | | SR |
| 75 | SR | HP | 13.1 | DUP | 4227 | 2100 | | SR |
| 75 | SR | HP | 13.7 | DUP | 4227 | 2200 | | SR |
| 75 | SR | HP | 14.3 | DUP | 4227 | 2300 | | SR |
| 75 | SR | HP | 14.9 | DUP | 4227 | 2400 | | SR |
| 75 | SR | HP | 15.6 | DUP | 4227 | 2500 | | SR |
| 87 | SR | SPZ | 10.6 | HER | 2400 | 1800 | | SR |
| 87 | SR | SPZ | 11.2 | HER | 2400 | 1900 | | SR |
| 87 | SR | SPZ | 11.8 | HER | 2400 | 2000 | | SR |
| 87 | SR | SPZ | 12.4 | HER | 2400 | 2100 | | SR |
| 87 | SR | SPZ | 13.0 | HER | 2400 | 2200 | | SR |
| 87 | SR | SPZ | 12.0 | DUP | 4227 | 1800 | | SR |
| 87 | SR | SPZ | 12.5 | DUP | 4227 | 1900 | | SR |
| 87 | SR | SPZ | 13.0 | DUP | 4227 | 2000 | | SR |
| 87 | SR | SPZ | 13.5 | DUP | 4227 | 2100 | | SR |
| 87 | SR | SPZ | 14.0 | DUP | 4227 | 2200 | | SR |

LOADS FOR .250/3000 SAVAGE .257

| BULLET | | | POWDER | | | | | |
|---|---|---|---|---|---|---|---|---|
| Wgt | Mfg | Type | GRN | Mfg | Type | VEL(fps) | OAL | Source |
| 75 | SR | HP | 31.2 | DUP | 3031 | 2900 | 2.515 | SR |
| 75 | SR | HP | 32.0 | DUP | 3031 | 3000 | | SR |
| 75 | SR | HP | 32.8 | DUP | 3031 | 3100 | | SR |
| 75 | SR | HP | 33.6 | DUP | 3031 | 3200 | | SR |
| 75 | SR | HP | 34.5 | DUP | 3031 | 3300 | | SR |
| 75 | SR | HP | 35.4 | DUP | 3031 | 3400 | | SR |
| 75 | SR | HP | 34.0 | DUP | 4064 | 2900 | | SR |
| 75 | SR | HP | 34.9 | DUP | 4064 | 3000 | | SR |
| 75 | SR | HP | 35.8 | DUP | 4064 | 3100 | | SR |
| 75 | SR | HP | 36.8 | DUP | 4064 | 3200 | | SR |
| 75 | SR | HP | 37.8 | DUP | 4064 | 3300 | | SR |
| 87 | SP | SP | 31.5 | DUP | 4064 | 2593 | | SP |
| 87 | SP | SP | 33.5 | DUP | 4064 | 2769 | | SP |
| 87 | SP | SP | 35.5 | DUP | 4064 | 2892 | | SP |
| 87&90 | SR | all | 29.7 | DUP | 3031 | 2700 | | SR |
| 87&90 | SR | all | 30.6 | DUP | 3031 | 2800 | | SR |
| 87&90 | SR | all | 31.5 | DUP | 3031 | 2900 | | SR |
| 87&90 | SR | all | 32.5 | DUP | 3031 | 3000 | | SR |
| 87&90 | SR | all | 33.5 | DUP | 3031 | 3100 | | SR |
| 87&90 | SR | all | 34.5 | DUP | 3031 | 3200 | | SR |
| 87&90 | SR | all | 31.8 | DUP | 4064 | 2700 | | SR |
| 87&90 | SR | all | 32.8 | DUP | 4064 | 2800 | | SR |
| 87&90 | SR | all | 33.8 | DUP | 4064 | 2900 | | SR |
| 87&90 | SR | all | 34.8 | DUP | 4064 | 3000 | | SR |
| 87&90 | SR | all | 35.9 | DUP | 4064 | 3100 | | SR |
| 87&90 | SR | all | 37.0 | DUP | 4064 | 3200 | | SR |
| 100 | SR | all | 28.2 | DUP | 3031 | 2500 | | SR |
| 100 | SR | all | 29.1 | DUP | 3031 | 2600 | | SR |
| 100 | SR | all | 30.0 | DUP | 3031 | 2700 | | SR |
| 100 | SR | all | 30.9 | DUP | 3031 | 2800 | | SR |
| 100 | SR | all | 31.9 | DUP | 3031 | 2900 | | SR |
| 100 | SR | all | 32.9 | DUP | 3031 | 3000 | | SR |
| 100 | SR | all | 30.2 | DUP | 4064 | 2500 | | SR |
| 100 | SR | all | 31.2 | DUP | 4064 | 2600 | | SR |
| 100 | SR | all | 32.2 | DUP | 4064 | 2700 | | SR |
| 100 | SR | all | 33.2 | DUP | 4064 | 2800 | | SR |
| 100 | SR | all | 34.2 | DUP | 4064 | 2900 | | SR |
| 100 | SR | all | 35.3 | DUP | 4064 | 3000 | | SR |
| 100 | SP | all | 29.0 | DUP | 4064 | 2361 | | SP |
| 100 | SP | all | 31.0 | DUP | 4064 | 2519 | | SP |
| 100 | SP | all | 33.0 | DUP | 4064 | 2679 | | SP |
| 100 | NO | SP | 30.0 | DUP | 4064 | 2443 | | NO |
| 100 | NO | SP | 32.0 | DUP | 4064 | 2615 | | NO |
| 100 | NO | SP | 34.0 | DUP | 4064 | 2721 | | NO |
| 115 | NO | SP | 28.0 | DUP | 4064 | 2202 | | NO |
| 115 | NO | SP | 30.0 | DUP | 4064 | 2360 | | NO |
| 115 | NO | SP | 32.0 | DUP | 4064 | 2509 | | NO |
| 117 | NO | SSP | 28.0 | DUP | 4064 | 2197 | | NO |
| 117 | NO | SSP | 30.0 | DUP | 4064 | 2356 | | NO |
| 117 | NO | SSP | 32.0 | DUP | 4064 | 2498 | | NO |
| 117 & 120 | SR | all | 26.1 | DUP | 3031 | 2300 | | SR |

LOADS FOR .250/3000 SAVAGE .257

| BULLET | | | POWDER | | | | | |
|---|---|---|---|---|---|---|---|---|
| Wgt | Mfg | Type | GRN | Mfg | Type | VEL(fps) | OAL | Source |
| 117 & 120 | SR | all | 27.4 | DUP | 3031 | 2400 | | SR |
| 117 & 120 | SR | all | 28.7 | DUP | 3031 | 2500 | | SR |
| 117 & 120 | SR | all | 30.0 | DUP | 3031 | 2600 | | SR |
| 117 & 120 | SR | all | 31.3 | DUP | 3031 | 2700 | | SR |
| 117 & 120 | SR | all | 27.7 | DUP | 4064 | 2300 | | SR |
| 117 & 120 | SR | all | 29.1 | DUP | 4064 | 2400 | | SR |
| 117 & 120 | SR | all | 30.5 | DUP | 4064 | 2500 | | SR |
| 117 & 120 | SR | all | 31.9 | DUP | 4064 | 2600 | | SR |
| 117 & 120 | SR | all | 33.4 | DUP | 4064 | 2700 | | SR |
| 120 | SP | all | 28.0 | DUP | 4064 | 2221 | | SP |
| 120 | SP | all | 30.0 | DUP | 4064 | 2361 | | SP |
| 120 | SP | all | 32.0 | DUP | 4064 | 2516 | | SP |
| 120 | NO | SP | 28.0 | DUP | 4064 | 2168 | | NO |
| 120 | NO | SP | 30.0 | DUP | 4064 | 2324 | | NO |
| 120 | NO | SP | 32.0 | DUP | 4064 | 2476 | | NO |

## LOADS FOR .257 ROBERTS .257

| BULLET | | | POWDER | | | | | |
|---|---|---|---|---|---|---|---|---|
| Wgt | Mfg | Type | GRN | Mfg | Type | VEL (fps) | OAL | Source |
| 75 | SR | HP | 36.3 | DUP | 3031 | 3000 | 2.750 | SR |
| 75 | SR | HP | 37.5 | DUP | 3031 | 3100 | | SR |
| 75 | SR | HP | 38.7 | DUP | 3031 | 3200 | | SR |
| 75 | SR | HP | 39.9 | DUP | 3031 | 3300 | | SR |
| 75 | SR | HP | 39.9 | DUP | 4064 | 3000 | | SR |
| 75 | SR | HP | 40.7 | DUP | 4064 | 3100 | | SR |
| 75 | SR | HP | 41.5 | DUP | 4064 | 3200 | | SR |
| 75 | SR | HP | 42.3 | DUP | 4064 | 3300 | | SR |
| 75 | SR | HP | 43.1 | DUP | 4064 | 3400 | | SR |
| 75 | SR | HP | 44.0 | DUP | 4064 | 3500 | | SR |
| 87 | SP | SP | 47.0 | DUP | 4831 | 2982 | | SP |
| 87 | SP | SP | 49.0 | DUP | 4831 | 3091 | | SP |
| 87 | SP | SP | 51.0 | DUP | 4831 | 3213 | | SP |
| 87&90 | SR | all | 34.0 | DUP | 3031 | 2800 | | SR |
| 87&90 | SR | all | 35.5 | DUP | 3031 | 2900 | | SR |
| 87&90 | SR | all | 37.0 | DUP | 3031 | 3000 | | SR |
| 87&90 | SR | all | 38.5 | DUP | 3031 | 3100 | | SR |
| 87&90 | SR | all | 40.0 | DUP | 3031 | 3200 | | SR |
| 87&90 | SR | all | 37.7 | DUP | 4064 | 2800 | | SR |
| 87&90 | SR | all | 38.7 | DUP | 4064 | 2900 | | SR |
| 87&90 | SR | all | 39.7 | DUP | 4064 | 3000 | | SR |
| 87&90 | SR | all | 40.7 | DUP | 4064 | 3100 | | SR |
| 87&90 | SR | all | 41.8 | DUP | 4064 | 3200 | | SR |
| 87&90 | SR | all | 42.9 | DUP | 4064 | 3300 | | SR |
| 100 | SR | SPZ | 31.0 | DUP | 3031 | 2500 | 2.775 | SR |
| 100 | SR | SPZ | 32.4 | DUP | 3031 | 2600 | | SR |
| 100 | SR | SPZ | 33.8 | DUP | 3031 | 2700 | | SR |
| 100 | SR | SPZ | 35.2 | DUP | 3031 | 2800 | | SR |
| 100 | SR | SPZ | 36.7 | DUP | 3031 | 2900 | | SR |
| 100 | SR | SPZ | 38.2 | DUP | 3031 | 3000 | | SR |
| 100 | SR | SPZ | 34.4 | DUP | 4064 | 2500 | | SR |
| 100 | SR | SPZ | 35.6 | DUP | 4064 | 2600 | | SR |
| 100 | SR | SPZ | 36.8 | DUP | 4064 | 2700 | | SR |
| 100 | SR | SPZ | 38.0 | DUP | 4064 | 2800 | | SR |
| 100 | SR | SPZ | 39.3 | DUP | 4064 | 2900 | | SR |
| 100 | SP | all | 44.0 | DUP | 4831 | 2797 | | SP |
| 100 | SP | all | 46.0 | DUP | 4831 | 2914 | | SP |
| 100 | SP | all | 48.0 | DUP | 4831 | 3042 | | SP |
| 100 | NO | SP | 41.5 | DUP | 4831 | 2801 | | NO |
| 100 | NO | SP | 43.5 | DUP | 4831 | 2928 | | NO |
| 100 | NO | SP | 45.5 | DUP | 4831 | 3094 | | NO |
| 100 | NO | SPP | 41.5 | DUP | 4831 | 2798 | | NO |
| 100 | NO | SPP | 43.5 | DUP | 4831 | 2938 | | NO |
| 100 | NO | SPP | 45.5 | DUP | 4831 | 3092 | | NO |
| 115 | NO | SPP | 38.5 | DUP | 4831 | 2581 | | NO |
| 115 | NO | SPP | 40.5 | DUP | 4831 | 2700 | | NO |
| 115 | NO | SPP | 42.5 | DUP | 4831 | 2828 | | NO |
| 117 | NO | SSPP | 38.5 | DUP | 4831 | 2570 | | NO |
| 117 | NO | SSPP | 40.5 | DUP | 4831 | 2693 | | NO |
| 117 | NO | SSPP | 42.5 | DUP | 4831 | 2822 | | NO |
| 117 | SR | SPBT | 47.0C | DUP | 7828 | 2720 | | DUP |
| 117 & 120 | SR | all | 29.4 | DUP | 3031 | 2300 | | SR |

LOADS FOR .257 ROBERTS .257

| BULLET | | | POWDER | | | | | |
|---|---|---|---|---|---|---|---|---|
| Wgt | Mfg | Type | GRN | Mfg | Type | VEL(fps) | OAL | Source |
| 117 & 120 | SR | all | 30.8 | DUP | 3031 | 2400 | | SR |
| 117 & 120 | SR | all | 32.2 | DUP | 3031 | 2500 | | SR |
| 117 & 120 | SR | all | 33.7 | DUP | 3031 | 2600 | | SR |
| 117 & 120 | SR | all | 31.3 | DUP | 4064 | 2300 | | SR |
| 117 & 120 | SR | all | 32.8 | DUP | 4064 | 2400 | | SR |
| 117 & 120 | SR | all | 34.3 | DUP | 4064 | 2500 | | SR |
| 117 & 120 | SR | all | 35.9 | DUP | 4064 | 2600 | | SR |
| 120 | SP | all | 41.0 | DUP | 4831 | 2547 | | SP |
| 120 | SP | all | 43.0 | DUP | 4831 | 2673 | | SP |
| 120 | SP | all | 45.0 | DUP | 4831 | 2786 | | SP |
| 120 | SP | SP | 47.0C | DUP | 7828 | 2745 | | DUP |

Letter C indicates compressed powder charge.

LOADS FOR .25/06 REMINGTON .257

| BULLET | | | POWDER | | | | | |
|---|---|---|---|---|---|---|---|---|
| Wgt | Mfg | Type | GRN | Mfg | Type | VEL(fps) | OAL | Source |
| 75 | SR | HP | 41.5 | DUP | 3031 | 3200 | 3.250 | SR |
| 75 | SR | HP | 43.0 | DUP | 3031 | 3300 | | SR |
| 75 | SR | HP | 44.5 | DUP | 3031 | 3400 | | SR |
| 75 | SR | HP | 46.0 | DUP | 3031 | 3500 | | SR |
| 75 | SR | HP | 47.5 | DUP | 3031 | 3600 | | SR |
| 75 | SR | HP | 43.7 | DUP | 4064 | 3200 | | SR |
| 75 | SR | HP | 45.0 | DUP | 4064 | 3300 | | SR |
| 75 | SR | HP | 46.3 | DUP | 4064 | 3400 | | SR |
| 75 | SR | HP | 47.6 | DUP | 4064 | 3500 | | SR |
| 75 | SR | HP | 49.0 | DUP | 4064 | 3600 | | SR |
| 75 | SR | HP | 50.4 | DUP | 4064 | 3700 | | SR |
| 87 | SP | SP | 55.0 | DUP | 4831 | 3395 | | SP |
| 87 | SP | SP | 57.0 | DUP | 4831 | 3518 | | SP |
| 87 | SP | SP | 59.0 | DUP | 4831 | 3395 | | SP |
| 87&90 | SR | all | 40.3 | DUP | 3031 | 3000 | | SR |
| 87&90 | SR | all | 41.8 | DUP | 3031 | 3100 | | SR |
| 87&90 | SR | all | 43.3 | DUP | 3031 | 3200 | | SR |
| 87&90 | SR | all | 44.8 | DUP | 3031 | 3300 | | SR |
| 87&90 | SR | all | 46.3 | DUP | 3031 | 3400 | | SR |
| 87&90 | SR | all | 41.7 | DUP | 4064 | 3000 | | SR |
| 87&90 | SR | all | 43.2 | DUP | 4064 | 3100 | | SR |
| 87&90 | SR | all | 44.7 | DUP | 4064 | 3200 | | SR |
| 87&90 | SR | all | 46.2 | DUP | 4064 | 3300 | | SR |
| 87&90 | SR | all | 47.7 | DUP | 4064 | 3400 | | SR |
| 100 | SR | SPZ | 39.2 | DUP | 4064 | 2800 | | SR |
| 100 | SR | SPZ | 40.6 | DUP | 4064 | 2900 | | SR |
| 100 | SR | SPZ | 42.1 | DUP | 4064 | 3000 | | SR |
| 100 | SR | SPZ | 43.6 | DUP | 4064 | 3100 | | SR |
| 100 | SR | SPZ | 50.5 | DUP | 4831 | 2900 | | SR |
| 100 | SR | SPZ | 51.5 | DUP | 4831 | 3000 | | SR |
| 100 | SR | SPZ | 52.5 | DUP | 4831 | 3100 | | SR |
| 100 | SR | SPZ | 53.6 | DUP | 4831 | 3200 | | SR |
| 100 | SR | SPZ | 54.6 | DUP | 4831 | 3300 | | SR |
| 100 | SP | all | 52.0 | DUP | 4831 | 3231 | | SP |
| 100 | SP | all | 54.0 | DUP | 4831 | 3379 | | SP |
| 100 | SP | all | 56.0 | DUP | 4831 | 3425 | | SP |
| 100 | NO | SPP | 50.0 | DUP | 4831 | 3202 | | NO |
| 100 | NO | SPP | 52.0 | DUP | 4831 | 3321 | | NO |
| 100 | NO | SPP | 54.0 | DUP | 4831 | 3488 | | NO |
| 100 | NO | SP | 50.0 | DUP | 4831 | 3210 | | NO |
| 100 | NO | SP | 52.0 | DUP | 4831 | 3325 | | NO |
| 100 | NO | SP | 54.0 | DUP | 4831 | 3492 | | NO |
| 115 | NO | SPP | 48.0 | DUP | 4831 | 2931 | | NO |
| 115 | NO | SPP | 50.0 | DUP | 4831 | 3028 | | NO |
| 115 | NO | SPP | 52.0 | DUP | 4831 | 3114 | | NO |
| 117 | NO | SSPP | 48.0 | DUP | 4831 | 2936 | | NO |
| 117 | NO | SSPP | 50.0 | DUP | 4831 | 3020 | | NO |
| 117 | NO | SSPP | 52.0 | DUP | 4831 | 3113 | | NO |
| 117 | SR | SPBT | 55.0 | DUP | 7828 | 3130 | | DUP |
| 117 & 120 | SR | all | 46.8 | DUP | 4831 | 2700 | | SR |

LOADS FOR .25/06 REMINGTON .257

| BULLET | | | POWDER | | | | | |
|---|---|---|---|---|---|---|---|---|
| Wgt | Mfg | Type | GRN | Mfg | Type | VEL (fps) | OAL | Source |
| 117 & 120 | SR | all | 48.6 | DUP | 4831 | 2800 | | SR |
| 117 & 120 | SR | all | 50.5 | DUP | 4831 | 2900 | | SR |
| 117 & 120 | SR | all | 52.3 | DUP | 4831 | 3000 | | SR |
| 120 | SP | all | 46.0 | DUP | 4831 | 2769 | | SP |
| 120 | SP | all | 48.0 | DUP | 4831 | 2863 | | SP |
| 120 | SP | all | 50.0 | DUP | 4831 | 2980 | | SP |
| 120 | SP | SP | 55.0 | DUP | 7828 | 3105 | | DUP |

LOADS FOR .257 WEATHERBY MAGNUM .257

| BULLET | | | POWDER | | | | | |
|---|---|---|---|---|---|---|---|---|
| Wgt | Mfg | Type | GRN | Mfg | Type | VEL(fps) | OAL | Source |
| 75 | SR | HP | 57.1 | DUP | 4064 | 3500 | 3.250 | SR |
| 75 | SR | HP | 58.7 | DUP | 4064 | 3600 | | SR |
| 75 | SR | HP | 60.3 | DUP | 4064 | 3700 | | SR |
| 75 | SR | HP | 67.2 | DUP | 4831 | 3500 | | SR |
| 75 | SR | HP | 68.2 | DUP | 4831 | 3600 | | SR |
| 75 | SR | HP | 69.2 | DUP | 4831 | 3700 | | SR |
| 75 | SR | HP | 70.3 | DUP | 4831 | 3800 | | SR |
| 75 | SR | HP | 71.5 | DUP | 4831 | 3900 | | SR |
| 87 | SP | SP | 67.0 | DUP | 4831 | 3459 | | SP |
| 87 | SP | SP | 69.0 | DUP | 4831 | 3583 | | SP |
| 87 | SP | SP | 71.0 | DUP | 4831 | 3676 | | SP |
| 87&90 | SR | all | 53.1 | DUP | 4064 | 3200 | | SR |
| 87&90 | SR | all | 54.6 | DUP | 4064 | 3300 | | SR |
| 87&90 | SR | all | 56.1 | DUP | 4064 | 3400 | | SR |
| 87&90 | SR | all | 57.7 | DUP | 4064 | 3500 | | SR |
| 87&90 | SR | all | 62.6 | DUP | 4831 | 3200 | | SR |
| 87&90 | SR | all | 64.1 | DUP | 4831 | 3300 | | SR |
| 87&90 | SR | all | 65.6 | DUP | 4831 | 3400 | | SR |
| 87&90 | SR | all | 67.2 | DUP | 4831 | 3500 | | SR |
| 87&90 | SR | all | 68.7 | DUP | 4831 | 3600 | | SR |
| 87&90 | SR | all | 70.2 | DUP | 4831 | 3700 | | SR |
| 100 | SR | SPZ | 51.6 | DUP | 4064 | 3000 | | SR |
| 100 | SR | SPZ | 53.1 | DUP | 4064 | 3100 | | SR |
| 100 | SR | SPZ | 54.7 | DUP | 4064 | 3200 | | SR |
| 100 | SR | SPZ | 56.3 | DUP | 4064 | 3300 | | SR |
| 100 | SR | SPZ | 59.9 | DUP | 4831 | 3000 | | SR |
| 100 | SR | SPZ | 61.6 | DUP | 4831 | 3100 | | SR |
| 100 | SR | SPZ | 63.2 | DUP | 4831 | 3200 | | SR |
| 100 | SR | SPZ | 64.9 | DUP | 4831 | 3300 | | SR |
| 100 | SR | SPZ | 66.6 | DUP | 4831 | 3400 | | SR |
| 100 | SP | all | 64.0 | DUP | 4831 | 3125 | | SP |
| 100 | SP | all | 66.0 | DUP | 4831 | 3238 | | SP |
| 100 | SP | all | 68.0 | DUP | 4831 | 3361 | | SP |
| 100 | NO | SPP | 64.0 | DUP | 4831 | 3392 | | NO |
| 100 | NO | SPP | 66.0 | DUP | 4831 | 3473 | | NO |
| 100 | NO | SPP | 68.0 | DUP | 4831 | 3572 | | NO |
| 100 | NO | SP | 64.0 | DUP | 4831 | 3381 | | NO |
| 100 | NO | SP | 66.0 | DUP | 4831 | 3470 | | NO |
| 100 | NO | SP | 68.0 | DUP | 4831 | 3575 | | NO |
| 115 | NO | SPP | 60.0 | DUP | 4831 | 3086 | | NO |
| 115 | NO | SPP | 62.0 | DUP | 4831 | 3184 | | NO |
| 115 | NO | SPP | 64.0 | DUP | 4831 | 3279 | | NO |
| 117 | NO | SSPP | 60.0 | DUP | 4831 | 3077 | | NO |
| 117 | NO | SSPP | 62.0 | DUP | 4831 | 3163 | | NO |
| 117 | NO | SSPP | 64.0 | DUP | 4831 | 3272 | | NO |
| 117 & 120 | SR | all | 47.3 | DUP | 4064 | 2700 | | SR |
| 117 & 120 | SR | all | 49.1 | DUP | 4064 | 2800 | | SR |
| 117 & 120 | SR | all | 50.9 | DUP | 4064 | 2900 | | SR |

LOADS FOR .257 WEATHERBY MAGNUM .257

| BULLET | | | POWDER | | | | | |
|---|---|---|---|---|---|---|---|---|
| Wgt | Mfg | Type | GRN | Mfg | Type | VEL(fps) | OAL | Source |
| 117 & 120 | SR | all | 52.8 | DUP | 4064 | 3000 | | SR |
| 117 & 120 | SR | all | 56.7 | DUP | 4831 | 2800 | | SR |
| 117 & 120 | SR | all | 58.4 | DUP | 4831 | 2900 | | SR |
| 117 & 120 | SR | all | 60.2 | DUP | 4831 | 3000 | | SR |
| 117 & 120 | SR | all | 61.9 | DUP | 4831 | 3100 | | SR |
| 117 & 120 | SR | all | 63.7 | DUP | 4831 | 3200 | | SR |
| 120 | SP | all | 61.0 | DUP | 4831 | 2968 | | SP |
| 120 | SP | all | 63.0 | DUP | 4831 | 3062 | | SP |
| 120 | SP | all | 65.0 | DUP | 4831 | 3199 | | SP |
| 120 | NO | SP | 60.0 | DUP | 4831 | 3050 | | NO |
| 120 | NO | SP | 62.0 | DUP | 4831 | 3142 | | NO |
| 120 | NO | SP | 64.0 | DUP | 4831 | 3050 | | NO |

LOADS FOR 6.5 REMINGTON MAGNUM .264

| BULLET | | | POWDER | | | | | |
|---|---|---|---|---|---|---|---|---|
| Wgt | Mfg | Type | GRN | Mfg | Type | VEL(fps) | OAL | Source |
| 85 | SR | HP | 45.9 | DUP | 4064 | 2700 | 2.800 | SR |
| 85 | SR | HP | 47.4 | DUP | 4064 | 2800 | | SR |
| 85 | SR | HP | 49.0 | DUP | 4064 | 2900 | | SR |
| 85 | SR | HP | 52.4 | DUP | 4350 | 2700 | | SR |
| 85 | SR | HP | 53.6 | DUP | 4350 | 2800 | | SR |
| 85 | SR | HP | 54.8 | DUP | 4350 | 2900 | | SR |
| 85 | SR | HP | 56.0 | DUP | 4350 | 3000 | | SR |
| 85 | SR | HP | 57.2 | DUP | 4350 | 3100 | | SR |
| 100 | SR | HP | 43.0 | DUP | 4064 | 2600 | | SR |
| 100 | SR | HP | 45.0 | DUP | 4064 | 2700 | | SR |
| 100 | SR | HP | 47.0 | DUP | 4064 | 2800 | | SR |
| 100 | SR | HP | 49.4 | DUP | 4350 | 2600 | | SR |
| 100 | SR | HP | 51.0 | DUP | 4350 | 2700 | | SR |
| 100 | SR | HP | 52.6 | DUP | 4350 | 2800 | | SR |
| 100 | SR | HP | 54.2 | DUP | 4350 | 2900 | | SR |
| 100 | SR | HP | 55.8 | DUP | 4350 | 3000 | | SR |
| 120 | SR | SPZ | 39.0 | DUP | 4064 | 2400 | | SR |
| 120 | SR | SPZ | 41.5 | DUP | 4064 | 2500 | | SR |
| 120 | SR | SPZ | 44.0 | DUP | 4064 | 2600 | | SR |
| 120 | SR | SPZ | 46.1 | DUP | 4350 | 2400 | | SR |
| 120 | SR | SPZ | 47.8 | DUP | 4350 | 2500 | | SR |
| 120 | SR | SPZ | 49.5 | DUP | 4350 | 2600 | | SR |
| 120 | SR | SPZ | 51.2 | DUP | 4350 | 2700 | | SR |
| 120 | SR | SPZ | 52.9 | DUP | 4350 | 2800 | | SR |
| 120 | SP | SP | 51.0 | DUP | 4831 | 2724 | | SP |
| 120 | SP | SP | 53.0 | DUP | 4831 | 2847 | | SP |
| 120 | SP | SP | 55.0 | DUP | 4831 | 2970 | | SP |
| 140 | SP | SP | 48.0 | DUP | 4831 | 2547 | | SP |
| 140 | SP | SP | 50.0 | DUP | 4831 | 2650 | | SP |
| 140 | SP | SP | 52.0 | DUP | 4831 | 2755 | | SP |

LOADS FOR .264 WINCHESTER MAGNUM .264

| BULLET | | | POWDER | | | | | |
|---|---|---|---|---|---|---|---|---|
| Wgt | Mfg | Type | GRN | Mfg | Type | VEL.(fps) | OAL | Source |
| 85 | SR | HP | 48.8 | DUP | 3031 | 3300 | 3.340 | SR |
| 85 | SR | HP | 50.7 | DUP | 3031 | 3400 | | SR |
| 85 | SR | HP | 52.6 | DUP | 3031 | 3500 | | SR |
| 85 | SR | HP | 59.4 | DUP | 4831 | 3300 | | SR |
| 85 | SR | HP | 61.4 | DUP | 4831 | 3400 | | SR |
| 85 | SR | HP | 63.4 | DUP | 4831 | 3500 | | SR |
| 85 | SR | HP | 65.5 | DUP | 4831 | 3600 | | SR |
| 100 | SR | HP | 48.9 | DUP | 3031 | 3200 | | SR |
| 100 | SR | HP | 50.9 | DUP | 3031 | 3300 | | SR |
| 100 | SR | HP | 57.2 | DUP | 4831 | 3200 | | SR |
| 100 | SR | HP | 59.5 | DUP | 4831 | 3300 | | SR |
| 100 | SR | HP | 61.9 | DUP | 4831 | 3400 | | SR |
| 100 | SR | HP | 64.4 | DUP | 4831 | 3500 | | SR |
| 120 | SR | SPZ | 46.7 | DUP | 3031 | 2900 | | SR |
| 120 | SR | SPZ | 49.4 | DUP | 3031 | 3000 | | SR |
| 120 | SR | SPZ | 55.1 | DUP | 4831 | 3100 | | SR |
| 120 | SR | SPZ | 58.6 | DUP | 4831 | 3200 | | SR |
| 120 | SR | SPZ | 62.1 | DUP | 4831 | 3300 | | SR |
| 120 | SP | SPZ | 55.0 | DUP | 4831 | 2816 | | SP |
| 120 | SP | SPZ | 57.0 | DUP | 4831 | 2954 | | SP |
| 120 | SP | SPZ | 59.0 | DUP | 4831 | 3082 | | SP |
| 120 | NO | SPZ | 58.0 | DUP | 4831 | 2953 | | NO |
| 120 | NO | SPZ | 60.o | DUP | 4831 | 3075 | | NO |
| 120 | NO | SPZ | 62.0 | DUP | 4831 | 3162 | | NO |
| 125 | NO | SSP | 58.0 | DUP | 4831 | 2925 | | NO |
| 125 | NO | SSP | 60.0 | DUP | 4831 | 3046 | | NO |
| 125 | NO | SSP | 62.0 | DUP | 4831 | 3138 | | NO |
| 140 | SR | all | 43.1 | DUP | 3031 | 2600 | | SR |
| 140 | SR | all | 45.5 | DUP | 3031 | 2700 | | SR |
| 140 | SR | all | 48.0 | DUP | 3031 | 2800 | | SR |
| 140 | SP | SPZ | 52.0 | DUP | 4831 | 2683 | | SP |
| 140 | SP | SPZ | 54.0 | DUP | 4831 | 2796 | | SP |
| 140 | SP | SPZ | 56.0 | DUP | 4831 | 2931 | | SP |
| 140 | NO | SPP | 55.0 | DUP | 4831 | 2770 | | NO |
| 140 | NO | SPP | 57.0 | DUP | 4831 | 2853 | | NO |
| 140 | NO | SPP | 59.0 | DUP | 4831 | 2968 | | NO |
| 140 | SR | SPBT | 65.0 | DUP | 7828 | 3115 | | DUP |

LOADS FOR .270 WINCHESTER .277

| BULLET | | | POWDER | | | | | |
|---|---|---|---|---|---|---|---|---|
| Wgt. | Mfg | Type | GRN | Mfg | Type | VEL(fps) | OAL | Source |
| 90 | SR | HP | 45.0 | DUP | 3031 | 3100 | 3.340 | SR |
| 90 | SR | HP | 46.3 | DUP | 3031 | 3200 | | SR |
| 90 | SR | HP | 47.6 | DUP | 3031 | 3300 | | SR |
| 90 | SR | HP | 49.0 | DUP | 3031 | 3400 | | SR |
| 90 | SR | HP | 50.4 | DUP | 3031 | 3500 | | SR |
| 90 | SR | HP | 47.2 | DUP | 4064 | 3100 | | SR |
| 90 | SR | HP | 48.6 | DUP | 4064 | 3200 | | SR |
| 90 | SR | HP | 50.0 | DUP | 4064 | 3300 | | SR |
| 90 | SR | HP | 51.4 | DUP | 4064 | 3400 | | SR |
| 90 | SR | HP | 52.8 | DUP | 4064 | 3500 | | SR |
| 100 | SP | all | 57.0 | DUP | 4831 | 3095 | | SP |
| 100 | SP | all | 59.0 | DUP | 4831 | 3214 | | SP |
| 100 | SP | all | 61.0 | DUP | 4831 | 3303 | | SP |
| 100 | NO | SPZ | 56.0 | DUP | 4831 | 3052 | | NO |
| 100 | NO | SPZ | 58.0 | DUP | 4831 | 3178 | | NO |
| 100 | NO | SPZ | 60.0 | DUP | 4831 | 3312 | | NO |
| 110 | SR | SPZ | 42.4 | DUP | 3031 | 2800 | | SR |
| 110 | SR | SPZ | 43.6 | DUP | 3031 | 2900 | | SR |
| 110 | SR | SPZ | 44.9 | DUP | 3031 | 3000 | | SR |
| 110 | SR | SPZ | 46.2 | DUP | 3031 | 3100 | | SR |
| 110 | SR | SPZ | 47.5 | DUP | 3031 | 3200 | | SR |
| 110 | SR | SPZ | 43.8 | DUP | 4064 | 2800 | | SR |
| 110 | SR | SPZ | 45.2 | DUP | 4064 | 2900 | | SR |
| 110 | SR | SPZ | 46.7 | DUP | 4064 | 3000 | | SR |
| 110 | SR | SPZ | 48.2 | DUP | 4064 | 3100 | | SR |
| 110 | SR | SPZ | 49.7 | DUP | 4064 | 3200 | | SR |
| 130 | SR | all | 41.3 | DUP | 3031 | 2700 | | SR |
| 130 | SR | all | 43.0 | DUP | 3031 | 2800 | | SR |
| 130 | SR | all | 44.7 | DUP | 3031 | 2900 | | SR |
| 130 | SR | all | 46.5 | DUP | 3031 | 3000 | | SR |
| 130 | SR | all | 43.1 | DUP | 4064 | 2700 | | SR |
| 130 | SR | all | 44.8 | DUP | 4064 | 2800 | | SR |
| 130 | SR | all | 46.5 | DUP | 4064 | 2900 | | SR |
| 130 | SR | all | 48.3 | DUP | 4064 | 3000 | | SR |
| 130 | SR | all | 53.7 | DUP | 4831 | 2900 | | SR |
| 130 | SR | all | 55.4 | DUP | 4831 | 3000 | | SR |
| 130 | SR | all | 57.1 | DUP | 4831 | 3100 | | SR |
| 130 | SP | all | 54.0 | DUP | 4831 | 2892 | | SP |
| 130 | SP | all | 56.0 | DUP | 4831 | 3001 | | SP |
| 130 | SP | all | 58.0 | DUP | 4831 | 3052 | | SP |
| 130 | NO | SPP | 53.0 | DUP | 4831 | 2786 | | NO |
| 130 | NO | SPP | 55.0 | DUP | 4831 | 2920 | | NO |
| 130 | NO | SPP | 57.0 | DUP | 4831 | 3053 | | NO |
| 130 | NO | SPZ | 53.0 | DUP | 4831 | 2794 | | NO |
| 130 | NO | SPZ | 55.0 | DUP | 4831 | 2932 | | NO |
| 130 | NO | SPZ | 57.0 | DUP | 4831 | 3065 | | NO |
| 150 | SR | all | 38.7 | DUP | 3031 | 2500 | | SR |
| 150 | SR | all | 40.8 | DUP | 3031 | 2600 | | SR |
| 150 | SR | all | 42.9 | DUP | 3031 | 2700 | | SR |
| 150 | SR | all | 45.0 | DUP | 3031 | 2800 | | SR |
| 150 | SR | all | 40.8 | DUP | 4064 | 2500 | | SR |
| 150 | SR | all | 42.7 | DUP | 4064 | 2600 | | SR |
| 150 | SR | all | 44.6 | DUP | 4064 | 2700 | | SR |

LOADS FOR .270 WINCHESTER .277

| BULLET | | | POWDER | | | | | |
|---|---|---|---|---|---|---|---|---|
| Wgt | Mfg | Type | GRN | Mfg | Type | VEL(fps) | OAL | Source |
| 150 | SR | all | 46.5 | DUP | 4064 | 2800 | | SR |
| 150 | SR | all | 48.7 | DUP | 4831 | 2600 | | SR |
| 150 | SR | all | 50.7 | DUP | 4831 | 2700 | | SR |
| 150 | SR | all | 52.6 | DUP | 4831 | 2800 | | SR |
| 150 | SR | all | 54.5 | DUP | 4831 | 2900 | | SR |
| 150 | SP | all | 51.0 | DUP | 4831 | 2632 | | SP |
| 150 | SP | all | 53.0 | DUP | 4831 | 2744 | | SP |
| 150 | SP | all | 55.0 | DUP | 4831 | 2827 | | SP |
| 150 | NO | SPP | 51.0 | DUP | 4831 | 2700 | | NO |
| 150 | NO | SPP | 53.0 | DUP | 4831 | 2802 | | NO |
| 150 | NO | SPP | 55.0 | DUP | 4831 | 2904 | | NO |
| 150 | NO | SPZ | 51.0 | DUP | 4831 | 2713 | | NO |
| 150 | NO | SPZ | 53.0 | DUP | 4831 | 2806 | | NO |
| 150 | NO | SPZ | 55.0 | DUP | 4831 | 2911 | | NO |
| 150 | SP | SP | 56.5C | DUP | 7828 | 2860 | | DUP |
| 160 | NO | SSPP | 50.0 | DUP | 4831 | 2601 | | NO |
| 160 | NO | SSPP | 52.0 | DUP | 4831 | 2709 | | NO |
| 160 | NO | SSPP | 54.0 | DUP | 4831 | 2806 | | NO |
| 160 | NO | SP | 56.5 | DUP | 7828 | 2780 | | DUP |

Letter C indicates compressed powder charge.

LOADS FOR .270 WEATHERBY MAGNUM .277

| BULLET | | | POWDER | | | | | |
|---|---|---|---|---|---|---|---|---|
| Wgt | Mfg | Type | GRN | Mfg | Type | VEL(fps) | OAL | Source |
| 90 | SR | HP | 58.7 | DUP | 4064 | 3300 | | SR |
| 90 | SR | HP | 60.0 | DUP | 4064 | 3400 | | SR |
| 90 | SR | HP | 61.3 | DUP | 4064 | 3500 | | SR |
| 90 | SR | HP | 69.1 | DUP | 4831 | 3300 | | SR |
| 90 | SR | HP | 70.6 | DUP | 4831 | 3400 | | SR |
| 90 | SR | HP | 72.1 | DUP | 4831 | 3500 | | SR |
| 90 | SR | HP | 73.5 | DUP | 4831 | 3600 | | SR |
| 90 | SR | HP | 74.9 | DUP | 4831 | 3700 | | SR |
| 90 | SR | HP | 76.3 | DUP | 4831 | 3800 | | SR |
| 100 | SP | all | 69.0 | DUP | 4831 | 3323 | | SP |
| 100 | SP | all | 71.0 | DUP | 4831 | 3418 | | SP |
| 100 | SP | all | 73.0 | DUP | 4831 | 3505 | | SP |
| 100 | NO | SPZ | 69.0 | DUP | 4831 | 3255 | | NO |
| 100 | NO | SPZ | 71.0 | DUP | 4831 | 3386 | | NO |
| 100 | NO | SPZ | 73.0 | DUP | 4831 | 3518 | | NO |
| 110 | SR | SPZ | 54.4 | DUP | 4064 | 3000 | | SR |
| 110 | SR | SPZ | 55.8 | DUP | 4064 | 3100 | | SR |
| 110 | SR | SPZ | 57.2 | DUP | 4064 | 3200 | | SR |
| 110 | SR | SPZ | 58.6 | DUP | 4064 | 3300 | | SR |
| 110 | SR | SPZ | 61.0 | DUP | 4064 | 3400 | | SR |
| 110 | SR | SPZ | 63.8 | DUP | 4831 | 3000 | | SR |
| 110 | SR | SPZ | 65.6 | DUP | 4831 | 3100 | | SR |
| 110 | SR | SPZ | 67.4 | DUP | 4831 | 3200 | | SR |
| 110 | SR | SPZ | 69.2 | DUP | 4831 | 3300 | | SR |
| 110 | SR | SPZ | 70.9 | DUP | 4831 | 3400 | | SR |
| 110 | SR | SPZ | 72.6 | DUP | 4831 | 3500 | | SR |
| 130 | SR | all | 52.2 | DUP | 4064 | 2800 | | SR |
| 130 | SR | all | 54.2 | DUP | 4064 | 2900 | | SR |
| 130 | SR | all | 56.2 | DUP | 4064 | 3000 | | SR |
| 130 | SR | all | 58.2 | DUP | 4064 | 3100 | | SR |
| 130 | SR | all | 60.3 | DUP | 4064 | 3200 | | SR |
| 130 | SR | all | 63.6 | DUP | 4831 | 3000 | | SR |
| 130 | SR | all | 65.8 | DUP | 4831 | 3100 | | SR |
| 130 | SR | all | 68.0 | DUP | 4831 | 3200 | | SR |
| 130 | SR | all | 70.2 | DUP | 4831 | 3300 | | SR |
| 130 | SP | all | 66.0 | DUP | 4831 | 3105 | | SP |
| 130 | SP | all | 68.0 | DUP | 4831 | 3199 | | SP |
| 130 | SP | all | 70.0 | DUP | 4831 | 3281 | | SP |
| 130 | NO | SPP | 65.0 | DUP | 4831 | 3019 | | NO |
| 130 | NO | SPP | 67.0 | DUP | 4831 | 3127 | | NO |
| 130 | NO | SPP | 69.0 | DUP | 4831 | 3239 | | NO |
| 130 | NO | SPZ | 65.0 | DUP | 4831 | 3021 | | NO |
| 130 | NO | SPZ | 67.0 | DUP | 4831 | 3132 | | NO |
| 130 | NO | SPZ | 69.0 | DUP | 4831 | 3243 | | NO |
| 150 | SR | all | 51.3 | DUP | 4064 | 2700 | | SR |
| 150 | SR | all | 53.5 | DUP | 4064 | 2800 | | SR |
| 150 | SR | all | 55.7 | DUP | 4064 | 2900 | | SR |
| 150 | SR | all | 57.9 | DUP | 4064 | 3000 | | SR |
| 150 | SR | all | 60.8 | DUP | 4831 | 2800 | | SR |
| 150 | SR | all | 62.8 | DUP | 4831 | 2900 | | SR |
| 150 | SR | all | 64.7 | DUP | 4831 | 3000 | | SR |
| 150 | SR | all | 66.6 | DUP | 4831 | 3100 | | SR |

LOADS FOR .270 WEATHERBY MAGNUM .277

| BULLET | | | POWDER | | | | | |
|---|---|---|---|---|---|---|---|---|
| Wgt | Mfg | Type | GRN | Mfg | Type | VEL(fps) | OAL | Source |
| 150 | SR | all | 68.5 | DUP | 4831 | 3200 | | SR |
| 150 | SP | all | 63.0 | DUP | 4831 | 2852 | | SP |
| 150 | SP | all | 65.0 | DUP | 4831 | 2946 | | SP |
| 150 | SP | all | 67.0 | DUP | 4831 | 3019 | | SP |
| 150 | NO | SPP | 63.0 | DUP | 4831 | 2839 | | NO |
| 150 | NO | SPP | 65.0 | DUP | 4831 | 2951 | | NO |
| 150 | NO | SPP | 67.0 | DUP | 4831 | 3057 | | NO |
| 150 | NO | SPP | 63.0 | DUP | 4831 | 2850 | | NO |
| 150 | NO | SPP | 65.0 | DUP | 4831 | 2955 | | NO |
| 150 | NO | SPP | 67.0 | DUP | 4831 | 3060 | | NO |
| 160 | NO | SSPP | 61.0 | DUP | 4831 | 2755 | | NO |
| 160 | NO | SSPP | 63.0 | DUP | 4831 | 2836 | | NO |
| 160 | NO | SSPP | 65.0 | DUP | 4831 | 2917 | | NO |

LOADS FOR 6.5 JAPANESE .264

| BULLET | | | POWDER | | | | | |
|---|---|---|---|---|---|---|---|---|
| Wgt | Mfg | Type | GRN | Mfg | Type | VEL(fps) | OAL | Source |
| 85 | SR | HP | 33.0 | DUP | 3031 | 2500 | 2.940 | SR |
| 85 | SR | HP | 34.0 | DUP | 3031 | 2600 | | SR |
| 85 | SR | HP | 35.0 | DUP | 3031 | 2700 | | SR |
| 85 | SR | HP | 36.0 | DUP | 3031 | 2800 | | SR |
| 85 | SR | HP | 37.0 | DUP | 3031 | 2900 | | SR |
| 85 | SR | HP | 36.0 | DUP | 4064 | 2500 | | SR |
| 85 | SR | HP | 37.1 | DUP | 4064 | 2600 | | SR |
| 85 | SR | HP | 38.2 | DUP | 4064 | 2700 | | SR |
| 85 | SR | HP | 39.3 | DUP | 4064 | 2800 | | SR |
| 100 | SR | HP | 31.4 | DUP | 3031 | 2400 | | SR |
| 100 | SR | HP | 32.7 | DUP | 3031 | 2500 | | SR |
| 100 | SR | HP | 34.0 | DUP | 3031 | 2600 | | SR |
| 100 | SR | HP | 35.3 | DUP | 3031 | 2700 | | SR |
| 100 | SR | HP | 34.8 | DUP | 4064 | 2400 | | SR |
| 100 | SR | HP | 36.0 | DUP | 4064 | 2500 | | SR |
| 100 | SR | HP | 37.2 | DUP | 4064 | 2600 | | SR |
| 100 | SR | HP | 38.4 | DUP | 4064 | 2700 | | SR |
| 120 | SR | SPZ | 30.2 | DUP | 3031 | 2300 | | SR |
| 120 | SR | SPZ | 31.6 | DUP | 3031 | 2400 | | SR |
| 120 | SR | SPZ | 33.0 | DUP | 3031 | 2500 | | SR |
| 120 | SR | SPZ | 34.5 | DUP | 3031 | 2600 | | SR |
| 120 | SR | SPZ | 33.3 | DUP | 4064 | 2300 | | SR |
| 120 | SR | SPZ | 34.5 | DUP | 4064 | 2400 | | SR |
| 120 | SR | SPZ | 35.7 | DUP | 4064 | 2500 | | SR |
| 120 | SR | SPZ | 37.0 | DUP | 4064 | 2600 | | SR |
| 140 | SR | all | 28.7 | DUP | 3031 | 2100 | | SR |
| 140 | SR | all | 30.1 | DUP | 3031 | 2200 | | SR |
| 140 | SR | all | 31.5 | DUP | 3031 | 2300 | | SR |
| 140 | SR | all | 32.9 | DUP | 3031 | 2400 | | SR |
| 140 | SR | all | 30.9 | DUP | 4064 | 2100 | | SR |
| 140 | SR | all | 32.2 | DUP | 4064 | 2200 | | SR |
| 140 | SR | all | 33.6 | DUP | 4064 | 2300 | | SR |
| 140 | SR | all | 35.0 | DUP | 4064 | 2400 | | SR |

## LOADS FOR 6.5 x 54mm MANNLICHER-SCHOENAUER .264

| BULLET | | | POWDER | | | | | |
|---|---|---|---|---|---|---|---|---|
| Wgt | Mfg | Type | GRN | Mfg | Type | VEL(fps) | OAL | Source |
| 85 | SR | HP | 32.6 | DUP | 3031 | 2500 | 3.010 | SR |
| 85 | SR | HP | 33.8 | DUP | 3031 | 2600 | | SR |
| 85 | SR | HP | 35.0 | DUP | 3031 | 2700 | | SR |
| 85 | SR | HP | 36.2 | DUP | 3031 | 2800 | | SR |
| 85 | SR | HP | 37.4 | DUP | 3031 | 2900 | | SR |
| 85 | SR | HP | 38.7 | DUP | 3031 | 3000 | | SR |
| 85 | SR | HP | 36.3 | DUP | 4064 | 2500 | | SR |
| 85 | SR | HP | 37.3 | DUP | 4064 | 2600 | | SR |
| 85 | SR | HP | 38.3 | DUP | 4064 | 2700 | | SR |
| 85 | SR | HP | 39.3 | DUP | 4064 | 2800 | | SR |
| 85 | SR | HP | 40.3 | DUP | 4064 | 2900 | | SR |
| 85 | SR | HP | 41.4 | DUP | 4064 | 3000 | | SR |
| 100 | SR | HP | 29.7 | DUP | 3031 | 2300 | | SR |
| 100 | SR | HP | 31.1 | DUP | 3031 | 2400 | | SR |
| 100 | SR | HP | 32.5 | DUP | 3031 | 2500 | | SR |
| 100 | SR | HP | 33.9 | DUP | 3031 | 2600 | | SR |
| 100 | SR | HP | 35.3 | DUP | 3031 | 2700 | | SR |
| 100 | SR | HP | 36.7 | DUP | 3031 | 2800 | | SR |
| 100 | SR | HP | 32.7 | DUP | 4064 | 2300 | | SR |
| 100 | SR | HP | 34.0 | DUP | 4064 | 2400 | | SR |
| 100 | SR | HP | 35.3 | DUP | 4064 | 2500 | | SR |
| 100 | SR | HP | 36.6 | DUP | 4064 | 2600 | | SR |
| 100 | SR | HP | 37.9 | DUP | 4064 | 2700 | | SR |
| 100 | SR | HP | 39.2 | DUP | 4064 | 2800 | | SR |
| 120 | SR | SPZ | 29.5 | DUP | 3031 | 2200 | | SR |
| 120 | SR | SPZ | 31.1 | DUP | 3031 | 2300 | | SR |
| 120 | SR | SPZ | 32.7 | DUP | 3031 | 2400 | | SR |
| 120 | SR | SPZ | 34.3 | DUP | 3031 | 2500 | | SR |
| 120 | SR | SPZ | 35.9 | DUP | 3031 | 2600 | | SR |
| 120 | SR | SPZ | 37.5 | DUP | 3031 | 2700 | | SR |
| 120 | SR | SPZ | 32.2 | DUP | 4064 | 2200 | | SR |
| 120 | SR | SPZ | 33.6 | DUP | 4064 | 2300 | | SR |
| 120 | SR | SPZ | 35.0 | DUP | 4064 | 2400 | | SR |
| 120 | SR | SPZ | 36.4 | DUP | 4064 | 2500 | | SR |
| 120 | SR | SPZ | 37.8 | DUP | 4064 | 2600 | | SR |
| 120 | SR | SPZ | 39.3 | DUP | 4064 | 2700 | | SR |
| 140 | SR | all | 27.1 | DUP | 3031 | 1900 | | SR |
| 140 | SR | all | 28.6 | DUP | 3031 | 2000 | | SR |
| 140 | SR | all | 30.1 | DUP | 3031 | 2100 | | SR |
| 140 | SR | all | 31.6 | DUP | 3031 | 2200 | | SR |
| 140 | SR | all | 33.1 | DUP | 3031 | 2300 | | SR |
| 140 | SR | all | 34.7 | DUP | 3031 | 2400 | | SR |
| 140 | SR | all | 28.7 | DUP | 4064 | 1900 | | SR |
| 140 | SR | all | 30.3 | DUP | 4064 | 2000 | | SR |
| 140 | SR | all | 31.9 | DUP | 4064 | 2100 | | SR |
| 140 | SR | all | 33.5 | DUP | 4064 | 2200 | | SR |
| 140 | SR | all | 35.1 | DUP | 4064 | 2300 | | SR |
| 140 | SR | all | 36.7 | DUP | 4064 | 2400 | | SR |

LOADS FOR 6.5 x 52mm MANNLICHER-CARCANO .264

| BULLET | | | POWDER | | | | | |
|---|---|---|---|---|---|---|---|---|
| Wgt | Mfg | Type | GRN | Mfg | Type | VEL(fps) | OAL | Source |
| 85 | SR | HP | 30.2 | DUP | 3031 | 2400 | 2.900 | SR |
| 85 | SR | HP | 32.5 | DUP | 3031 | 2500 | | SR |
| 85 | SR | HP | 34.9 | DUP | 3031 | 2600 | | SR |
| 85 | SR | HP | 34.8 | DUP | 4064 | 2400 | | SR |
| 85 | SR | HP | 36.1 | DUP | 4064 | 2500 | | SR |
| 85 | SR | HP | 37.5 | DUP | 4064 | 2600 | | SR |
| 85 | SR | HP | 38.9 | DUP | 4064 | 2700 | | SR |
| 100 | SR | HP | 30.1 | DUP | 3031 | 2200 | | SR |
| 100 | SR | HP | 31.1 | DUP | 3031 | 2300 | | SR |
| 100 | SR | HP | 32.2 | DUP | 3031 | 2400 | | SR |
| 100 | SR | HP | 33.3 | DUP | 3031 | 2500 | | SR |
| 100 | SR | HP | 33.6 | DUP | 4064 | 2200 | | SR |
| 100 | SR | HP | 34.7 | DUP | 4064 | 2300 | | SR |
| 100 | SR | HP | 35.8 | DUP | 4064 | 2400 | | SR |
| 100 | SR | HP | 37.0 | DUP | 4064 | 2500 | | SR |
| 120 | SR | SPZ | 27.3 | DUP | 3031 | 2100 | | SR |
| 120 | SR | SPZ | 28.9 | DUP | 3031 | 2200 | | SR |
| 120 | SR | SPZ | 30.6 | DUP | 3031 | 2300 | | SR |
| 120 | SR | SPZ | 32.3 | DUP | 3031 | 2400 | | SR |
| 120 | SR | SPZ | 30.5 | DUP | 4064 | 2100 | | SR |
| 120 | SR | SPZ | 32.1 | DUP | 4064 | 2200 | | SR |
| 120 | SR | SPZ | 33.7 | DUP | 4064 | 2300 | | SR |
| 120 | SR | SPZ | 35.3 | DUP | 4064 | 2400 | | SR |
| 120 | SR | SPZ | 36.9 | DUP | 4064 | 2500 | | SR |
| 140 | SR | all | 29.2 | DUP | 3031 | 2000 | | SR |
| 140 | SR | all | 30.3 | DUP | 3031 | 2100 | | SR |
| 140 | SR | all | 31.4 | DUP | 3031 | 2200 | | SR |
| 140 | SR | all | 30.3 | DUP | 4064 | 2000 | | SR |
| 140 | SR | all | 31.7 | DUP | 4064 | 2100 | | SR |
| 140 | SR | all | 33.1 | DUP | 4064 | 2200 | | SR |
| 140 | SR | all | 34.5 | DUP | 4064 | 2300 | | SR |
| 140 | SR | all | 35.9 | DUP | 4064 | 2400 | | SR |

LOADS FOR 6.5 x 55mm SWEDISH .264

| BULLET | | | POWDER | | | | | |
|---|---|---|---|---|---|---|---|---|
| Wgt | Mfg | Type | GRN | Mfg | Type | VEL(fps) | OAL | Source |
| 85 | SR | HP | 38.6 | DUP | 3031 | 2700 | 3.062 | SR |
| 85 | SR | HP | 39.7 | DUP | 3031 | 2800 | | SR |
| 85 | SR | HP | 40.9 | DUP | 3031 | 2900 | | SR |
| 85 | SR | HP | 42.1 | DUP | 3031 | 3000 | | SR |
| 85 | SR | HP | 43.3 | DUP | 3031 | 3100 | | SR |
| 85 | SR | HP | 41.8 | DUP | 4064 | 2700 | | SR |
| 85 | SR | HP | 42.9 | DUP | 4064 | 2800 | | SR |
| 85 | SR | HP | 44.0 | DUP | 4064 | 2900 | | SR |
| 85 | SR | HP | 45.1 | DUP | 4064 | 3000 | | SR |
| 85 | SR | HP | 46.3 | DUP | 4064 | 3100 | | SR |
| 100 | SR | HP | 35.6 | DUP | 3031 | 2500 | | SR |
| 100 | SR | HP | 37.0 | DUP | 3031 | 2600 | | SR |
| 100 | SR | HP | 38.4 | DUP | 3031 | 2700 | | SR |
| 100 | SR | HP | 39.8 | DUP | 3031 | 2800 | | SR |
| 100 | SR | HP | 41.3 | DUP | 3031 | 2900 | | SR |
| 100 | SR | HP | 38.9 | DUP | 4064 | 2500 | | SR |
| 100 | SR | HP | 40.1 | DUP | 4064 | 2600 | | SR |
| 100 | SR | HP | 41.4 | DUP | 4064 | 2700 | | SR |
| 100 | SR | HP | 42.7 | DUP | 4064 | 2800 | | SR |
| 100 | SR | HP | 44.0 | DUP | 4064 | 2900 | | SR |
| 120 | SR | SPZ | 34.0 | DUP | 3031 | 2300 | | SR |
| 120 | SR | SPZ | 35.5 | DUP | 3031 | 2400 | | SR |
| 120 | SR | SPZ | 37.0 | DUP | 3031 | 2500 | | SR |
| 120 | SR | SPZ | 38.5 | DUP | 3031 | 2600 | | SR |
| 120 | SR | SPZ | 40.0 | DUP | 3031 | 2700 | | SR |
| 120 | SR | SPZ | 36.0 | DUP | 4064 | 2300 | | SR |
| 120 | SR | SPZ | 37.5 | DUP | 4064 | 2400 | | SR |
| 120 | SR | SPZ | 39.0 | DUP | 4064 | 2500 | | SR |
| 120 | SR | SPZ | 40.5 | DUP | 4064 | 2600 | | SR |
| 120 | SR | SPZ | 42.0 | DUP | 4064 | 2700 | | SR |
| 140 | SR | all | 29.7 | DUP | 3031 | 2100 | | SR |
| 140 | SR | all | 32.0 | DUP | 3031 | 2200 | | SR |
| 140 | SR | all | 34.3 | DUP | 3031 | 2300 | | SR |
| 140 | SR | all | 36.7 | DUP | 3031 | 2400 | | SR |
| 140 | SR | all | 33.2 | DUP | 4064 | 2100 | | SR |
| 140 | SR | all | 34.8 | DUP | 4064 | 2200 | | SR |
| 140 | SR | all | 36.5 | DUP | 4064 | 2300 | | SR |
| 140 | SR | all | 38.2 | DUP | 4064 | 2400 | | SR |
| 140 | SR | all | 39.9 | DUP | 4064 | 2500 | | SR |

LOADS FOR 7 X 57mm MAUSER .284

| BULLET | | | POWDER | | | | | |
|---|---|---|---|---|---|---|---|---|
| Wgt | Mfg | Type | GRN | Mfg | Type | VEL(fps) | OAL | Source |
| 115 | SP | HP | 51.0 | DUP | 4831 | 2683 | | SP |
| 115 | SP | HP | 53.0 | DUP | 4831 | 2799 | | SP |
| 115 | SP | HP | 55.0 | DUP | 4831 | 2910 | | SP |
| 120 | SR | SPZ | 37.1 | DUP | 3031 | 2600 | 3.065 | SR |
| 120 | SR | SPZ | 38.7 | DUP | 3031 | 2700 | | SR |
| 120 | SR | SPZ | 40.3 | DUP | 3031 | 2800 | | SR |
| 120 | SR | SPZ | 42.0 | DUP | 3031 | 2900 | | SR |
| 120 | SR | SPZ | 38.6 | DUP | 4064 | 2600 | | SR |
| 120 | SR | SPZ | 40.0 | DUP | 4064 | 2700 | | SR |
| 120 | SR | SPZ | 41.0 | DUP | 4064 | 2800 | | SR |
| 120 | SR | SPZ | 42.8 | DUP | 4064 | 2900 | | SR |
| 120 | SR | SPZ | 44.3 | DUP | 4064 | 3000 | | SR |
| 120 | NO | SPZ | 48.0 | DUP | 4831 | 2674 | | NO |
| 120 | NO | SPZ | 50.0 | DUP | 4831 | 2750 | | NO |
| 120 | NO | SPZ | 52.0 | DUP | 4831 | 2834 | | NO |
| 130 | SP | SPZ | 49.0 | DUP | 4831 | 2586 | | SP |
| 130 | SP | SPZ | 51.0 | DUP | 4831 | 2691 | | SP |
| 130 | SP | SPZ | 53.0 | DUP | 4831 | 2814 | | SP |
| 140 | SR | SPZ | 35.6 | DUP | 3031 | 2400 | | SR |
| 140 | SR | SPZ | 37.1 | DUP | 3031 | 2500 | | SR |
| 140 | SR | SPZ | 38.6 | DUP | 3031 | 2600 | | SR |
| 140 | SR | SPZ | 40.1 | DUP | 3031 | 2700 | | SR |
| 140 | SR | SPZ | 36.7 | DUP | 4064 | 2400 | | SR |
| 140 | SR | SPZ | 38.1 | DUP | 4064 | 2500 | | SR |
| 140 | SR | SPZ | 39.6 | DUP | 4064 | 2600 | | SR |
| 140 | SR | SPZ | 41.1 | DUP | 4064 | 2700 | | SR |
| 140 | SR | SPZ | 42.6 | DUP | 4064 | 2800 | | SR |
| 140 | NO | SPP | 47.0 | DUP | 4831 | 2501 | | NO |
| 140 | NO | SPP | 49.0 | DUP | 4831 | 2614 | | NO |
| 140 | NO | SPP | 51.0 | DUP | 4831 | 2759 | | NO |
| 140 | NO | SPZ | 47.0 | DUP | 4831 | 2509 | | NO |
| 140 | NO | SPZ | 49.0 | DUP | 4831 | 2618 | | NO |
| 140 | NO | SPZ | 51.0 | DUP | 4831 | 2761 | | NO |
| 145 | SP | all | 47.0 | DUP | 4831 | 2523 | | SP |
| 145 | SP | all | 49.0 | DUP | 4831 | 2607 | | SP |
| 145 | SP | all | 51.0 | DUP | 4831 | 2713 | | SP |
| 150 | NO | SPP | 45.0 | DUP | 4831 | 2327 | | NO |
| 150 | NO | SPP | 47.0 | DUP | 4831 | 2491 | | NO |
| 150 | NO | SPP | 49.0 | DUP | 4831 | 2653 | | NO |
| 150 | NO | SPZ | 45.0 | DUP | 4831 | 2332 | | NO |
| 150 | NO | SPZ | 47.0 | DUP | 4831 | 2501 | | NO |
| 150 | NO | SPZ | 49.0 | DUP | 4831 | 2655 | | NO |
| 160 | SR | SBT | 32.6 | DUP | 3031 | 2200 | | SR |
| 160 | SR | SBT | 34.2 | DUP | 3031 | 2300 | | SR |
| 160 | SR | SBT | 35.8 | DUP | 3031 | 2400 | | SR |
| 160 | SR | SBT | 37.4 | DUP | 3031 | 2500 | | SR |
| 160 | SR | SBT | 34.1 | DUP | 4064 | 2200 | | SR |
| 160 | SR | SBT | 35.6 | DUP | 4064 | 2300 | | SR |
| 160 | SR | SBT | 37.1 | DUP | 4064 | 2400 | | SR |
| 160 | SR | SBT | 38.7 | DUP | 4064 | 2500 | | SR |
| 160 | SP | all | 45.0 | DUP | 4831 | 2412 | | SP |
| 160 | SP | all | 47.0 | DUP | 4831 | 2509 | | SP |

## LOADS FOR 7 X 57mm MAUSER .284

| BULLET | | | POWDER | | | | | |
|---|---|---|---|---|---|---|---|---|
| Wgt | Mfg | Type | GRN | Mfg | Type | VEL(fps) | OAL | Source |
| 160 | SP | all | 49.0 | DUP | 4831 | 2616 | | SP |
| 160 | NO | SPP | 45.0 | DUP | 4831 | 2398 | | NO |
| 160 | NO | SPP | 47.0 | DUP | 4831 | 2570 | | NO |
| 160 | NO | SPP | 49.0 | DUP | 4831 | 2650 | | NO |
| 168 | SR | HP | 32.9 | DUP | 3031 | 2200 | | SR |
| 168 | SR | HP | 34.5 | DUP | 3031 | 2300 | | SR |
| 168 | SR | HP | 36.1 | DUP | 3031 | 2400 | | SR |
| 168 | SR | HP | 37.7 | DUP | 3031 | 2500 | | SR |
| 168 | SR | HP | 34.4 | DUP | 4064 | 2200 | | SR |
| 168 | SR | HP | 35.9 | DUP | 4064 | 2300 | | SR |
| 168 | SR | HP | 37.3 | DUP | 4064 | 2400 | | SR |
| 168 | SR | HP | 39.0 | DUP | 4064 | 2500 | | SR |
| 170 | SR | RN | 32.0 | DUP | 3031 | 2200 | | SR |
| 170 | SR | RN | 34.2 | DUP | 3031 | 2300 | | SR |
| 170 | SR | RN | 36.5 | DUP | 3031 | 2400 | | SR |
| 170 | SR | RN | 38.8 | DUP | 3031 | 2500 | | SR |
| 170 | SR | RN | 35.3 | DUP | 4064 | 2200 | | SR |
| 170 | SR | RN | 36.9 | DUP | 4064 | 2300 | | SR |
| 170 | SR | RN | 38.4 | DUP | 4064 | 2400 | | SR |
| 170 | SR | RN | 39.9 | DUP | 4064 | 2500 | | SR |
| 175 | SP | all | 44.0 | DUP | 4831 | 2366 | | SP |
| 175 | SP | all | 46.0 | DUP | 4831 | 2471 | | SP |
| 175 | SP | all | 48.0 | DUP | 4831 | 2578 | | SP |
| 175 | NO | SSPP | 44.0 | DUP | 4831 | 2307 | | NO |
| 175 | NO | SSPP | 46.0 | DUP | 4831 | 2445 | | NO |
| 175 | NO | SSPP | 48.0 | DUP | 4831 | 2563 | | NO |

LOADS FOR .284 WINCHESTER .284

| BULLET | | | POWDER | | | | | |
|---|---|---|---|---|---|---|---|---|
| Wgt | Mfg | Type | GRN | Mfg | Type | VEL(fps) | OAL | Source |
| 115 | SP | HP | 56.0 | DUP | 4831 | 2919 | | SP |
| 115 | SP | HP | 58.0 | DUP | 4831 | 3036 | | SP |
| 115 | SP | HP | 60.0 | DUP | 4831 | 3143 | | SP |
| 120 | SR | SPZ | 40.4 | DUP | 3031 | 2600 | 2.800 | SR |
| 120 | SR | SPZ | 41.8 | DUP | 3031 | 2700 | | SR |
| 120 | SR | SPZ | 43.2 | DUP | 3031 | 2800 | | SR |
| 120 | SR | SPZ | 44.6 | DUP | 3031 | 2900 | | SR |
| 120 | SR | SPZ | 43.1 | DUP | 4064 | 2600 | | SR |
| 120 | SR | SPZ | 44.5 | DUP | 4064 | 2700 | | SR |
| 120 | SR | SPZ | 46.0 | DUP | 4064 | 2800 | | SR |
| 120 | SR | SPZ | 47.5 | DUP | 4064 | 2900 | | SR |
| 120 | SR | SPZ | 49.0 | DUP | 4064 | 3000 | | SR |
| 130 | SP | SPZ | 55.0 | DUP | 4831 | 2846 | | SP |
| 130 | SP | SPZ | 57.0 | DUP | 4831 | 2924 | | SP |
| 130 | SP | SPZ | 59.0 | DUP | 4831 | 3039 | | SP |
| 140 | SR | SPZ | 38.0 | DUP | 3031 | 2400 | | SR |
| 140 | SR | SPZ | 39.6 | DUP | 3031 | 2500 | | SR |
| 140 | SR | SPZ | 41.2 | DUP | 3031 | 2600 | | SR |
| 140 | SR | SPZ | 42.9 | DUP | 3031 | 2700 | | SR |
| 140 | SR | SPZ | 44.6 | DUP | 3031 | 2800 | | SR |
| 140 | SR | SPZ | 40.2 | DUP | 4064 | 2400 | | SR |
| 140 | SR | SPZ | 41.9 | DUP | 4064 | 2500 | | SR |
| 140 | SR | SPZ | 43.6 | DUP | 4064 | 2600 | | SR |
| 140 | SR | SPZ | 45.3 | DUP | 4064 | 2700 | | SR |
| 140 | SR | SPZ | 47.0 | DUP | 4064 | 2800 | | SR |
| 140 | SR | SPZ | 48.7 | DUP | 4064 | 2900 | | SR |
| 145 | SP | all | 53.0 | DUP | 4831 | 2720 | | SP |
| 145 | SP | all | 55.0 | DUP | 4831 | 2815 | | SP |
| 145 | SP | all | 57.0 | DUP | 4831 | 2907 | | SP |
| 160 | SR | SBT | 36.2 | DUP | 3031 | 2200 | | SR |
| 160 | SR | SBT | 37.9 | DUP | 3031 | 2300 | | SR |
| 160 | SR | SBT | 39.7 | DUP | 3031 | 2400 | | SR |
| 160 | SR | SBT | 41.5 | DUP | 3031 | 2500 | | SR |
| 160 | SR | SBT | 43.3 | DUP | 3031 | 2600 | | SR |
| 160 | SR | SBT | 37.6 | DUP | 4064 | 2200 | | SR |
| 160 | SR | SBT | 39.3 | DUP | 4064 | 2300 | | SR |
| 160 | SR | SBT | 41.0 | DUP | 4064 | 2400 | | SR |
| 160 | SR | SBT | 42.8 | DUP | 4064 | 2500 | | SR |
| 160 | SR | SBT | 44.6 | DUP | 4064 | 2600 | | SR |
| 160 | SR | SBT | 46.4 | DUP | 4064 | 2700 | | SR |
| 160 | SP | all | 51.0 | DUP | 4831 | 2580 | | SP |
| 160 | SP | all | 53.0 | DUP | 4831 | 2673 | | SP |
| 160 | SP | all | 55.0 | DUP | 4831 | 2765 | | SP |
| 168 | SR | HP | 36.5 | DUP | 3031 | 2200 | | SR |
| 168 | SR | HP | 38.2 | DUP | 3031 | 2300 | | SR |
| 168 | SR | HP | 40.0 | DUP | 3031 | 2400 | | SR |
| 168 | SR | HP | 41.8 | DUP | 3031 | 2500 | | SR |
| 168 | SR | HP | 43.6 | DUP | 3031 | 2600 | | SR |
| 168 | SR | HP | 37.9 | DUP | 4064 | 2200 | | SR |
| 168 | SR | HP | 39.6 | DUP | 4064 | 2300 | | SR |
| 168 | SR | HP | 41.3 | DUP | 4064 | 2400 | | SR |

LOADS FOR .284 WINCHESTER .284

| BULLET | | | POWDER | | | | | |
|---|---|---|---|---|---|---|---|---|
| Wgt | Mfg | Type | GRN | Mfg | Type | VEL(fps) | OAL | Source |
| 168 | SR | HP | 43.1 | DUP | 4064 | 2500 | | SR |
| 168 | SR | HP | 44.9 | DUP | 4064 | 2600 | | SR |
| 170 | SR | RN | 36.4 | DUP | 3031 | 2200 | | SR |
| 170 | SR | RN | 38.0 | DUP | 3031 | 2300 | | SR |
| 170 | SR | RN | 39.6 | DUP | 3031 | 2400 | | SR |
| 170 | SR | RN | 38.9 | DUP | 4064 | 2200 | | SR |
| 170 | SR | RN | 40.7 | DUP | 4064 | 2300 | | SR |
| 170 | SR | RN | 42.5 | DUP | 4064 | 2400 | | SR |
| 170 | SR | RN | 44.3 | DUP | 4064 | 2500 | | SR |
| 175 | SP | all | 49.0 | DUP | 4831 | 2451 | | SP |
| 175 | SP | all | 51.0 | DUP | 4831 | 2523 | | SP |
| 175 | SP | all | 53.0 | DUP | 4831 | 2638 | | SP |

LOADS FOR .280 REMINGTON .284
ALSO FOR 7mm-06 &
7mm EXPRESS REMINGTON

| BULLET | | | POWDER | | | | | |
|---|---|---|---|---|---|---|---|---|
| Wgt | Mfg | Type | GRN | Mfg | Type | VEL.(fps) | OAL | Source |
| 115 | SP | HP | 56.0 | DUP | 4831 | 3020 | | SP |
| 115 | SP | HP | 58.0 | DUP | 4831 | 3135 | | SP |
| 115 | SP | HP | 60.0 | DUP | 4831 | 3246 | | SP |
| 120 | SR | SPZ | 38.5 | DUP | 3031 | 2600 | 3.330 | SR |
| 120 | SR | SPZ | 40.6 | DUP | 3031 | 2700 | | SR |
| 120 | SR | SPZ | 42.7 | DUP | 3031 | 2800 | | SR |
| 120 | SR | SPZ | 44.9 | DUP | 3031 | 2900 | | SR |
| 120 | SR | SPZ | 42.6 | DUP | 4064 | 2600 | | SR |
| 120 | SR | SPZ | 44.3 | DUP | 4064 | 2700 | | SR |
| 120 | SR | SPZ | 46.0 | DUP | 4064 | 2800 | | SR |
| 120 | SR | SPZ | 47.7 | DUP | 4064 | 2900 | | SR |
| 120 | NO | SPZ | 53.0 | DUP | 4831 | 2777 | | NO |
| 120 | NO | SPZ | 55.0 | DUP | 4831 | 2904 | | NO |
| 120 | NO | SPZ | 57.0 | DUP | 4831 | 3078 | | NO |
| 130 | SP | SPZ | 54.0 | DUP | 4831 | 2865 | | SP |
| 130 | SP | SPZ | 56.0 | DUP | 4831 | 2960 | | SP |
| 130 | SP | SPZ | 58.0 | DUP | 4831 | 3077 | | SP |
| 140 | SR | SPZ | 39.6 | DUP | 3031 | 2500 | | SR |
| 140 | SR | SPZ | 41.4 | DUP | 3031 | 2600 | | SR |
| 140 | SR | SPZ | 43.2 | DUP | 3031 | 2700 | | SR |
| 140 | SR | SPZ | 45.0 | DUP | 3031 | 2800 | | SR |
| 140 | SR | SPZ | 42.5 | DUP | 4064 | 2500 | | SR |
| 140 | SR | SPZ | 44.1 | DUP | 4064 | 2600 | | SR |
| 140 | SR | SPZ | 45.8 | DUP | 4064 | 2700 | | SR |
| 140 | SR | SPZ | 47.5 | DUP | 4064 | 2800 | | SR |
| 140 | NO | SPP | 52.0 | DUP | 4831 | 2724 | | NO |
| 140 | NO | SPP | 54.0 | DUP | 4831 | 2843 | | NO |
| 140 | NO | SPP | 56.0 | DUP | 4831 | 2980 | | NO |
| 140 | NO | SPZ | 52.0 | DUP | 4831 | 2713 | | NO |
| 140 | NO | SPZ | 54.0 | DUP | 4831 | 2840 | | NO |
| 140 | NO | SPZ | 56.0 | DUP | 4831 | 2982 | | NO |
| 145 | SP | SPZ | 52.5 | DUP | 4831 | 2765 | | SP |
| 145 | SP | SPZ | 54.5 | DUP | 4831 | 2870 | | SP |
| 145 | SP | SPZ | 56.5 | DUP | 4831 | 2976 | | SP |
| 150 | NO | SPP | 51.0 | DUP | 4831 | 2587 | | NO |
| 150 | NO | SPP | 53.0 | DUP | 4831 | 2743 | | NO |
| 150 | NO | SPP | 55.0 | DUP | 4831 | 2863 | | NO |
| 150 | NO | SPZ | 51.0 | DUP | 4831 | 2593 | | NO |
| 150 | NO | SPZ | 53.0 | DUP | 4831 | 2751 | | NO |
| 150 | NO | SPZ | 55.0 | DUP | 4831 | 2868 | | NO |
| 160 | SR | SBT | 37.4 | DUP | 3031 | 2300 | | SR |
| 160 | SR | SBT | 39.3 | DUP | 3031 | 2400 | | SR |
| 160 | SR | SBT | 41.3 | DUP | 3031 | 2500 | | SR |
| 160 | SR | SBT | 40.2 | DUP | 4064 | 2300 | | SR |
| 160 | SR | SBT | 41.8 | DUP | 4064 | 2400 | | SR |
| 160 | SR | SBT | 43.5 | DUP | 4064 | 2500 | | SR |
| 160 | SR | SBT | 45.2 | DUP | 4064 | 2600 | | SR |
| 160 | SP | all | 51.0 | DUP | 4831 | 2671 | | SP |
| 160 | SP | all | 53.0 | DUP | 4831 | 2795 | | SP |
| 160 | SP | all | 55.0 | DUP | 4831 | 2893 | | SP |

LOADS FOR .280 REMINGTON .284
ALSO FOR 7mm-06 &
7mm EXPRESS REMINGTON

| BULLET | | | POWDER | | | | | |
|---|---|---|---|---|---|---|---|---|
| Wgt | Mfg | Type | GRN | Mfg | Type | VEL.(fps) | OAL | Source |
| 160 | NO | SPP | 50.0 | DUP | 4831 | 2640 | | NO |
| 160 | NO | SPP | 52.0 | DUP | 4831 | 2719 | | NO |
| 160 | NO | SPP | 54.0 | DUP | 4831 | 2812 | | NO |
| 160 | SR | SPBT | 59.0C | DUP | 7828 | 2775 | | DUP |
| 168 | SR | HP | 37.7 | DUP | 3031 | 2300 | | SR |
| 168 | SR | HP | 39.6 | DUP | 3031 | 2400 | | SR |
| 168 | SR | HP | 41.6 | DUP | 3031 | 2500 | | SR |
| 168 | SR | HP | 40.5 | DUP | 4064 | 2300 | | SR |
| 168 | SR | HP | 42.1 | DUP | 4064 | 2400 | | SR |
| 168 | SR | HP | 43.8 | DUP | 4064 | 2500 | | SR |
| 168 | SR | HP | 45.5 | DUP | 4064 | 2600 | | SR |
| 170 | SR | RN | 36.2 | DUP | 3031 | 2200 | | SR |
| 170 | SR | RN | 38.3 | DUP | 3031 | 2300 | | SR |
| 170 | SR | RN | 40.4 | DUP | 3031 | 2400 | | SR |
| 170 | SR | RN | 39.0 | DUP | 4064 | 2200 | | SR |
| 170 | SR | RN | 40.9 | DUP | 4064 | 2300 | | SR |
| 170 | SR | RN | 42.8 | DUP | 4064 | 2400 | | SR |
| 170 | SR | RN | 44.7 | DUP | 4064 | 2500 | | SR |
| 175 | SP | all | 49.0 | DUP | 4831 | 2561 | | SP |
| 175 | SP | all | 51.0 | DUP | 4831 | 2650 | | SP |
| 175 | SP | all | 53.0 | DUP | 4831 | 2767 | | SP |
| 175 | NO | SSPP | 48.5 | DUP | 4831 | 2461 | | NO |
| 175 | NO | SSPP | 50.5 | DUP | 4831 | 2569 | | NO |
| 175 | NO | SSPP | 52.5 | DUP | 4831 | 2678 | | NO |

Letter C indicates compressed powder charge.

LOADS FOR 7MM-08 REMINGTON .284

| BULLET | | | POWDER | | | | | |
|---|---|---|---|---|---|---|---|---|
| Wgt | Mfg | Type | GRN | Mfg | Type | VEL(fps) | OAL | Source |
| 120 | SR | SPZ | 35.6 | DUP | 3031 | 2600 | 2.800 | SR |
| 120 | SR | SPZ | 37.1 | DUP | 3031 | 2700 | | SR |
| 120 | SR | SPZ | 38.6 | DUP | 3031 | 2800 | | SR |
| 120 | SR | SPZ | 40.1 | DUP | 3031 | 2900 | | SR |
| 120 | SR | SPZ | 41.5 | DUP | 3031 | 3000 | | SR |
| 120 | SR | SPZ | 40.5 | DUP | 4064 | 2600 | | SR |
| 120 | SR | SPZ | 41.8 | DUP | 4064 | 2700 | | SR |
| 120 | SR | SPZ | 43.1 | DUP | 4064 | 2800 | | SR |
| 120 | SR | SPZ | 44.4C | DUP | 4064 | 2900 | | SR |
| 120 | SR | SPZ | 45.7C | DUP | 4064 | 3000 | | SR |
| 140 | SR | ALL | 33.8 | DUP | 3031 | 2400 | | SR |
| 140 | SR | ALL | 35.5 | DUP | 3031 | 2500 | | SR |
| 140 | SR | ALL | 37.2 | DUP | 3031 | 2600 | | SR |
| 140 | SR | ALL | 38.9 | DUP | 3031 | 2700 | | SR |
| 140 | SR | ALL | 40.5 | DUP | 3031 | 2800 | | SR |
| 140 | SR | ALL | 37.8 | DUP | 4064 | 2400 | | SR |
| 140 | SR | ALL | 39.4 | DUP | 4064 | 2500 | | SR |
| 140 | SR | ALL | 40.9 | DUP | 4064 | 2600 | | SR |
| 140 | SR | ALL | 42.4 | DUP | 4064 | 2700 | | SR |
| 140 | SR | All | 43.9C | DUP | 4064 | 2800 | | SR |
| 140 | SR | ALL | 45.4 | DUP | 4064 | 2900 | | SR |
| 150 | SR | HP | 32.8 | DUP | 3031 | 2300 | | SR |
| 150 | SR | HP | 34.4 | DUP | 3031 | 2400 | | SR |
| 150 | SR | HP | 36.0 | DUP | 3031 | 2500 | | SR |
| 150 | SR | HP | 37.6 | DUP | 3031 | 2600 | | SR |
| 150 | SR | HP | 39.1 | DUP | 3031 | 2700 | | SR |
| 150 | SR | HP | 36.1 | DUP | 4064 | 2300 | | SR |
| 150 | SR | HP | 37.4 | DUP | 4064 | 2400 | | SR |
| 150 | SR | HP | 38.8 | DUP | 4064 | 2500 | | SR |
| 150 | SR | HP | 40.1 | DUP | 4064 | 2600 | | SR |
| 150 | SR | HP | 41.4C | DUP | 4064 | 2700 | | SR |
| 150 | SR | HP | 42.1C | DUP | 4064 | 2750 | | SR |
| 160 | SR | SPBT | 31.1 | DUP | 3031 | 2200 | | SR |
| 160 | SR | SPBT | 32.8 | DUP | 3031 | 2300 | | SR |
| 160 | SR | SPBT | 34.5 | DUP | 3031 | 2400 | | SR |
| 160 | SR | SPBT | 36.1 | DUP | 3031 | 2500 | | SR |
| 160 | SR | SPBT | 37.7 | DUP | 3031 | 2600 | | SR |
| 160 | SR | SPBT | 39.3 | DUP | 3031 | 2700 | | SR |
| 160 | SR | SPBT | 34.6 | DUP | 4064 | 2200 | | SR |
| 160 | SR | SPBT | 36.1 | DUP | 4064 | 2300 | | SR |
| 160 | SR | SPBT | 37.5 | DUP | 4064 | 2400 | | SR |
| 160 | SR | SPBT | 38.9 | DUP | 4064 | 2500 | | SR |
| 160 | SR | SPBT | 40.3C | DUP | 4064 | 2600 | | SR |
| 160 | SR | SPBT | 41.7C | DUP | 4064 | 2700 | | SR |
| 168 | SR | HP | 31.4 | DUP | 3031 | 2200 | | SR |
| 168 | SR | HP | 33.0 | DUP | 3031 | 2300 | | SR |
| 168 | SR | HP | 34.6 | DUP | 3031 | 2400 | | SR |
| 168 | SR | HP | 36.2 | DUP | 3031 | 2500 | | SR |
| 168 | SR | HP | 37.8 | DUP | 3031 | 2600 | | SR |
| 168 | SR | HP | 32.2 | DUP | 4064 | 2200 | | SR |
| 168 | SR | HP | 34.2 | DUP | 4064 | 2300 | | SR |
| 168 | SR | HP | 36.2 | DUP | 4064 | 2400 | | SR |

LOADS FOR 7MM-08 REMINGTON .284

| BULLET | | | POWDER | | | | | |
|---|---|---|---|---|---|---|---|---|
| Wgt | Mfg | Type | GRN | Mfg | Type | VEL(fps) | OAL | Source |
| 168 | SR | HP | 38.2 | DUP | 4064 | 2500 | | SR |
| 168 | SR | HP | 40.1C | DUP | 4064 | 2600 | | SR |
| 168 | SR | HP | 42.0C | DUP | 4064 | 2700 | | SR |
| 170 | SR | RN | 33.0 | DUP | 3031 | 2200 | | SR |
| 170 | SR | RN | 34.7 | DUP | 3031 | 2300 | | SR |
| 170 | SR | RN | 36.3 | DUP | 3031 | 2400 | | SR |
| 170 | SR | RN | 37.9 | DUP | 3031 | 2500 | | SR |
| 170 | SR | RN | 39.5 | DUP | 3031 | 2600 | | SR |
| 170 | SR | RN | 36.0 | DUP | 4064 | 2200 | | SR |
| 170 | SR | RN | 37.6 | DUP | 4064 | 2300 | | SR |
| 170 | SR | RN | 39.1 | DUP | 4064 | 2400 | | SR |
| 170 | SR | RN | 40.6 | DUP | 4064 | 2500 | | SR |
| 170 | SR | RN | 42.1 | DUP | 4064 | 2600 | | SR |
| 175 | SR | SPBT | 31.8 | DUP | 3031 | 2200 | | SR |
| 175 | SR | SPBT | 33.6 | DUP | 3031 | 2300 | | SR |
| 175 | SR | SPBT | 35.3 | DUP | 3031 | 2400 | | SR |
| 175 | SR | SPBT | 37.0 | DUP | 3031 | 2500 | | SR |
| 175 | SR | SPBT | 34.5 | DUP | 4064 | 2200 | | SR |
| 175 | SR | SPBT | 36.3 | DUP | 4064 | 2300 | | SR |
| 175 | SR | SPBT | 38.0 | DUP | 4064 | 2400 | | SR |
| 175 | SR | SPBT | 39.7C | DUP | 4064 | 2500 | | SR |
| 175 | SR | SPBT | 41.4C | DUP | 4064 | 2600 | | SR |

Letter C indicates compressed powder charge.

LOADS FOR 7 x 61 SHARPE & HART .284

| BULLET | | | | POWDER | | | | |
|---|---|---|---|---|---|---|---|---|
| Wgt | Mfg | Type | GRN | Mfg | Type | VEL(fps) | OAL | Source |
| 120 | SR | SPZ | 51.1 | DUP | 4064 | 2900 | 3.190 | SR |
| 120 | SR | SPZ | 52.5 | DUP | 4064 | 3000 | | SR |
| 120 | SR | SPZ | 54.0 | DUP | 4064 | 3100 | | SR |
| 120 | SR | SPZ | 55.5 | DUP | 4064 | 3200 | | SR |
| 120 | SR | SPZ | 57.0 | DUP | 4064 | 3300 | | SR |
| 120 | SR | SPZ | 58.7 | DUP | 4831 | 2900 | | SR |
| 120 | SR | SPZ | 60.7 | DUP | 4831 | 3000 | | SR |
| 120 | SR | SPZ | 62.6 | DUP | 4831 | 3100 | | SR |
| 140 | SR | SPZ | 50.8 | DUP | 4064 | 2800 | | SR |
| 140 | SR | SPZ | 52.3 | DUP | 4064 | 2900 | | SR |
| 140 | SR | SPZ | 53.9 | DUP | 4064 | 3000 | | SR |
| 140 | SR | SPZ | 55.5 | DUP | 4064 | 3100 | | SR |
| 140 | SR | SPZ | 57.3 | DUP | 4831 | 2800 | | SR |
| 140 | SR | SPZ | 59.3 | DUP | 4831 | 2900 | | SR |
| 140 | SR | SPZ | 61.3 | DUP | 4831 | 3000 | | SR |
| 140 | SR | SPZ | 63.3 | DUP | 4831 | 3100 | | SR |
| 160 | SR | SBT | 48.0 | DUP | 4064 | 2600 | | SR |
| 160 | SR | SBT | 50.0 | DUP | 4064 | 2700 | | SR |
| 160 | SR | SBT | 52.1 | DUP | 4064 | 2800 | | SR |
| 160 | SR | SBT | 54.2 | DUP | 4064 | 2900 | | SR |
| 160 | SR | SBT | 54.9 | DUP | 4831 | 2600 | | SR |
| 160 | SR | SBT | 57.1 | DUP | 4831 | 2700 | | SR |
| 160 | SR | SBT | 59.3 | DUP | 4831 | 2800 | | SR |
| 160 | SR | SBT | 61.4 | DUP | 4831 | 2900 | | SR |
| 160 | SR | SBT | 63.5 | DUP | 4831 | 3000 | | SR |
| 168 | SR | HP | 48.2 | DUP | 4064 | 2600 | | SR |
| 168 | SR | HP | 50.3 | DUP | 4064 | 2700 | | SR |
| 168 | SR | HP | 52.4 | DUP | 4064 | 2800 | | SR |
| 168 | SR | HP | 55.2 | DUP | 4831 | 2600 | | SR |
| 168 | SR | HP | 57.4 | DUP | 4831 | 2700 | | SR |
| 168 | SR | HP | 59.6 | DUP | 4831 | 2800 | | SR |
| 168 | SR | HP | 61.8 | DUP | 4831 | 2900 | | SR |
| 168 | SR | HP | 64.0 | DUP | 4831 | 3000 | | SR |
| 170 | SR | RN | 47.2 | DUP | 4064 | 2500 | | SR |
| 170 | SR | RN | 49.1 | DUP | 4064 | 2600 | | SR |
| 170 | SR | RN | 51.0 | DUP | 4064 | 2700 | | SR |
| 170 | SR | RN | 52.8 | DUP | 4064 | 2800 | | SR |
| 170 | SR | RN | 53.3 | DUP | 4831 | 2500 | | SR |
| 170 | SR | RN | 55.5 | DUP | 4831 | 2600 | | SR |
| 170 | SR | RN | 57.7 | DUP | 4831 | 2700 | | SR |
| 170 | SR | RN | 59.8 | DUP | 4831 | 2800 | | SR |
| 170 | SR | RN | 61.9 | DUP | 4831 | 2900 | | SR |

LOADS FOR 7MM REMINGTON MAGNUM .284

| BULLET | | | POWDER | | | | | |
|---|---|---|---|---|---|---|---|---|
| Wgt | Mfg | Type | GRN | Mfg | Type | VEL(fps) | OAL | Source |
| 115 | SP | HP | 67.0 | DUP | 4831 | 3208 | | SP |
| 115 | SP | HP | 69.0 | DUP | 4831 | 3320 | | SP |
| 115 | SP | HP | 71.0 | DUP | 4831 | 3412 | | SP |
| 120 | SR | SPZ | 48.3 | DUP | 3031 | 2900 | 3.290 | SR |
| 120 | SR | SPZ | 50.3 | DUP | 3031 | 3000 | | SR |
| 120 | SR | SPZ | 52.4 | DUP | 3031 | 3100 | | SR |
| 120 | SR | SPZ | 54.5 | DUP | 3031 | 3200 | | SR |
| 120 | SR | SPZ | 61.7 | DUP | 4831 | 2900 | | SR |
| 120 | SR | SPZ | 63.6 | DUP | 4831 | 3000 | | SR |
| 120 | SR | SPZ | 65.5 | DUP | 4831 | 3100 | | SR |
| 120 | SR | SPZ | 67.4 | DUP | 4831 | 3200 | | SR |
| 120 | SR | SPZ | 69.3 | DUP | 4831 | 3300 | | SR |
| 120 | SR | SPZ | 71.1 | DUP | 4831 | 3400 | | SR |
| 120 | NO | SPZ | 64.0 | DUP | 4831 | 3127 | | NO |
| 120 | NO | SPZ | 66.0 | DUP | 4831 | 3229 | | NO |
| 120 | NO | SPZ | 68.0 | DUP | 4831 | 3366 | | NO |
| 130 | SP | SPZ | 61.0 | DUP | 4831 | 3044 | | SP |
| 130 | SP | SPZ | 63.0 | DUP | 4831 | 3153 | | SP |
| 130 | SP | SPZ | 65.0 | DUP | 4831 | 3243 | | SP |
| 140 | SR | SPZ | 47.5 | DUP | 4064 | 2600 | | SR |
| 140 | SR | SPZ | 49.8 | DUP | 4064 | 2700 | | SR |
| 140 | SR | SPZ | 52.1 | DUP | 4064 | 2800 | | SR |
| 140 | SR | SPZ | 54.4 | DUP | 4064 | 2900 | | SR |
| 140 | SR | SPZ | 56.7 | DUP | 4064 | 3000 | | SR |
| 140 | SR | SPZ | 59.0 | DUP | 4064 | 3100 | | SR |
| 140 | SR | SPZ | 55.2 | DUP | 4831 | 2700 | | SR |
| 140 | SR | SPZ | 58.0 | DUP | 4831 | 2800 | | SR |
| 140 | SR | SPZ | 60.8 | DUP | 4831 | 2900 | | SR |
| 140 | SR | SPZ | 63.5 | DUP | 4831 | 3000 | | SR |
| 140 | SR | SPZ | 66.2 | DUP | 4831 | 3100 | | SR |
| 140 | NO | SPP | 62.0 | DUP | 4831 | 3021 | | NO |
| 140 | NO | SPP | 64.0 | DUP | 4831 | 3124 | | NO |
| 140 | NO | SPP | 66.0 | DUP | 4831 | 3218 | | NO |
| 140 | NO | SPZ | 62.0 | DUP | 4831 | 3021 | | NO |
| 140 | NO | SPZ | 64.0 | DUP | 4831 | 3127 | | NO |
| 140 | NO | SPZ | 66.0 | DUP | 4831 | 3221 | | NO |
| 145 | SP | all | 59.0 | DUP | 4831 | 2920 | | SP |
| 145 | SP | all | 61.0 | DUP | 4831 | 3006 | | SP |
| 145 | SP | all | 63.0 | DUP | 4831 | 3081 | | SP |
| 150 | NO | SPP | 61.0 | DUP | 4831 | 3033 | | NO |
| 150 | NO | SPP | 63.0 | DUP | 4831 | 3118 | | NO |
| 150 | NO | SPP | 65.0 | DUP | 4831 | 3235 | | NO |
| 150 | NO | SPZ | 61.0 | DUP | 4831 | 3017 | | NO |
| 150 | NO | SPZ | 63.0 | DUP | 4831 | 3133 | | NO |
| 150 | NO | SPZ | 65.0 | DUP | 4831 | 3242 | | NO |
| 160 | SR | SBT | 43.2 | DUP | 4064 | 2500 | | SR |
| 160 | SR | SBT | 45.8 | DUP | 4064 | 2600 | | SR |
| 160 | SR | SBT | 48.4 | DUP | 4064 | 2700 | | SR |
| 160 | SR | SBT | 51.0 | DUP | 4064 | 2800 | | SR |
| 160 | SR | SBT | 53.7 | DUP | 4064 | 2900 | | SR |
| 160 | SR | SBT | 51.7 | DUP | 4831 | 2500 | | SR |
| 160 | SR | SBT | 54.0 | DUP | 4831 | 2600 | | SR |
| 160 | SR | SBT | 56.2 | DUP | 4831 | 2700 | | SR |

## LOADS FOR 7MM REMINGTON MAGNUM .284

| BULLET | | | POWDER | | | | | |
|---|---|---|---|---|---|---|---|---|
| Wgt | Mfg | Type | GRN | Mfg | Type | VEL(fps) | OAL | Source |
| 160 | SR | SBT | 58.4 | DUP | 4831 | 2800 | | SR |
| 160 | SR | SBT | 60.6 | DUP | 4831 | 2900 | | SR |
| 160 | SR | SBT | 62.8 | DUP | 4831 | 3000 | | SR |
| 160 | SP | all | 58.0 | DUP | 4831 | 2801 | | SP |
| 160 | SP | all | 60.0 | DUP | 4831 | 2892 | | SP |
| 160 | SP | all | 62.0 | DUP | 4831 | 2979 | | SP |
| 160 | NO | SPP | 59.0 | DUP | 4831 | 2855 | | NO |
| 160 | NO | SPP | 61.0 | DUP | 4831 | 2960 | | NO |
| 160 | NO | SPP | 63.0 | DUP | 4831 | 3011 | | NO |
| 168 | SR | HP | 45.5 | DUP | 4064 | 2500 | | SR |
| 168 | SR | HP | 47.7 | DUP | 4064 | 2600 | | SR |
| 168 | SR | HP | 49.9 | DUP | 4064 | 2700 | | SR |
| 168 | SR | HP | 52.1 | DUP | 4064 | 2800 | | SR |
| 168 | SR | HP | 52.9 | DUP | 4831 | 2500 | | SR |
| 168 | SR | HP | 55.3 | DUP | 4831 | 2600 | | SR |
| 168 | SR | HP | 57.6 | DUP | 4831 | 2700 | | SR |
| 168 | SR | HP | 59.9 | DUP | 4831 | 2800 | | SR |
| 168 | SR | HP | 62.2 | DUP | 4831 | 2900 | | SR |
| 170 | SR | all | 46.6 | DUP | 4064 | 2500 | | SR |
| 170 | SR | all | 49.0 | DUP | 4064 | 2600 | | SR |
| 170 | SR | all | 51.3 | DUP | 4064 | 2700 | | SR |
| 170 | SR | all | 54.6 | DUP | 4831 | 2500 | | SR |
| 170 | SR | all | 56.7 | DUP | 4831 | 2600 | | SR |
| 170 | SR | all | 58.8 | DUP | 4831 | 2700 | | SR |
| 170 | SR | all | 60.8 | DUP | 4831 | 2800 | | SR |
| 170 | SR | all | 62.8 | DUP | 4831 | 2900 | | SR |
| 175 | SP | all | 56.0 | DUP | 4831 | 2688 | | SP |
| 175 | SP | all | 58.0 | DUP | 4831 | 2783 | | SP |
| 175 | SP | all | 60.0 | DUP | 4831 | 2865 | | SP |
| 175 | NO | SSPP | 56.0 | DUP | 4831 | 2692 | | NO |
| 175 | NO | SSPP | 58.0 | DUP | 4831 | 2784 | | NO |
| 175 | NO | SSPP | 60.0 | DUP | 4831 | 2866 | | NO |
| 175 | NO | SSP | 66.0 | DUP | 7828 | 2910 | | DUP |
| 175 | SP | MAGTIP | 66.0 | DUP | 7828 | 2910 | | DUP |

## LOADS FOR 7MM WEATHERBY MAGNUM .284

| BULLET | | | POWDER | | | | | |
|---|---|---|---|---|---|---|---|---|
| Wgt | Mfg | Type | GRN | Mfg | Type | VEL(fps) | OAL | Source |
| 115 | SP | HP | 72.0 | DUP | 4831 | 3381 | | SP |
| 115 | SP | HP | 74.0 | DUP | 4831 | 3464 | | SP |
| 115 | SP | HP | 76.0 | DUP | 4831 | 3580 | | SP |
| 120 | SR | SPZ | 56.8 | DUP | 4064 | 3000 | 3.250 | SR |
| 120 | SR | SPZ | 58.4 | DUP | 4064 | 3100 | | SR |
| 120 | SR | SPZ | 60.1 | DUP | 4064 | 3200 | | SR |
| 120 | SR | SPZ | 61.8 | DUP | 4064 | 3300 | | SR |
| 120 | SR | SPZ | 66.1 | DUP | 4831 | 3000 | | SR |
| 120 | SR | SPZ | 67.7 | DUP | 4831 | 3100 | | SR |
| 120 | SR | SPZ | 69.3 | DUP | 4831 | 3200 | | SR |
| 120 | SR | SPZ | 70.9 | DUP | 4831 | 3300 | | SR |
| 120 | SR | SPZ | 72.4 | DUP | 4831 | 3400 | | SR |
| 120 | NO | SPZ | 68.0 | DUP | 4831 | 3257 | | NO |
| 120 | NO | SPZ | 70.0 | DUP | 4831 | 3430 | | NO |
| 120 | NO | SPZ | 72.0 | DUP | 4831 | 3542 | | NO |
| 130 | SP | SPZ | 69.0 | DUP | 4831 | 3193 | | SP |
| 130 | SP | SPZ | 71.0 | DUP | 4831 | 3295 | | SP |
| 130 | SP | SPZ | 73.0 | DUP | 4831 | 3370 | | SP |
| 140 | SR | SPZ | 56.6 | DUP | 4064 | 2900 | | SR |
| 140 | SR | SPZ | 58.5 | DUP | 4064 | 3000 | | SR |
| 140 | SR | SPZ | 60.4 | DUP | 4064 | 3100 | | SR |
| 140 | SR | SPZ | 64.7 | DUP | 4831 | 2900 | | SR |
| 140 | SR | SPZ | 66.5 | DUP | 4831 | 3000 | | SR |
| 140 | SR | SPZ | 68.3 | DUP | 4831 | 3100 | | SR |
| 140 | SR | SPZ | 70.1 | DUP | 4831 | 3200 | | SR |
| 140 | NO | SPP | 66.0 | DUP | 4831 | 3182 | | NO |
| 140 | NO | SPP | 68.0 | DUP | 4831 | 3246 | | NO |
| 140 | NO | SPP | 70.0 | DUP | 4831 | 3289 | | NO |
| 140 | NO | SPZ | 66.0 | DUP | 4831 | 3181 | | NO |
| 140 | NO | SPZ | 68.0 | DUP | 4831 | 3253 | | NO |
| 140 | NO | SPZ | 70.0 | DUP | 4831 | 3291 | | NO |
| 145 | SP | all | 67.0 | DUP | 4831 | 3052 | | SP |
| 145 | SP | all | 69.0 | DUP | 4831 | 3139 | | SP |
| 145 | SP | all | 71.0 | DUP | 4831 | 3228 | | SP |
| 150 | NO | SPP | 64.0 | DUP | 4831 | 2989 | | NO |
| 150 | NO | SPP | 66.0 | DUP | 4831 | 3091 | | NO |
| 150 | NO | SPP | 68.0 | DUP | 4831 | 3172 | | NO |
| 150 | NO | SPZ | 64.0 | DUP | 4831 | 2997 | | NO |
| 150 | NO | SPZ | 66.0 | DUP | 4831 | 3099 | | NO |
| 150 | NO | SPZ | 68.0 | DUP | 4831 | 3194 | | NO |
| 160 | SR | SBT | 54.0 | DUP | 4064 | 2700 | | SR |
| 160 | SR | SBT | 55.9 | DUP | 4064 | 2800 | | SR |
| 160 | SR | SBT | 57.8 | DUP | 4064 | 2900 | | SR |
| 160 | SR | SBT | 60.9 | DUP | 4831 | 2700 | | SR |
| 160 | SR | SBT | 63.1 | DUP | 4831 | 2800 | | SR |
| 160 | SR | SBT | 65.3 | DUP | 4831 | 2900 | | SR |
| 160 | SR | SBT | 67.4 | DUP | 4831 | 3000 | | SR |
| 160 | SP | all | 66.0 | DUP | 4831 | 2922 | | SP |
| 160 | SP | all | 68.0 | DUP | 4831 | 2991 | | SP |
| 160 | SP | all | 70.0 | DUP | 4831 | 3099 | | SP |
| 160 | NO | SPP | 62.0 | DUP | 4831 | 2878 | | NO |
| 160 | NO | SPP | 64.0 | DUP | 4831 | 2966 | | NO |
| 160 | NO | SPP | 66.0 | DUP | 4831 | 3075 | | NO |

LOADS FOR 7MM WEATHERBY MAGNUM .284

| BULLET | | | POWDER | | | | | |
|---|---|---|---|---|---|---|---|---|
| Wgt | Mfg | Type | GRN | Mfg | Type | VEL(fps) | OAL | Source |
| 168 | SR | HP | 54.3 | DUP | 4064 | 2700 | | SR |
| 168 | SR | HP | 56.2 | DUP | 4064 | 2800 | | SR |
| 168 | SR | HP | 58.1 | DUP | 4064 | 2900 | | SR |
| 168 | SR | HP | 61.2 | DUP | 4831 | 2700 | | SR |
| 168 | SR | HP | 63.4 | DUP | 4831 | 2800 | | SR |
| 168 | SR | HP | 65.6 | DUP | 4831 | 2900 | | SR |
| 168 | SR | HP | 67.7 | DUP | 4831 | 3000 | | SR |
| 170 | SR | RN | 50.2 | DUP | 4064 | 2500 | | SR |
| 170 | SR | RN | 52.7 | DUP | 4064 | 2600 | | SR |
| 170 | SR | RN | 55.2 | DUP | 4064 | 2700 | | SR |
| 170 | SR | RN | 59.5 | DUP | 4831 | 2600 | | SR |
| 170 | SR | RN | 61.7 | DUP | 4831 | 2700 | | SR |
| 170 | SR | RN | 63.8 | DUP | 4831 | 2800 | | SR |
| 170 | SR | RN | 65.9 | DUP | 4831 | 2900 | | SR |
| 175 | SP | all | 64.0 | DUP | 4831 | 2748 | | SP |
| 175 | SP | all | 66.0 | DUP | 4831 | 2855 | | SP |
| 175 | SP | all | 68.0 | DUP | 4831 | 2935 | | SP |
| 175 | NO | SSPP | 60.0 | DUP | 4831 | 2653 | | NO |
| 175 | NO | SSPP | 62.0 | DUP | 4831 | 2814 | | NO |
| 175 | NO | SSPP | 64.0 | DUP | 4831 | 2923 | | NO |

LOADS FOR .30 M1 CARBINE .308

| BULLET | | | POWDER | | | | | |
|---|---|---|---|---|---|---|---|---|
| Wgt | Mfg | Type | GRN | Mfg | Type | VEL (fps) | OAL | Source |
| 100 | SP | PL | 13.5 | DUP | 4198 | 1459 | | SP |
| 100 | SP | PL | 14.5 | DUP | 4198 | 1581 | | SP |
| 100 | SP | PL | 15.5 | DUP | 4198 | 1685 | | SP |
| 110 | SR | RN | 11.0 | HER | 2400 | 1600 | 1.680 | SR |
| 110 | SR | RN | 11.9 | HER | 2400 | 1700 | | SR |
| 110 | SR | RN | 12.8 | HER | 2400 | 1800 | | SR |
| 110 | SR | RN | 13.6 | HER | 2400 | 1900 | | SR |
| 110 | SR | RN | 13.2 | DUP | 4227 | 1600 | | SR |
| 110 | SR | RN | 14.0 | DUP | 4227 | 1700 | | SR |
| 110 | SR | RN | 14.7 | DUP | 4227 | 1800 | | SR |
| 110 | SR | RN | 15.4 | DUP | 4227 | 1900 | | SR |
| 110 | SP | all | 13.0 | DUP | 4198 | 1428 | | SP |
| 110 | SP | all | 14.0 | DUP | 4198 | 1512 | | SP |
| 110 | SP | all | 15.0 | DUP | 4198 | 1631 | | SP |

## LOADS FOR 7.62MM x 39MM .308 & .311

| BULLET | | | POWDER | | | | | |
|---|---|---|---|---|---|---|---|---|
| Wgt | Mfg | Type | GRN | Mfg | Type | VEL(fps) | OAL | Source |
| 110 | SR | HP | 19.3 | DUP | 4227 | 2200 | 2.200 | SR |
| 110 | SR | HP | 20.7 | DUP | 4227 | 2300 | | SR |
| 110 | SR | HP | 22.1 | DUP | 4227 | 2400 | | SR |
| 110 | SR | HP | 23.5 | DUP | 4227 | 2500 | | SR |
| 110 | SR | HP | 24.8 | DUP | 4227 | 2600 | | SR |
| 110 | SR | HP | 23.0 | DUP | 4198 | 2200 | | SR |
| 110 | SR | HP | 24.3 | DUP | 4198 | 2300 | | SR |
| 110 | SR | HP | 25.6 | DUP | 4198 | 2400 | | SR |
| 110 | SR | HP | 26.8 | DUP | 4198 | 2500 | | SR |
| 125 | SR | SPZ | 20.6 | DUP | 4227 | 2200 | | SR |
| 125 | SR | SPZ | 21.9 | DUP | 4227 | 2300 | | SR |
| 125 | SR | SPZ | 23.1 | DUP | 4227 | 2400 | | SR |
| 125 | SR | SPZ | 24.3 | DUP | 4227 | 2500 | | SR |
| 125 | SR | SPZ | 23.7 | DUP | 4198 | 2200 | | SR |
| 125 | SR | SPZ | 25.0 | DUP | 4198 | 2300 | | SR |
| 125 | SR | SPZ | 26.3 | DUP | 4198 | 2400 | | SR |
| 150 | SR | SPZ .311 | 17.2 | DUP | 4227 | 1900 | | SR |
| 150 | SR | SPZ | 18.8 | DUP | 4227 | 2000 | | SR |
| 150 | SR | SPZ | 20.4 | DUP | 4227 | 2100 | | SR |
| 150 | SR | SPZ | 22.0 | DUP | 4227 | 2200 | | SR |
| 150 | SR | SPZ | 20.6 | DUP | 4198 | 1900 | | SR |
| 150 | SR | SPZ | 22.0 | DUP | 4198 | 2000 | | SR |
| 150 | SR | SPZ | 23.3 | DUP | 4198 | 2100 | | SR |
| 150 | SR | SPZ | 24.6 | DUP | 4198 | 2200 | | SR |

LOADS FOR .30-30 WINCHESTER .308 & .307

| BULLET | | | POWDER | | | | | |
|---|---|---|---|---|---|---|---|---|
| Wgt | Mfg | Type | GRN | Mfg | Type | VEL(fps) | OAL | Source |
| 100 | SP | PL | 31.0 | DUP | 4895 | 2254 | | SP |
| 100 | SP | PL | 33.0 | DUP | 4895 | 2375 | | SP |
| 100 | SP | PL | 35.0 | DUP | 4895 | 2468 | | SP |
| 110 | SR | all .308 | 27.4 | DUP | 4198 | 2400 | 2.550 | SR |
| 110 | SR | all | 28.4 | DUP | 4198 | 2500 | | SR |
| 110 | SR | all | 29.4 | DUP | 4198 | 2600 | | SR |
| 110 | SR | all | 26.8 | HER | RE-7 | 2400 | | SR |
| 110 | SR | all | 27.9 | HER | RE-7 | 2500 | | SR |
| 110 | SR | all | 29.1 | HER | RE-7 | 2600 | | SR |
| 110 | SR | all | 30.3 | HER | RE-7 | 2700 | | SR |
| 110 | SR | all | 31.5 | HER | RE-7 | 2800 | | SR |
| 110 | SP | all | 31.0 | DUP | 4895 | 2258 | | SP |
| 110 | SP | all | 33.0 | DUP | 4895 | 2361 | | SP |
| 110 | SP | all | 35.0 | DUP | 4895 | 2434 | | SP |
| 125 | SR | HPFN .307 | 24.8 | DUP | 4198 | 2100 | | SR |
| 125 | SR | HPFN | 26.1 | DUP | 4198 | 2200 | | SR |
| 125 | SR | HPFN | 27.4 | DUP | 4198 | 2300 | | SR |
| 125 | SR | HPFN | 28.7 | DUP | 4198 | 2400 | | SR |
| 125 | SR | HPFN | 30.0 | DUP | 4198 | 2500 | | SR |
| 125 | SR | HPFN | 25.5 | HER | RE-7 | 2100 | | SR |
| 125 | SR | HPFN | 27.2 | HER | RE-7 | 2200 | | SR |
| 125 | SR | HPFN | 28.8 | HER | RE-7 | 2300 | | SR |
| 125 | SR | HPFN | 30.4 | HER | RE-7 | 2400 | | SR |
| 125 | SR | HPFN | 32.0 | HER | RE-7 | 2500 | | SR |
| 130 | SP | all | 35.5 | DUP | 4831 | 2000 | | SP |
| 130 | SP | all | 37.5 | DUP | 4831 | 2132 | | SP |
| 130 | SP | all | 39.5 | DUP | 4831 | 2243 | | SP |
| 150 | SR | FN .307 | 21.9 | DUP | 4198 | 1900 | | SR |
| 150 | SR | FN | 23.1 | DUP | 4198 | 2000 | | SR |
| 150 | SR | FN | 24.3 | DUP | 4198 | 2100 | | SR |
| 150 | SR | FN | 25.6 | DUP | 4198 | 2200 | | SR |
| 150 | SR | FN | 21.1 | HER | RE-7 | 1900 | | SR |
| 150 | SR | FN | 22.6 | HER | RE-7 | 2000 | | SR |
| 150 | SR | FN | 24.1 | HER | RE-7 | 2100 | | SR |
| 150 | SR | FN | 25.6 | HER | RE-7 | 2200 | | SR |
| 150 | SR | FN | 27.2 | HER | RE-7 | 2300 | | SR |
| 150 | SP | all | 25.0 | HER | RE-7 | 1902 | | SP |
| 150 | SP | all | 27.0 | HER | RE-7 | 2108 | | SP |
| 150 | SP | all | 29.0 | HER | RE-7 | 2319 | | SP |
| 150 | NO | FP | 22.0 | DUP | 4198 | 1963 | | NO |
| 150 | NO | FP | 24.0 | DUP | 4198 | 2102 | | NO |
| 150 | NO | FP | 26.0 | DUP | 4198 | 2257 | | NO |
| 170 | SR | FN .307 | 20.1 | HER | RE-7 | 1800 | | SR |
| 170 | SR | FN | 21.9 | HER | RE-7 | 1900 | | SR |
| 170 | SR | FN | 23.7 | HER | RE-7 | 2000 | | SR |
| 170 | SR | FN | 25.5 | HER | RE-7 | 2100 | | SR |
| 170 | SR | FN | 26.5 | DUP | 4064 | 1800 | | SR |
| 170 | SR | FN | 27.8 | DUP | 4064 | 1900 | | SR |
| 170 | SR | FN | 28.7 | DUP | 4064 | 2000 | | SR |
| 170 | SR | FN | 29.9 | DUP | 4064 | 2100 | | SR |
| 170 | SP | FN | 27.0 | DUP | 4064 | 1726 | | SP |

LOADS FOR .30-30 WINCHESTER .308 & .307

| BULLET | | | POWDER | | | | | |
|---|---|---|---|---|---|---|---|---|
| Wgt | Mfg | Type | GRN | Mfg | Type | VEL(fps) | OAL | Source |
| 170 | SP | FN | 29.0 | DUP | 4064 | 1895 | | SP |
| 170 | SP | FN | 31.0 | DUP | 4064 | 2056 | | SP |
| 170 | NO | FP | 21.0 | DUP | 4198 | 1809 | | NO |
| 170 | NO | FP | 23.0 | DUP | 4198 | 1947 | | NO |
| 170 | NO | FP | 25.0 | DUP | 4198 | 2092 | | NO |

LOADS FOR .30/40 KRAG .308

| BULLET | | | POWDER | | | | | |
|---|---|---|---|---|---|---|---|---|
| Wgt | Mfg | Type | GRN | Mfg | Type | VEL(fps) | OAL | Source |
| 100 | SP | PL | 42.0 | DUP | 4064 | 2551 | | SP |
| 100 | SP | PL | 44.0 | DUP | 4064 | 2694 | | SP |
| 100 | SP | PL | 46.0 | DUP | 4064 | 2811 | | SP |
| 110 | SR | HP | 36.7 | DUP | 3031 | 2400 | 3.089 | SR |
| 110 | SR | HP | 38.8 | DUP | 3031 | 2500 | | SR |
| 110 | SR | HP | 40.9 | DUP | 3031 | 2600 | | SR |
| 110 | SR | HP | 43.0 | DUP | 3031 | 2700 | | SR |
| 110 | SR | HP | 44.2 | DUP | 3031 | 2800 | | SR |
| 110 | SR | HP | 41.5 | DUP | 4895 | 2400 | | SR |
| 110 | SR | HP | 42.9 | DUP | 4895 | 2500 | | SR |
| 110 | SR | HP | 44.3 | DUP | 4895 | 2600 | | SR |
| 110 | SR | HP | 45.7 | DUP | 4895 | 2700 | | SR |
| 110 | SR | HP | 47.1 | DUP | 4895 | 2800 | | SR |
| 110 | SP | VAR | 41.0 | DUP | 4064 | 2413 | | SP |
| 110 | SP | VAR | 43.0 | DUP | 4064 | 2552 | | SP |
| 110 | SP | VAR | 45.0 | DUP | 4064 | 2667 | | SP |
| 110 | SP | SP | 42.0 | DUP | 4064 | 2599 | | SP |
| 110 | SP | SP | 44.0 | DUP | 4064 | 2734 | | SP |
| 110 | SP | SP | 46.0 | DUP | 4064 | 2843 | | SP |
| 125 | SR | SPZ | 36.3 | DUP | 3031 | 2200 | | SR |
| 125 | SR | SPZ | 37.6 | DUP | 3031 | 2300 | | SR |
| 125 | SR | SPZ | 39.0 | DUP | 3031 | 2400 | | SR |
| 125 | SR | SPZ | 40.4 | DUP | 3031 | 2500 | | SR |
| 125 | SR | SPZ | 41.8 | DUP | 3031 | 2600 | | SR |
| 125 | SR | SPZ | 37.3 | DUP | 4895 | 2200 | | SR |
| 125 | SR | SPZ | 39.1 | DUP | 4895 | 2300 | | SR |
| 125 | SR | SPZ | 41.0 | DUP | 4895 | 2400 | | SR |
| 125 | SR | SPZ | 42.9 | DUP | 4895 | 2500 | | SR |
| 125 | SR | SPZ | 44.8 | DUP | 4895 | 2600 | | SR |
| 130 | SP | all | 42.0 | DUP | 4064 | 2404 | | SP |
| 130 | SP | all | 44.0 | DUP | 4064 | 2539 | | SP |
| 130 | SP | all | 46.0 | DUP | 4064 | 2647 | | SP |
| 150 | SR | all | 35.7 | DUP | 3031 | 2200 | | SR |
| 150 | SR | all | 37.4 | DUP | 3031 | 2300 | | SR |
| 150 | SR | all | 39.2 | DUP | 3031 | 2400 | | SR |
| 150 | SR | all | 41.0 | DUP | 3031 | 2500 | | SR |
| 150 | SR | all | 36.1 | DUP | 4895 | 2200 | | SR |
| 150 | SR | all | 38.4 | DUP | 4895 | 2300 | | SR |
| 150 | SR | all | 40.7 | DUP | 4895 | 2400 | | SR |
| 150 | SR | all | 43.0 | DUP | 4895 | 2500 | | SR |
| 150 | SP | all | 38.0 | DUP | 4064 | 2204 | | SP |
| 150 | SP | all | 40.0 | DUP | 4064 | 2321 | | SP |
| 150 | SP | all | 42.0 | DUP | 4064 | 2452 | | SP |
| 150 | NO | SPP | 47.0 | DUP | 4831 | 2304 | | NO |
| 150 | NO | SPP | 49.0 | DUP | 4831 | 2438 | | NO |
| 150 | NO | SPP | 51.0 | DUP | 4831 | 2574 | | NO |
| 150 | NO | SPZ | 47.0 | DUP | 4831 | 2319 | | NO |
| 150 | NO | SPZ | 49.0 | DUP | 4831 | 2450 | | NO |
| 150 | NO | SPZ | 51.0 | DUP | 4831 | 2581 | | NO |
| 165 | SR | all | 34.8 | DUP | 3031 | 2100 | | SR |
| 165 | SR | all | 36.4 | DUP | 3031 | 2200 | | SR |
| 165 | SR | all | 38.0 | DUP | 3031 | 2300 | | SR |
| 165 | SR | all | 35.9 | DUP | 4895 | 2100 | | SR |

LOADS FOR .30/40 KRAG .308

| BULLET | | | POWDER | | | | | |
|---|---|---|---|---|---|---|---|---|
| Wgt | Mfg | Type | GRN | Mfg | Type | VEL(fps) | OAL | Source |
| 165 | SR | all | 37.9 | DUP | 4895 | 2200 | | SR |
| 165 | SR | all | 39.9 | DUP | 4895 | 2300 | | SR |
| 165 | SR | all | 41.9 | DUP | 4895 | 2400 | | SR |
| 165 | SP | all | 44.0 | DUP | 4831 | 2096 | | SP |
| 165 | SP | all | 46.0 | DUP | 4831 | 2177 | | SP |
| 165 | SP | all | 48.0 | DUP | 4831 | 2283 | | SP |
| 165 | NO | SPP | 45.0 | DUP | 4831 | 2160 | | NO |
| 165 | NO | SPP | 47.0 | DUP | 4831 | 2277 | | NO |
| 165 | NO | SPP | 49.0 | DUP | 4831 | 2421 | | NO |
| 165 | NO | SPZ | 45.0 | DUP | 4831 | 2169 | | NO |
| 165 | NO | SPZ | 47.0 | DUP | 4831 | 2299 | | NO |
| 165 | NO | SPZ | 49.0 | DUP | 4831 | 2429 | | NO |
| 180 | SR | all | 32.4 | DUP | 3031 | 1900 | | SR |
| 180 | SR | all | 33.9 | DUP | 3031 | 2000 | | SR |
| 180 | SR | all | 35.5 | DUP | 3031 | 2100 | | SR |
| 180 | SR | all | 37.1 | DUP | 3031 | 2200 | | SR |
| 180 | SR | all | 34.2 | DUP | 4895 | 1900 | | SR |
| 180 | SR | all | 35.8 | DUP | 4895 | 2000 | | SR |
| 180 | SR | all | 37.4 | DUP | 4895 | 2100 | | SR |
| 180 | SR | all | 39.1 | DUP | 4895 | 2200 | | SR |
| 180 | SR | all | 40.8 | DUP | 4895 | 2300 | | SR |
| 180 | SP | all | 43.0 | DUP | 4831 | 1983 | | SP |
| 180 | SP | all | 45.0 | DUP | 4831 | 2075 | | SP |
| 180 | SP | all | 47.0 | DUP | 4831 | 2162 | | SP |
| 180 | NO | SPP | 43.0 | DUP | 4831 | 2087 | | NO |
| 180 | NO | SPP | 45.0 | DUP | 4831 | 2215 | | NO |
| 180 | NO | SPP | 47.0 | DUP | 4831 | 2356 | | NO |
| 180 | NO | PPP | 43.0 | DUP | 4831 | 2089 | | NO |
| 180 | NO | PPP | 45.0 | DUP | 4831 | 2231 | | NO |
| 180 | NO | PPP | 47.0 | DUP | 4831 | 2367 | | NO |
| 180 | NO | SPZ | 43.0 | DUP | 4831 | 2094 | | NO |
| 180 | NO | SPZ | 45.0 | DUP | 4831 | 2233 | | NO |
| 180 | NO | SPZ | 47.0 | DUP | 4831 | 2371 | | NO |
| 200 | SP | SPZ | 43.0 | DUP | 4831 | 1898 | | SP |
| 200 | SP | SPZ | 45.0 | DUP | 4831 | 1962 | | SP |
| 200 | SP | SPZ | 47.0 | DUP | 4831 | 2002 | | SP |
| 200 | NO | SPP | 41.0 | DUP | 4831 | 2013 | | NO |
| 200 | NO | SPP | 43.0 | DUP | 4831 | 2124 | | NO |
| 200 | NO | SPP | 45.0 | DUP | 4831 | 2235 | | NO |
| 220 | SR | all | 31.5 | DUP | 3031 | 1800 | | SR |
| 220 | SR | all | 33.1 | DUP | 3031 | 1900 | | SR |
| 220 | SR | all | 34.8 | DUP | 3031 | 2000 | | SR |
| 220 | SR | all | 32.0 | DUP | 4895 | 1800 | | SR |
| 220 | SR | all | 34.1 | DUP | 4895 | 1900 | | SR |
| 220 | SR | all | 36.2 | DUP | 4895 | 2000 | | SR |
| 220 | SR | all | 38.3 | DUP | 4895 | 2100 | | SR |

LOADS FOR 7.5 X 55 SWISS (Schmidt-Rubin) .308

| BULLET | | | POWDER | | | | | |
|---|---|---|---|---|---|---|---|---|
| Wgt | Mfg | Type | GRN | Mfg | Type | VEL(fps) | OAL | Source |
| 110 | SR | HP | 38.7 | DUP | 3031 | 2600 | 2.810 | SR |
| 110 | SR | HP | 40.2 | DUP | 3031 | 2700 | | SR |
| 110 | SR | HP | 41.7 | DUP | 3031 | 2800 | | SR |
| 110 | SR | HP | 43.2 | DUP | 3031 | 2900 | | SR |
| 110 | SR | HP | 44.7 | DUP | 3031 | 3000 | | SR |
| 110 | SR | HP | 41.5 | DUP | 4895 | 2700 | | SR |
| 110 | SR | HP | 43.2 | DUP | 4895 | 2800 | | SR |
| 110 | SR | HP | 44.9 | DUP | 4895 | 2900 | | SR |
| 110 | SR | HP | 46.6 | DUP | 4895 | 3000 | | SR |
| 125 | SR | SPZ | 36.0 | DUP | 3031 | 2500 | 2.900 | SR |
| 125 | SR | SPZ | 38.0 | DUP | 3031 | 2600 | | SR |
| 125 | SR | SPZ | 40.0 | DUP | 3031 | 2700 | | SR |
| 125 | SR | SPZ | 42.0 | DUP | 3031 | 2800 | | SR |
| 125 | SR | SPZ | 44.0 | DUP | 3031 | 2900 | | SR |
| 125 | SR | SPZ | 38.5 | DUP | 4895 | 2600 | | SR |
| 125 | SR | SPZ | 40.5 | DUP | 4895 | 2700 | | SR |
| 125 | SR | SPZ | 42.5 | DUP | 4895 | 2800 | | SR |
| 125 | SR | SPZ | 44.5 | DUP | 4895 | 2900 | | SR |
| 150 | SR | all | 37.6 | DUP | 3031 | 2400 | 2.925 | SR |
| 150 | SR | all | 39.3 | DUP | 3031 | 2500 | | SR |
| 150 | SR | all | 41.0 | DUP | 3031 | 2600 | | SR |
| 150 | SR | all | 42.7 | DUP | 3031 | 2700 | | SR |
| 150 | SR | all | 44.4 | DUP | 3031 | 2800 | | SR |
| 150 | SR | all | 38.5 | DUP | 4895 | 2400 | | SR |
| 150 | SR | all | 40.5 | DUP | 4895 | 2500 | | SR |
| 150 | SR | all | 42.5 | DUP | 4895 | 2600 | | SR |
| 150 | SR | all | 44.5 | DUP | 4895 | 2700 | | SR |
| 150 | SR | all | 46.5 | DUP | 4895 | 2800 | | SR |
| 165 | SR | all | 35.8 | DUP | 3031 | 2300 | 2.935 | SR |
| 165 | SR | all | 37.6 | DUP | 3031 | 2400 | | SR |
| 165 | SR | all | 39.4 | DUP | 3031 | 2500 | | SR |
| 165 | SR | all | 41.2 | DUP | 3031 | 2600 | | SR |
| 165 | SR | all | 43.0 | DUP | 3031 | 2700 | | SR |
| 165 | SR | all | 38.4 | DUP | 4895 | 2300 | | SR |
| 165 | SR | all | 40.2 | DUP | 4895 | 2400 | | SR |
| 165 | SR | all | 42.0 | DUP | 4895 | 2500 | | SR |
| 165 | SR | all | 43.8 | DUP | 4895 | 2600 | | SR |
| 165 | SR | all | 45.6 | DUP | 4895 | 2700 | | SR |
| 180 | SR | all | 32.5 | DUP | 3031 | 2100 | 3.020 | SR |
| 180 | SR | all | 34.5 | DUP | 3031 | 2200 | | SR |
| 180 | SR | all | 36.5 | DUP | 3031 | 2300 | | SR |
| 180 | SR | all | 38.5 | DUP | 3031 | 2400 | | SR |
| 180 | SR | all | 34.0 | DUP | 4895 | 2100 | | SR |
| 180 | SR | all | 36.1 | DUP | 4895 | 2200 | | SR |
| 180 | SR | all | 38.2 | DUP | 4895 | 2300 | | SR |
| 180 | SR | all | 40.3 | DUP | 4895 | 2400 | | SR |
| 200 | SR | all | 33.8 | DUP | 3031 | 2100 | 3.018 | SR |
| 200 | SR | all | 36.0 | DUP | 3031 | 2200 | | SR |
| 200 | SR | all | 38.2 | DUP | 3031 | 2300 | | SR |
| 200 | SR | all | 35.9 | DUP | 4895 | 2100 | | SR |
| 200 | SR | all | 38.1 | DUP | 4895 | 2200 | | SR |
| 200 | SR | all | 40.3 | DUP | 4895 | 2300 | | SR |

LOADS FOR .300 SAVAGE .308

| BULLET | | | POWDER | | | | | |
|---|---|---|---|---|---|---|---|---|
| Wgt | Mfg | Type | GRN | Mfg | Type | VEL(fps) | OAL | Source |
| 100 | SP | PL | 42.0 | DUP | 4064 | 2723 | | SP |
| 100 | SP | PL | 44.0 | DUP | 4064 | 2864 | | SP |
| 100 | SP | PL | 46.0 | DUP | 4064 | 3003 | | SP |
| 110 | SR | HP | 37.0 | DUP | 3031 | 2600 | 2.600 | SR |
| 110 | SR | HP | 38.3 | DUP | 3031 | 2700 | | SR |
| 110 | SR | HP | 39.7 | DUP | 3031 | 2800 | | SR |
| 110 | SR | HP | 41.1 | DUP | 3031 | 2900 | | SR |
| 110 | SR | HP | 40.2 | DUP | 4895 | 2600 | | SR |
| 110 | SR | HP | 41.5 | DUP | 4895 | 2700 | | SR |
| 110 | SR | HP | 42.8 | DUP | 4895 | 2800 | | SR |
| 110 | SR | HP | 44.1 | DUP | 4895 | 2900 | | SR |
| 110 | SP | VAR | 39.0 | DUP | 4064 | 2542 | | SP |
| 110 | SP | VAR | 41.0 | DUP | 4064 | 2677 | | SP |
| 110 | SP | VAR | 43.0 | DUP | 4064 | 2793 | | SP |
| 110 | SP | RN&SP | 41.0 | DUP | 4064 | 2586 | | SP |
| 110 | SP | RN&SP | 43.0 | DUP | 4064 | 2734 | | SP |
| 110 | SP | RN&SP | 45.0 | DUP | 4064 | 2849 | | SP |
| 125 | SR | SPZ | 35.0 | DUP | 3031 | 2400 | | SR |
| 125 | SR | SPZ | 36.4 | DUP | 3031 | 2500 | | SR |
| 125 | SR | SPZ | 37.8 | DUP | 3031 | 2600 | | SR |
| 125 | SR | SPZ | 39.2 | DUP | 3031 | 2700 | | SR |
| 125 | SR | SPZ | 40.7 | DUP | 3031 | 2800 | | SR |
| 125 | SR | SPZ | 38.7 | DUP | 4895 | 2400 | | SR |
| 125 | SR | SPZ | 39.9 | DUP | 4895 | 2500 | | SR |
| 125 | SR | SPZ | 41.2 | DUP | 4895 | 2600 | | SR |
| 125 | SR | SPZ | 42.5 | DUP | 4895 | 2700 | | SR |
| 125 | SR | SPZ | 43.8 | DUP | 4895 | 2800 | | SR |
| 130 | SP | all | 39.0 | DUP | 4064 | 2458 | | SP |
| 130 | SP | all | 41.0 | DUP | 4064 | 2586 | | SP |
| 130 | SP | all | 43.0 | DUP | 4064 | 2717 | | SP |
| 150 | SR | all | 32.4 | DUP | 3031 | 2200 | | SR |
| 150 | SR | all | 33.9 | DUP | 3031 | 2300 | | SR |
| 150 | SR | all | 35.4 | DUP | 3031 | 2400 | | SR |
| 150 | SR | all | 36.9 | DUP | 3031 | 2500 | | SR |
| 150 | SR | all | 38.5 | DUP | 3031 | 2600 | | SR |
| 150 | SR | all | 35.3 | DUP | 4895 | 2200 | | SR |
| 150 | SR | all | 36.6 | DUP | 4895 | 2300 | | SR |
| 150 | SR | all | 38.0 | DUP | 4895 | 2400 | | SR |
| 150 | SR | all | 39.4 | DUP | 4895 | 2500 | | SR |
| 150 | SR | all | 40.8 | DUP | 4895 | 2600 | | SR |
| 150 | SP | all | 37.5 | DUP | 4064 | 2334 | | SP |
| 150 | SP | all | 39.5 | DUP | 4064 | 2472 | | SP |
| 150 | SP | all | 41.5 | DUP | 4064 | 2590 | | SP |
| 150 | NO | SPP | 36.0 | DUP | 4064 | 2335 | | NO |
| 150 | NO | SPP | 38.0 | DUP | 4064 | 2442 | | NO |
| 150 | NO | SPP | 40.0 | DUP | 4064 | 2547 | | NO |
| 150 | NO | SPZ | 36.0 | DUP | 4064 | 2356 | | NO |
| 150 | NO | SPZ | 38.0 | DUP | 4064 | 2461 | | NO |
| 150 | NO | SPZ | 40.0 | DUP | 4064 | 2563 | | NO |
| 150 | NO | HPBT | 34.0 | DUP | 4064 | 2204 | | NO |
| 150 | NO | HPBT | 36.0 | DUP | 4064 | 2267 | | NO |
| 150 | NO | HPBT | 38.0 | DUP | 4064 | 2431 | | NO |

LOADS FOR .300 SAVAGE .308

| BULLET | | | POWDER | | | | | |
|---|---|---|---|---|---|---|---|---|
| Wgt | Mfg | Type | GRN | Mfg | Type | VEL(fps) | OAL | Source |
| 165 | SR | all | 31.6 | DUP | 3031 | 2100 | | SR |
| 165 | SR | all | 33.2 | DUP | 3031 | 2200 | | SR |
| 165 | SR | all | 34.8 | DUP | 3031 | 2300 | | SR |
| 165 | SR | all | 36.4 | DUP | 3031 | 2400 | | SR |
| 165 | SR | all | 38.0 | DUP | 3031 | 2500 | | SR |
| 165 | SR | all | 34.0 | DUP | 4895 | 2100 | | SR |
| 165 | SR | all | 35.6 | DUP | 4895 | 2200 | | SR |
| 165 | SR | all | 37.2 | DUP | 4895 | 2300 | | SR |
| 165 | SR | all | 38.8 | DUP | 4895 | 2400 | | SR |
| 165 | SR | all | 40.4 | DUP | 4895 | 2500 | | SR |
| 165 | SP | all | 36.0 | DUP | 4064 | 2197 | | SP |
| 165 | SP | all | 38.0 | DUP | 4064 | 2331 | | SP |
| 165 | SP | all | 40.0 | DUP | 4064 | 2445 | | SP |
| 165 | NO | SPP | 34.0 | DUP | 4064 | 2155 | | NO |
| 165 | NO | SPP | 36.0 | DUP | 4064 | 2256 | | NO |
| 165 | NO | SPP | 38.0 | DUP | 4064 | 2358 | | NO |
| 165 | NO | SPZ | 34.0 | DUP | 4064 | 2173 | | NO |
| 165 | NO | SPZ | 36.0 | DUP | 4064 | 2265 | | NO |
| 165 | NO | SPZ | 38.0 | DUP | 4064 | 2372 | | NO |
| 168 | NO | HPBT | 32.0 | DUP | 4064 | 2065 | | NO |
| 168 | NO | HPBT | 34.0 | DUP | 4064 | 2172 | | NO |
| 168 | NO | HPBT | 36.0 | DUP | 4064 | 2301 | | NO |
| 180 | SR | all | 32.2 | DUP | 3031 | 2100 | | SR |
| 180 | SR | all | 34.0 | DUP | 3031 | 2200 | | SR |
| 180 | SR | all | 35.8 | DUP | 3031 | 2300 | | SR |
| 180 | SR | all | 34.3 | DUP | 4895 | 2100 | | SR |
| 180 | SR | all | 36.0 | DUP | 4895 | 2200 | | SR |
| 180 | SR | all | 37.7 | DUP | 4895 | 2300 | | SR |
| 180 | SR | all | 39.5 | DUP | 4895 | 2400 | | SR |
| 180 | SP | all | 35.0 | DUP | 4064 | 2094 | | SP |
| 180 | SP | all | 37.0 | DUP | 4064 | 2187 | | SP |
| 180 | SP | all | 39.0 | DUP | 4064 | 2315 | | SP |
| 180 | NO | SPP | 32.0 | DUP | 4064 | 2046 | | NO |
| 180 | NO | SPP | 34.0 | DUP | 4064 | 2153 | | NO |
| 180 | NO | SPP | 36.0 | DUP | 4064 | 2261 | | NO |
| 180 | NO | PPP | 32.0 | DUP | 4064 | 2038 | | NO |
| 180 | NO | PPP | 34.0 | DUP | 4064 | 2159 | | NO |
| 180 | NO | PPP | 36.0 | DUP | 4064 | 2272 | | NO |
| 180 | NO | SPZ | 32.0 | DUP | 4064 | 2056 | | NO |
| 180 | NO | SPZ | 34.0 | DUP | 4064 | 2163 | | NO |
| 180 | NO | SPZ | 36.0 | DUP | 4064 | 2274 | | NO |
| 200 | SP | SPZ | 32.0 | DUP | 4064 | 1899 | | SP |
| 200 | SP | SPZ | 34.0 | DUP | 4064 | 2018 | | SP |
| 200 | SP | SPZ | 36.0 | DUP | 4064 | 2137 | | SP |
| 200 | NO | SPP | 30.0 | DUP | 4064 | 1923 | | NO |
| 200 | NO | SPP | 32.0 | DUP | 4064 | 2010 | | NO |
| 200 | NO | SPP | 34.0 | DUP | 4064 | 2103 | | NO |

LOADS FOR 7.62 RUSSIAN .308

| BULLET | | | POWDER | | | | | |
|---|---|---|---|---|---|---|---|---|
| Wgt | Mfg | Type | GRN | Mfg | Type | VEL(fps) | OAL | Source |
| 110 | SR | HP | 45.0 | DUP | 3031 | 2800 | 3.037 | SR |
| 110 | SR | HP | 46.5 | DUP | 3031 | 2900 | | SR |
| 110 | SR | HP | 48.0 | DUP | 3031 | 3000 | | SR |
| 110 | SR | HP | 49.5 | DUP | 3031 | 3100 | | SR |
| 110 | SR | HP | 46.7 | DUP | 4895 | 2700 | | SR |
| 110 | SR | HP | 48.2 | DUP | 4895 | 2800 | | SR |
| 110 | SR | HP | 49.6 | DUP | 4895 | 2900 | | SR |
| 110 | SR | HP | 51.0 | DUP | 4895 | 3000 | | SR |
| 125 | SR | SPZ | 42.3 | DUP | 3031 | 2600 | | SR |
| 125 | SR | SPZ | 43.9 | DUP | 3031 | 2700 | | SR |
| 125 | SR | SPZ | 45.5 | DUP | 3031 | 2800 | | SR |
| 125 | SR | SPZ | 47.1 | DUP | 3031 | 2900 | | SR |
| 125 | SR | SPZ | 45.6 | DUP | 4895 | 2600 | | SR |
| 125 | SR | SPZ | 47.2 | DUP | 4895 | 2700 | | SR |
| 125 | SR | SPZ | 48.7 | DUP | 4895 | 2800 | | SR |
| 150 | SR | SPZ | 38.9 | DUP | 3031 | 2400 | | SR |
| 150 | SR | SPZ | 40.7 | DUP | 3031 | 2500 | | SR |
| 150 | SR | SPZ | 42.6 | DUP | 3031 | 2600 | | SR |
| 150 | SR | SPZ | 44.4 | DUP | 3031 | 2700 | | SR |
| 150 | SR | SPZ | 41.8 | DUP | 4895 | 2400 | | SR |
| 150 | SR | SPZ | 43.6 | DUP | 4895 | 2500 | | SR |
| 150 | SR | SPZ | 45.3 | DUP | 4895 | 2600 | | SR |
| 150 | SR | SPZ | 47.0 | DUP | 4895 | 2700 | | SR |
| 165 | SR | all | 38.5 | DUP | 3031 | 2300 | | SR |
| 165 | SR | all | 40.3 | DUP | 3031 | 2400 | | SR |
| 165 | SR | all | 42.0 | DUP | 3031 | 2500 | | SR |
| 165 | SR | all | 43.7 | DUP | 3031 | 2600 | | SR |
| 165 | SR | all | 40.7 | DUP | 4895 | 2300 | | SR |
| 165 | SR | all | 42.5 | DUP | 4895 | 2400 | | SR |
| 165 | SR | all | 44.3 | DUP | 4895 | 2500 | | SR |
| 165 | SR | all | 46.0 | DUP | 4895 | 2600 | | SR |
| 180 | SR | all | 39.7 | DUP | 3031 | 2300 | | SR |
| 180 | SR | all | 41.3 | DUP | 3031 | 2400 | | SR |
| 180 | SR | all | 42.9 | DUP | 3031 | 2500 | | SR |
| 180 | SR | all | 41.4 | DUP | 4895 | 2300 | | SR |
| 180 | SR | all | 43.3 | DUP | 4895 | 2400 | | SR |
| 180 | SR | all | 45.1 | DUP | 4895 | 2500 | | SR |
| 200 | SR | all | 36.6 | DUP | 3031 | 2100 | | SR |
| 200 | SR | all | 38.3 | DUP | 3031 | 2200 | | SR |
| 200 | SR | all | 40.0 | DUP | 3031 | 2300 | | SR |
| 200 | SR | all | 38.3 | DUP | 4895 | 2100 | | SR |
| 200 | SR | all | 40.1 | DUP | 4895 | 2200 | | SR |
| 200 | SR | all | 41.9 | DUP | 4895 | 2300 | | SR |
| 220 | SR | RN | 35.7 | DUP | 3031 | 2000 | | SR |
| 220 | SR | RN | 37.5 | DUP | 3031 | 2100 | | SR |
| 220 | SR | RN | 37.8 | DUP | 4895 | 2000 | | SR |
| 220 | SR | RN | 39.3 | DUP | 4895 | 2100 | | SR |
| 220 | SR | RN | 40.7 | DUP | 4895 | 2200 | | SR |

LOADS FOR .308 WINCHESTER .308
(7.62mm NATO)

| BULLET | | | POWDER | | | | | |
|---|---|---|---|---|---|---|---|---|
| Wgt | Mfg | Type | GRN | Mfg | Type | VEL(fps) | OAL | Source |
| 100 | SP | PL | 48.0 | DUP | 4064 | 3021 | | SP |
| 100 | SP | PL | 50.0 | DUP | 4064 | 3160 | | SP |
| 100 | SP | PL | 52.0 | DUP | 4064 | 3278 | | SP |
| 110 | SR | HP | 36.0 | HER | RE-7 | 2800 | 2.800 | SR |
| 110 | SR | HP | 37.5 | HER | RE-7 | 2900 | | SR |
| 110 | SR | HP | 39.0 | HER | RE-7 | 3000 | | SR |
| 110 | SR | HP | 40.6 | HER | RE-7 | 3100 | | SR |
| 110 | SR | HP | 41.3 | DUP | 3031 | 2800 | | SR |
| 110 | SR | HP | 42.7 | DUP | 3031 | 2900 | | SR |
| 110 | SR | HP | 44.2 | DUP | 3031 | 3000 | | SR |
| 110 | SR | HP | 45.7 | DUP | 3031 | 3100 | | SR |
| 110 | SR | HP | 47.2 | DUP | 3031 | 3200 | | SR |
| 110 | SP | VAR | 41.5 | DUP | 4064 | 2614 | | SP |
| 110 | SP | VAR | 43.5 | DUP | 4064 | 2743 | | SP |
| 110 | SP | VAR | 45.5 | DUP | 4064 | 2860 | | SP |
| 110 | SP | RN&SP | 48.0 | DUP | 4064 | 3046 | | SP |
| 110 | SP | RN&SP | 50.0 | DUP | 4064 | 3190 | | SP |
| 110 | SP | RN&SP | 52.0 | DUP | 4064 | 3311 | | SP |
| 125 | SR | SPZ | 36.0 | HER | RE-7 | 2700 | | SR |
| 125 | SR | SPZ | 37.3 | HER | RE-7 | 2800 | | SR |
| 125 | SR | SPZ | 38.6 | HER | RE-7 | 2900 | | SR |
| 125 | SR | SPZ | 39.9 | HER | RE-7 | 3000 | | SR |
| 125 | SR | SPZ | 40.8 | DUP | 3031 | 2700 | | SR |
| 125 | SR | SPZ | 42.0 | DUP | 3031 | 2800 | | SR |
| 125 | SR | SPZ | 43.2 | DUP | 3031 | 2900 | | SR |
| 125 | SR | SPZ | 44.4 | DUP | 3031 | 3000 | | SR |
| 130 | SP | all | 45.0 | DUP | 4064 | 2800 | | SP |
| 130 | SP | all | 47.0 | DUP | 4064 | 2943 | | SP |
| 130 | SP | all | 49.0 | DUP | 4064 | 3056 | | SP |
| 150 | SR | all | 35.4 | HER | RE-7 | 2500 | | SR |
| 150 | SR | all | 36.9 | HER | RE-7 | 2600 | | SR |
| 150 | SR | all | 38.4 | HER | RE-7 | 2700 | | SR |
| 150 | SR | all | 39.4 | DUP | 3031 | 2500 | | SR |
| 150 | SR | all | 40.8 | DUP | 3031 | 2600 | | SR |
| 150 | SR | all | 42.3 | DUP | 3031 | 2700 | | SR |
| 150 | SR | all | 43.8 | DUP | 3031 | 2800 | | SR |
| 150 | SP | all | 43.0 | DUP | 4064 | 2645 | | SP |
| 150 | SP | all | 45.0 | DUP | 4064 | 2757 | | SP |
| 150 | SP | all | 47.0 | DUP | 4064 | 2882 | | SP |
| 150 | NO | SPP | 44.0 | DUP | 4064 | 2728 | | NO |
| 150 | NO | SPP | 46.0 | DUP | 4064 | 2852 | | NO |
| 150 | NO | SPP | 48.0 | DUP | 4064 | 2912 | | NO |
| 150 | NO | SPZ | 44.0 | DUP | 4064 | 2736 | | NO |
| 150 | NO | SPZ | 46.0 | DUP | 4064 | 2861 | | NO |
| 150 | NO | SPZ | 48.0 | DUP | 4064 | 2918 | | NO |
| 150 | NO | HPBT | 43.0 | DUP | 4064 | 2603 | | NO |
| 150 | NO | HPBT | 45.0 | DUP | 4064 | 2754 | | NO |
| 150 | NO | HPBT | 47.0 | DUP | 4064 | 2875 | | NO |
| 165 | SR | all | 33.2 | HER | RE-7 | 2400 | | SR |
| 165 | SR | all | 35.1 | HER | RE-7 | 2500 | | SR |
| 165 | SR | all | 38.1 | DUP | 3031 | 2400 | | SR |

LOADS FOR .308 WINCHESTER .308
(7.62mm NATO)

| BULLET | | | POWDER | | | | | |
|---|---|---|---|---|---|---|---|---|
| Wgt | Mfg | Type | GRN | Mfg | Type | VEL (fps) | OAL | Source |
| 165 | SR | all | 39.6 | DUP | 3031 | 2500 | | SR |
| 165 | SR | all | 41.2 | DUP | 3031 | 2600 | | SR |
| 165 | SP | all | 41.0 | DUP | 4064 | 2391 | | SP |
| 165 | SP | all | 43.0 | DUP | 4064 | 2520 | | SP |
| 165 | SP | all | 45.0 | DUP | 4064 | 2634 | | SP |
| 165 | NO | SPP | 40.0 | DUP | 4064 | 2553 | | NO |
| 165 | NO | SPP | 42.0 | DUP | 4064 | 2612 | | NO |
| 165 | NO | SPP | 44.0 | DUP | 4064 | 2680 | | NO |
| 165 | NO | SPZ | 40.0 | DUP | 4064 | 2561 | | NO |
| 165 | NO | SPZ | 42.0 | DUP | 4064 | 2619 | | NO |
| 165 | NO | SPZ | 44.0 | DUP | 4064 | 2695 | | NO |
| 168 | NO | HPBT | 46.0 | DUP | 4350 | 2438 | | NO |
| 168 | NO | HPBT | 48.0 | DUP | 4350 | 2547 | | NO |
| 168 | NO | HPBT | 50.0 | DUP | 4350 | 2691 | | NO |
| 168 | SR | HP | 33.3 | HER | RE-7 | 2400 | | SR |
| 168 | SR | HP | 35.0 | HER | RE-7 | 2500 | | SR |
| 168 | SR | HP | 38.0 | DUP | 3031 | 2400 | | SR |
| 168 | SR | HP | 39.6 | DUP | 3031 | 2500 | | SR |
| 168 | SR | HP | 41.3 | DUP | 3031 | 2600 | | SR |
| 180 | SR | all | 35.3 | DUP | 3031 | 2200 | | SR |
| 180 | SR | all | 36.8 | DUP | 3031 | 2300 | | SR |
| 180 | SR | all | 38.3 | DUP | 3031 | 2400 | | SR |
| 180 | SR | all | 39.9 | DUP | 3031 | 2500 | | SR |
| 180 | SR | all | 35.9 | DUP | 4064 | 2200 | | SR |
| 180 | SR | all | 38.0 | DUP | 4064 | 2300 | | SR |
| 180 | SR | all | 40.1 | DUP | 4064 | 2400 | | SR |
| 180 | SR | all | 42.2 | DUP | 4064 | 2500 | | SR |
| 180 | SR | all | 44.4 | DUP | 4064 | 2600 | | SR |
| 180 | SP | all | 39.5 | DUP | 4064 | 2260 | | SP |
| 180 | SP | all | 41.5 | DUP | 4064 | 2376 | | SP |
| 180 | SP | all | 43.5 | DUP | 4064 | 2494 | | SP |
| 180 | NO | SPP | 40.0 | DUP | 4064 | 2422 | | NO |
| 180 | NO | SPP | 42.0 | DUP | 4064 | 2564 | | NO |
| 180 | NO | SPP | 44.0 | DUP | 4064 | 2705 | | NO |
| 180 | NO | SPZ | 40.0 | DUP | 4064 | 2431 | | NO |
| 180 | NO | SPZ | 42.0 | DUP | 4064 | 2569 | | NO |
| 180 | NO | SPZ | 44.0 | DUP | 4064 | 2718 | | NO |
| 180 | NO | PPP | 40.0 | DUP | 4064 | 2424 | | NO |
| 180 | NO | PPP | 42.0 | DUP | 4064 | 2566 | | NO |
| 180 | NO | PPP | 44.0 | DUP | 4064 | 2711 | | NO |
| 190 | SR | HP | 35.7 | DUP | 3031 | 2200 | | SR |
| 190 | SR | HP | 37.1 | DUP | 3031 | 2300 | | SR |
| 190 | SR | HP | 38.5 | DUP | 3031 | 2400 | | SR |
| 190 | SR | HP | 36.7 | DUP | 4895 | 2200 | | SR |
| 190 | SR | HP | 38.3 | DUP | 4895 | 2300 | | SR |
| 190 | SR | HP | 39.9 | DUP | 4895 | 2400 | | SR |
| 190 | SR | HP | 41.5 | DUP | 4895 | 2500 | | SR |
| 200 | SR | all | 34.9 | DUP | 3031 | 2100 | | SR |
| 200 | SR | all | 36.5 | DUP | 3031 | 2200 | | SR |
| 200 | SR | all | 38.1 | DUP | 3031 | 2300 | | SR |
| 200 | SR | all | 39.8 | DUP | 3031 | 2400 | | SR |

LOADS FOR .308 WINCHESTER .308
(7.62mm NATO)

| BULLET | | | POWDER | | | | | |
|---|---|---|---|---|---|---|---|---|
| Wgt | Mfg | Type | GRN | Mfg | Type | VEL(fps) | OAL | Source |
| 200 | SR | all | 35.7 | DUP | 4064 | 2100 | | SR |
| 200 | SR | all | 37.6 | DUP | 4064 | 2200 | | SR |
| 200 | SR | all | 39.6 | DUP | 4064 | 2300 | | SR |
| 200 | SR | all | 41.6 | DUP | 4064 | 2400 | | SR |
| 200 | SP | SPZ | 36.5 | DUP | 4064 | 2082 | | SP |
| 200 | SP | SPZ | 38.5 | DUP | 4064 | 2198 | | SP |
| 200 | SP | SPZ | 40.5 | DUP | 4064 | 2304 | | SP |
| 200 | NO | SPP | 44.0 | DUP | 4350 | 2262 | | NO |
| 200 | NO | SPP | 46.0 | DUP | 4350 | 2371 | | NO |
| 200 | NO | SPP | 48.0 | DUP | 4350 | 2463 | | NO |

LOADS FOR .30-06 .308

| BULLET | | | POWDER | | | | | |
|---|---|---|---|---|---|---|---|---|
| Wgt | Mfg | Type | GRN | Mfg | Type | VEL.(fps) | OAL | Source |
| 100 | SP | PL | 55.0 | DUP | 4064 | 3152 | | SP |
| 100 | SP | PL | 57.0 | DUP | 4064 | 3275 | | SP |
| 100 | SP | PL | 59.0 | DUP | 4064 | 3390 | | SP |
| 110 | SR | HP | 48.0 | DUP | 3031 | 3000 | 3.340 | SR |
| 110 | SR | HP | 49.5 | DUP | 3031 | 3100 | | SR |
| 110 | SR | HP | 51.1 | DUP | 3031 | 3200 | | SR |
| 110 | SR | HP | 49.5 | DUP | 4064 | 3000 | | SR |
| 110 | SR | HP | 51.1 | DUP | 4064 | 3100 | | SR |
| 110 | SR | HP | 52.8 | DUP | 4064 | 3200 | | SR |
| 110 | SR | HP | 54.5 | DUP | 4064 | 3300 | | SR |
| 110 | SP | VAR | 54.0 | DUP | 4350 | 2621 | | SP |
| 110 | SP | VAR | 56.0 | DUP | 4350 | 2719 | | SP |
| 110 | SP | VAR | 58.0 | DUP | 4350 | 2835 | | SP |
| 110 | SP | RN&SP | 54.0 | DUP | 4064 | 3110 | | SP |
| 110 | SP | RN&SP | 56.0 | DUP | 4064 | 3239 | | SP |
| 110 | SP | RN&SP | 58.0 | DUP | 4064 | 3355 | | SP |
| 125 | SR | SPZ | 45.9 | DUP | 3031 | 2800 | | SR |
| 125 | SR | SPZ | 47.3 | DUP | 3031 | 2900 | | SR |
| 125 | SR | SPZ | 48.7 | DUP | 3031 | 3000 | | SR |
| 125 | SR | SPZ | 50.1 | DUP | 3031 | 3100 | | SR |
| 125 | SR | SPZ | 48.6 | DUP | 4064 | 2800 | | SR |
| 125 | SR | SPZ | 50.1 | DUP | 4064 | 2900 | | SR |
| 125 | SR | SPZ | 51.7 | DUP | 4064 | 3000 | | SR |
| 125 | SR | SPZ | 53.2 | DUP | 4064 | 3100 | | SR |
| 125 | SR | SPZ | 54.8 | DUP | 4064 | 3200 | | SR |
| 130 | SP | all | 51.0 | DUP | 4064 | 2864 | | SP |
| 130 | SP | all | 53.0 | DUP | 4064 | 3001 | | SP |
| 130 | SP | all | 55.0 | DUP | 4064 | 3110 | | SP |
| 150 | SR | all | 44.4 | DUP | 3031 | 2600 | | SR |
| 150 | SR | all | 46.1 | DUP | 3031 | 2700 | | SR |
| 150 | SR | all | 47.8 | DUP | 3031 | 2800 | | SR |
| 150 | SR | all | 49.5 | DUP | 3031 | 2900 | | SR |
| 150 | SR | all | 46.5 | DUP | 4064 | 2600 | | SR |
| 150 | SR | all | 48.0 | DUP | 4064 | 2700 | | SR |
| 150 | SR | all | 49.6 | DUP | 4064 | 2800 | | SR |
| 150 | SR | all | 51.2 | DUP | 4064 | 2900 | | SR |
| 150 | SR | all | 52.8 | DUP | 4064 | 3000 | | SR |
| 150 | SP | all | 48.0 | DUP | 4064 | 2700 | | SP |
| 150 | SP | all | 50.0 | DUP | 4064 | 2830 | | SP |
| 150 | SP | all | 52.0 | DUP | 4064 | 2941 | | SP |
| 150 | NO | SPP | 48.0 | DUP | 4064 | 2678 | | NO |
| 150 | NO | SPP | 50.0 | DUP | 4064 | 2790 | | NO |
| 150 | NO | SPP | 52.0 | DUP | 4064 | 2906 | | NO |
| 150 | NO | SPZ | 48.0 | DUP | 4064 | 2684 | | NO |
| 150 | NO | SPZ | 50.0 | DUP | 4064 | 2796 | | NO |
| 150 | NO | SPZ | 52.0 | DUP | 4064 | 2921 | | NO |
| 150 | NO | HPBT | 53.0 | DUP | 4350 | 2619 | | NO |
| 150 | NO | HPBT | 55.0 | DUP | 4350 | 2746 | | NO |
| 150 | NO | HPBT | 57.0 | DUP | 4350 | 2872 | | NO |
| 165 | SR | all | 40.5 | DUP | 3031 | 2400 | | SR |
| 165 | SR | all | 42.3 | DUP | 3031 | 2500 | | SR |
| 165 | SR | all | 44.2 | DUP | 3031 | 2600 | | SR |
| 165 | SR | all | 46.1 | DUP | 3031 | 2700 | | SR |

LOADS FOR .30-06 .308

| BULLET | | | POWDER | | | | | |
|---|---|---|---|---|---|---|---|---|
| Wgt | Mfg | Type | GRN | Mfg | Type | VEL (fps) | OAL | Source |
| 165 | SR | all | 48.0 | DUP | 3031 | 2800 | | SR |
| 165 | SR | all | 42.8 | DUP | 4064 | 2400 | | SR |
| 165 | SR | all | 44.5 | DUP | 4064 | 2500 | | SR |
| 165 | SR | all | 46.2 | DUP | 4064 | 2600 | | SR |
| 165 | SR | all | 47.9 | DUP | 4064 | 2700 | | SR |
| 165 | SR | all | 49.6 | DUP | 4064 | 2800 | | SR |
| 165 | SP | all | 46.5 | DUP | 4064 | 2545 | | SP |
| 165 | SP | all | 48.5 | DUP | 4064 | 2650 | | SP |
| 165 | SP | all | 50.5 | DUP | 4064 | 2772 | | SP |
| 165 | NO | SPP | 46.0 | DUP | 4064 | 2561 | | NO |
| 165 | NO | SPP | 48.0 | DUP | 4064 | 2674 | | NO |
| 165 | NO | SPP | 50.0 | DUP | 4064 | 2746 | | NO |
| 165 | NO | SPZ | 46.0 | DUP | 4064 | 2574 | | NO |
| 165 | NO | SPZ | 48.0 | DUP | 4064 | 2683 | | NO |
| 165 | NO | SPZ | 50.0 | DUP | 4064 | 2750 | | NO |
| 168 | SR | HP | 40.9 | DUP | 3031 | 2400 | | SR |
| 168 | SR | HP | 42.6 | DUP | 3031 | 2500 | | SR |
| 168 | SR | HP | 44.4 | DUP | 3031 | 2600 | | SR |
| 168 | SR | HP | 46.1 | DUP | 3031 | 2700 | | SR |
| 168 | SR | HP | 47.9 | DUP | 3031 | 2800 | | SR |
| 168 | SR | HP | 42.1 | DUP | 4064 | 2400 | | SR |
| 168 | SR | HP | 44.0 | DUP | 4064 | 2500 | | SR |
| 168 | SR | HP | 45.9 | DUP | 4064 | 2600 | | SR |
| 168 | SR | HP | 47.8 | DUP | 4064 | 2700 | | SR |
| 168 | SR | HP | 49.7 | DUP | 4064 | 2800 | | SR |
| 168 | NO | HPBT | 51.0 | DUP | 4350 | 2511 | | NO |
| 168 | NO | HPBT | 53.0 | DUP | 4350 | 2603 | | NO |
| 168 | NO | HPBT | 55.0 | DUP | 4350 | 2706 | | NO |
| 180 | SR | all | 41.3 | DUP | 3031 | 2400 | | SR |
| 180 | SR | all | 42.9 | DUP | 3031 | 2500 | | SR |
| 180 | SR | all | 44.5 | DUP | 3031 | 2600 | | SR |
| 180 | SR | all | 46.2 | DUP | 3031 | 2700 | | SR |
| 180 | SR | all | 42.9 | DUP | 4064 | 2400 | | SR |
| 180 | SR | all | 44.8 | DUP | 4064 | 2500 | | SR |
| 180 | SR | all | 46.8 | DUP | 4064 | 2600 | | SR |
| 180 | SR | all | 48.8 | DUP | 4064 | 2700 | | SR |
| 180 | SP | all | 44.5 | DUP | 4064 | 2489 | | SP |
| 180 | SP | all | 46.5 | DUP | 4064 | 2580 | | SP |
| 180 | SP | all | 48.5 | DUP | 4064 | 2679 | | SP |
| 180 | NO | SPP | 43.0 | DUP | 4064 | 2362 | | NO |
| 180 | NO | SPP | 45.0 | DUP | 4064 | 2481 | | NO |
| 180 | NO | SPP | 47.0 | DUP | 4064 | 2563 | | NO |
| 180 | NO | PPP | 43.0 | DUP | 4064 | 2370 | | NO |
| 180 | NO | PPP | 45.0 | DUP | 4064 | 2483 | | NO |
| 180 | NO | PPP | 47.0 | DUP | 4064 | 2568 | | NO |
| 180 | NO | SPZ | 43.0 | DUP | 4064 | 2376 | | NO |
| 180 | NO | SPZ | 45.0 | DUP | 4064 | 2499 | | NO |
| 180 | NO | SPZ | 47.0 | DUP | 4064 | 2581 | | NO |
| 190 | SR | HP | 39.8 | DUP | 3031 | 2300 | | SR |
| 190 | SR | HP | 41.6 | DUP | 3031 | 2400 | | SR |
| 190 | SR | HP | 43.4 | DUP | 3031 | 2500 | | SR |
| 190 | SR | HP | 45.2 | DUP | 3031 | 2600 | | SR |

## LOADS FOR .30-06 .308

| BULLET | | | POWDER | | | | | |
|---|---|---|---|---|---|---|---|---|
| Wgt | Mfg | Type | GRN | Mfg | Type | VEL (fps) | OAL | Source |
| 190 | SR | HP | 41.8 | DUP | 4064 | 2300 | | SR |
| 190 | SR | HP | 43.5 | DUP | 4064 | 2400 | | SR |
| 190 | SR | HP | 45.3 | DUP | 4064 | 2500 | | SR |
| 190 | SR | HP | 47.1 | DUP | 4064 | 2600 | | SR |
| 200 | SR | all | 38.3 | DUP | 3031 | 2200 | | SR |
| 200 | SR | all | 40.5 | DUP | 3031 | 2300 | | SR |
| 200 | SR | all | 42.7 | DUP | 3031 | 2400 | | SR |
| 200 | SR | all | 39.8 | DUP | 4064 | 2200 | | SR |
| 200 | SR | all | 41.8 | DUP | 4064 | 2300 | | SR |
| 200 | SR | all | 43.8 | DUP | 4064 | 2400 | | SR |
| 200 | SR | all | 45.9 | DUP | 4064 | 2500 | | SR |
| 200 | SP | SPZ | 51.0 | DUP | 4350 | 2413 | | SP |
| 200 | SP | SPZ | 53.0 | DUP | 4350 | 2534 | | SP |
| 200 | SP | SPZ | 55.0 | DUP | 4350 | 2618 | | SP |
| 200 | NO | SPP | 49.0 | DUP | 4350 | 2359 | | NO |
| 200 | NO | SPP | 51.0 | DUP | 4350 | 2444 | | NO |
| 200 | NO | SPP | 53.0 | DUP | 4350 | 2531 | | NO |
| 200 | SP | SP | 55.0C | DUP | 7828 | 2385 | | DUP |
| 220 | SR | all | 38.3 | DUP | 3031 | 2100 | | SR |
| 220 | SR | all | 40.3 | DUP | 3031 | 2200 | | SR |
| 220 | SR | all | 42.3 | DUP | 3031 | 2300 | | SR |
| 220 | SR | all | 39.2 | DUP | 4064 | 2100 | | SR |
| 220 | SR | all | 41.3 | DUP | 4064 | 2200 | | SR |
| 220 | SR | all | 43.5 | DUP | 4064 | 2300 | | SR |
| 220 | SR | all | 45.7 | DUP | 4064 | 2400 | | SR |
| 220 | SR | RN | 55.0C | DUP | 7828 | 2285 | | DUP |

Letter C indicates compressed powder charge.

LOADS FOR .300 H & H MAGNUM .308

| BULLET | | | POWDER | | | | | |
|---|---|---|---|---|---|---|---|---|
| Wgt | Mfg | Type | GRN | Mfg | Type | VEL(fps) | OAL | Source |
| 100 | SP | PL | 52.0 | DUP | 4320 | 2786 | | SP |
| 100 | SP | PL | 54.0 | DUP | 4320 | 2879 | | SP |
| 100 | SP | PL | 56.0 | DUP | 4320 | 2994 | | SP |
| 110 | SR | HP | 55.2 | DUP | 3031 | 3300 | 3.655 | SR |
| 110 | SR | HP | 57.3 | DUP | 3031 | 3400 | | SR |
| 110 | SR | HP | 59.4 | DUP | 3031 | 3500 | | SR |
| 110 | SR | HP | 61.6 | DUP | 3031 | 3600 | | SR |
| 110 | SR | HP | 63.8 | DUP | 3031 | 3700 | | SR |
| 110 | SR | HP | 60.6 | DUP | 4064 | 3300 | | SR |
| 110 | SR | HP | 62.4 | DUP | 4064 | 3400 | | SR |
| 110 | SR | HP | 64.2 | DUP | 4064 | 3500 | | SR |
| 110 | SR | HP | 66.0 | DUP | 4064 | 3600 | | SR |
| 110 | SR | HP | 67.9 | DUP | 4064 | 3700 | | SR |
| 110 | SP | VAR | 48.0 | DUP | 4064 | 2663 | | SP |
| 110 | SP | VAR | 50.0 | DUP | 4064 | 2780 | | SP |
| 110 | SP | VAR | 52.0 | DUP | 4064 | 2874 | | SP |
| 110 | SP | SP | 72.0 | DUP | 4831 | 3402 | | SP |
| 110 | SP | SP | 74.0 | DUP | 4831 | 3499 | | SP |
| 110 | SP | SP | 76.0 | DUP | 4831 | 3584 | | SP |
| 125 | SR | SPZ | 57.0 | DUP | 3031 | 3200 | | SR |
| 125 | SR | SPZ | 59.3 | DUP | 3031 | 3300 | | SR |
| 125 | SR | SPZ | 61.7 | DUP | 3031 | 3400 | | SR |
| 125 | SR | SPZ | 60.7 | DUP | 4064 | 3200 | | SR |
| 125 | SR | SPZ | 62.7 | DUP | 4064 | 3300 | | SR |
| 125 | SR | SPZ | 64.8 | DUP | 4064 | 3400 | | SR |
| 125 | SR | SPZ | 66.9 | DUP | 4064 | 3500 | | SR |
| 130 | SP | HP | 70.0 | DUP | 4831 | 3197 | | SP |
| 130 | SP | HP | 72.0 | DUP | 4831 | 3308 | | SP |
| 130 | SP | HP | 74.0 | DUP | 4831 | 3390 | | SP |
| 150 | SR | all | 54.6 | DUP | 3031 | 3000 | | SR |
| 150 | SR | all | 57.6 | DUP | 3031 | 3100 | | SR |
| 150 | SR | all | 60.6 | DUP | 3031 | 3200 | | SR |
| 150 | SR | all | 58.7 | DUP | 4064 | 3000 | | SR |
| 150 | SR | all | 61.0 | DUP | 4064 | 3100 | | SR |
| 150 | SR | all | 63.4 | DUP | 4064 | 3200 | | SR |
| 150 | SP | all | 69.0 | DUP | 4831 | 3133 | | SP |
| 150 | SP | all | 71.0 | DUP | 4831 | 3240 | | SP |
| 150 | SP | all | 73.0 | DUP | 4831 | 3322 | | SP |
| 150 | NO | SPP | 69.0 | DUP | 4831 | 3105 | | NO |
| 150 | NO | SPP | 71.0 | DUP | 4831 | 3185 | | NO |
| 150 | NO | SPP | 73.0 | DUP | 4831 | 3306 | | NO |
| 150 | NO | SPZ | 69.0 | DUP | 4831 | 3115 | | NO |
| 150 | NO | SPZ | 71.0 | DUP | 4831 | 3196 | | NO |
| 150 | NO | SPZ | 73.0 | DUP | 4831 | 3316 | | NO |
| 150 | NO | HPBT | 65.0 | DUP | 4831 | 3000 | | NO |
| 150 | NO | HPBT | 67.0 | DUP | 4831 | 3105 | | NO |
| 150 | NO | HPBT | 69.0 | DUP | 4831 | 3156 | | NO |
| 165 | SR | all | 49.9 | DUP | 3031 | 2700 | | SR |
| 165 | SR | all | 52.6 | DUP | 3031 | 2800 | | SR |
| 165 | SR | all | 55.4 | DUP | 3031 | 2900 | | SR |
| 165 | SR | all | 53.3 | DUP | 4064 | 2700 | | SR |
| 165 | SR | all | 55.6 | DUP | 4064 | 2800 | | SR |

LOADS FOR .300 H & H MAGNUM .308

| BULLET | | | POWDER | | | | | |
|---|---|---|---|---|---|---|---|---|
| Wgt | Mfg | Type | GRN | Mfg | Type | VEL(fps) | OAL | Source |
| 165 | SR | all | 57.9 | DUP | 4064 | 2900 | | SR |
| 165 | SR | all | 60.2 | DUP | 4064 | 3000 | | SR |
| 165 | SP | all | 67.0 | DUP | 4831 | 2996 | | SP |
| 165 | SP | all | 69.0 | DUP | 4831 | 3103 | | SP |
| 165 | SP | all | 71.0 | DUP | 4831 | 3185 | | SP |
| 165 | NO | SPP | 66.0 | DUP | 4831 | 3012 | | NO |
| 165 | NO | SPP | 68.0 | DUP | 4831 | 3059 | | NO |
| 165 | NO | SPP | 70.0 | DUP | 4831 | 3138 | | NO |
| 165 | NO | SPZ | 66.0 | DUP | 4831 | 3022 | | NO |
| 165 | NO | SPZ | 68.0 | DUP | 4831 | 3072 | | NO |
| 165 | NO | SPZ | 70.0 | DUP | 4831 | 3147 | | NO |
| 168 | NO | HPBT | 63.0 | DUP | 4831 | 2949 | | NO |
| 168 | NO | HPBT | 65.0 | DUP | 4831 | 2972 | | NO |
| 168 | NO | HPBT | 67.0 | DUP | 4831 | 3028 | | NO |
| 180 | SR | all | 46.7 | DUP | 3031 | 2500 | | SR |
| 180 | SR | all | 49.4 | DUP | 3031 | 2600 | | SR |
| 180 | SR | all | 52.1 | DUP | 3031 | 2700 | | SR |
| 180 | SR | all | 54.8 | DUP | 3031 | 2800 | | SR |
| 180 | SR | all | 50.2 | DUP | 4064 | 2500 | | SR |
| 180 | SR | all | 52.4 | DUP | 4064 | 2600 | | SR |
| 180 | SR | all | 54.6 | DUP | 4064 | 2700 | | SR |
| 180 | SR | all | 56.8 | DUP | 4064 | 2800 | | SR |
| 180 | SR | all | 59.1 | DUP | 4064 | 2900 | | SR |
| 180 | SP | all | 66.0 | DUP | 4831 | 2853 | | SP |
| 180 | SP | all | 68.0 | DUP | 4831 | 2961 | | SP |
| 180 | SP | all | 70.0 | DUP | 4831 | 3039 | | SP |
| 180 | NO | SPP | 64.0 | DUP | 4831 | 2870 | | NO |
| 180 | NO | SPP | 66.0 | DUP | 4831 | 2906 | | NO |
| 180 | NO | SPP | 68.0 | DUP | 4831 | 2991 | | NO |
| 180 | NO | PPP | 64.0 | DUP | 4831 | 2876 | | NO |
| 180 | NO | PPP | 66.0 | DUP | 4831 | 2904 | | NO |
| 180 | NO | PPP | 68.0 | DUP | 4831 | 2990 | | NO |
| 180 | NO | SPZ | 64.0 | DUP | 4831 | 2884 | | NO |
| 180 | NO | SPZ | 66.0 | DUP | 4831 | 2912 | | NO |
| 180 | NO | SPZ | 68.0 | DUP | 4831 | 3001 | | NO |
| 190 | SR | HP | 47.9 | DUP | 3031 | 2500 | | SR |
| 190 | SR | HP | 50.5 | DUP | 3031 | 2600 | | SR |
| 190 | SR | HP | 53.1 | DUP | 3031 | 2700 | | SR |
| 190 | SR | HP | 50.8 | DUP | 4064 | 2500 | | SR |
| 190 | SR | HP | 53.2 | DUP | 4064 | 2600 | | SR |
| 190 | SR | HP | 55.6 | DUP | 4064 | 2700 | | SR |
| 190 | SR | HP | 58.0 | DUP | 4064 | 2800 | | SR |
| 200 | SR | all | 47.2 | DUP | 3031 | 2400 | | SR |
| 200 | SR | all | 49.7 | DUP | 3031 | 2500 | | SR |
| 200 | SR | all | 52.2 | DUP | 3031 | 2600 | | SR |
| 200 | SR | all | 50.0 | DUP | 4064 | 2400 | | SR |
| 200 | SR | all | 52.4 | DUP | 4064 | 2500 | | SR |
| 200 | SR | all | 54.9 | DUP | 4064 | 2600 | | SR |
| 200 | SR | all | 57.4 | DUP | 4064 | 2700 | | SR |
| 200 | SP | SP | 63.0 | DUP | 4831 | 2660 | | SP |
| 200 | SP | SP | 65.0 | DUP | 4831 | 2771 | | SP |
| 200 | SP | SP | 67.0 | DUP | 4831 | 2849 | | SP |

LOADS FOR .300 H & H MAGNUM .308

| BULLET | | | POWDER | | | | | |
|---|---|---|---|---|---|---|---|---|
| Wgt | Mfg | Type | GRN | Mfg | Type | VEL(fps) | OAL | Source |
| 200 | NO | SPP | 61.0 | DUP | 4831 | 2674 | | NO |
| 200 | NO | SPP | 63.0 | DUP | 4831 | 2785 | | NO |
| 200 | NO | SPP | 65.0 | DUP | 4831 | 2825 | | NO |
| 220 | SR | all | 46.0 | DUP | 3031 | 2300 | | SR |
| 220 | SR | all | 48.5 | DUP | 3031 | 2400 | | SR |
| 220 | SR | all | 51.0 | DUP | 3031 | 2500 | | SR |
| 220 | SR | all | 48.3 | DUP | 4064 | 2300 | | SR |
| 220 | SR | all | 50.5 | DUP | 4064 | 2400 | | SR |
| 220 | SR | all | 52.7 | DUP | 4064 | 2500 | | SR |
| 220 | SR | all | 55.0 | DUP | 4064 | 2600 | | SR |

LOADS FOR .30-338 - Bullet Diameter .308"

| BULLET | | | POWDER | | | | | |
|---|---|---|---|---|---|---|---|---|
| Wgt | Mfg | Type | GRN | Mfg | Type | VEL(fps) | OAL | Source |
| 110 | SR | ALL | 62.1 | DUP | 3031 | 3500 | 3.100 | SR |
| 110 | SR | ALL | 63.7 | DUP | 3031 | 3600 | | SR |
| 110 | SR | ALL | 65.2 | DUP | 3031 | 3700 | | SR |
| 110 | SR | ALL | 66.7 | DUP | 3031 | 3800 | | SR |
| 110 | SR | ALL | 63.5 | DUP | 4064 | 3500 | | SR |
| 110 | SR | ALL | 65.9 | DUP | 4064 | 3600 | | SR |
| 110 | SR | ALL | 68.2 | DUP | 4064 | 3700 | | SR |
| 110 | SR | ALL | 70.5 | DUP | 4064 | 3800 | | SR |
| 125 | SR | SPZ | 58.7 | DUP | 4064 | 3200 | 3.200 | SR |
| 125 | SR | SPZ | 61.1 | DUP | 4064 | 3300 | | SR |
| 125 | SR | SPZ | 63.4 | DUP | 4064 | 3400 | | SR |
| 125 | SR | SPZ | 65.7 | DUP | 4064 | 3500 | | SR |
| 125 | SR | SPZ | 72.3 | DUP | 4831 | 3200 | | SR |
| 125 | SR | SPZ | 74.3 | DUP | 4831 | 3300 | | SR |
| 125 | SR | SPZ | 76.3 | DUP | 4831 | 3400 | | SR |
| 150 | SR | ALL | 58.1 | DUP | 4064 | 3000 | 3.290 | SR |
| 150 | SR | ALL | 60.1 | DUP | 4064 | 3100 | | SR |
| 150 | SR | ALL | 62.2 | DUP | 4064 | 3200 | | SR |
| 150 | SR | ALL | 64.3 | DUP | 4064 | 3300 | | SR |
| 150 | SR | ALL | 67.6 | DUP | 4831 | 3000 | | SR |
| 150 | SR | ALL | 69.7 | DUP | 4831 | 3100 | | SR |
| 150 | SR | ALL | 71.8 | DUP | 4831 | 3200 | | SR |
| 150 | SR | ALL | 73.9 | DUP | 4831 | 3300 | | SR |
| 150 | SR | ALL | 76.0 | DUP | 4831 | 3400 | | SR |
| 165 | SR | ALL | 53.1 | DUP | 4064 | 2700 | | SR |
| 165 | SR | ALL | 55.5 | DUP | 4064 | 2800 | | SR |
| 165 | SR | ALL | 57.9 | DUP | 4064 | 2900 | | SR |
| 165 | SR | ALL | 60.3 | DUP | 4064 | 3000 | | SR |
| 165 | SR | ALL | 64.5 | DUP | 4831 | 2800 | | SR |
| 165 | SR | ALL | 66.5 | DUP | 4831 | 2900 | | SR |
| 165 | SR | ALL | 68.5 | DUP | 4831 | 3000 | | SR |
| 165 | SR | ALL | 70.5 | DUP | 4831 | 3100 | | SR |
| 180 | SR | ALL | 56.5 | DUP | 4064 | 2800 | | SR |
| 180 | SR | ALL | 59.0 | DUP | 4064 | 2900 | | SR |
| 180 | SR | ALL | 61.5 | DUP | 4064 | 3000 | | SR |
| 180 | SR | ALL | 65.0 | DUP | 4831 | 2800 | | SR |
| 180 | SR | ALL | 67.2 | DUP | 4831 | 2900 | | SR |
| 180 | SR | ALL | 69.3 | DUP | 4831 | 3000 | | SR |
| 180 | SR | ALL | 71.4 | DUP | 4831 | 3100 | | SR |
| 190 | SR | HP | 52.0 | DUP | 4064 | 2600 | 3.390 | SR |
| 190 | SR | HP | 54.3 | DUP | 4064 | 2700 | | SR |
| 190 | SR | HP | 56.6 | DUP | 4064 | 2800 | | SR |
| 190 | SR | HP | 62.0 | DUP | 4831 | 2700 | | SR |
| 190 | SR | HP | 64.3 | DUP | 4831 | 2800 | | SR |
| 190 | SR | HP | 66.5 | DUP | 4831 | 2900 | | SR |
| 190 | SR | HP | 68.7 | DUP | 4831 | 3000 | | SR |
| 200 | SR | ALL | 53.8 | DUP | 4064 | 2600 | 3.372 | SR |
| 200 | SR | ALL | 56.3 | DUP | 4064 | 2700 | | SR |
| 200 | SR | ALL | 58.8 | DUP | 4064 | 2800 | | SR |
| 200 | SR | ALL | 63.0 | DUP | 4831 | 2700 | | SR |
| 200 | SR | ALL | 65.4 | DUP | 4831 | 2800 | | SR |
| 200 | SR | ALL | 67.8 | DUP | 4831 | 2900 | | SR |

LOADS FOR .30-338 - Bullet Diameter .308"

| BULLET | | | POWDER | | | | | |
|---|---|---|---|---|---|---|---|---|
| Wgt | Mfg | Type | GRN | Mfg | Type | VEL(fps) | OAL | Source |
| 200 | SR | ALL | 70.2 | DUP | 4831 | 3000 | | SR |
| 220 | SR | ALL | 52.2 | DUP | 4320 | 2400 | 3.210 | SR |
| 220 | SR | ALL | 54.4 | DUP | 4320 | 2500 | | SR |
| 220 | SR | ALL | 56.5 | DUP | 4320 | 2600 | | SR |
| 220 | SR | ALL | 58.6 | DUP | 4320 | 2700 | | SR |
| 220 | SR | ALL | 60.2 | DUP | 4831 | 2500 | | SR |
| 220 | SR | ALL | 62.7 | DUP | 4831 | 2600 | | SR |
| 220 | SR | ALL | 65.2 | DUP | 4831 | 2700 | | SR |
| 220 | SR | ALL | 67.7 | DUP | 4831 | 2800 | | SR |

## LOADS FOR .300 WINCHESTER MAGNUM .308

| BULLET | | | POWDER | | | | | |
|---|---|---|---|---|---|---|---|---|
| Wgt | Mfg | Type | GRN | Mfg | Type | VEL (fps) | OAL | Source |
| 100 | SP | PL | 65.0 | DUP | 4064 | 3231 | | SP |
| 100 | SP | PL | 67.0 | DUP | 4064 | 3340 | | SP |
| 100 | SP | PL | 69.0 | DUP | 4064 | 3425 | | SP |
| 110 | SR | HP | 59.1 | DUP | 3031 | 3200 | 3.340 | SR |
| 110 | SR | HP | 61.1 | DUP | 3031 | 3300 | | SR |
| 110 | SR | HP | 63.1 | DUP | 3031 | 3400 | | SR |
| 110 | SR | HP | 65.1 | DUP | 3031 | 3500 | | SR |
| 110 | SR | HP | 63.5 | DUP | 4064 | 3200 | | SR |
| 110 | SR | HP | 65.3 | DUP | 4064 | 3300 | | SR |
| 110 | SR | HP | 67.2 | DUP | 4064 | 3400 | | SR |
| 110 | SR | HP | 69.1 | DUP | 4064 | 3500 | | SR |
| 110 | SP | VAR | 50.0 | DUP | 4064 | 2630 | | SP |
| 110 | SP | VAR | 52.0 | DUP | 4064 | 2756 | | SP |
| 110 | SP | VAR | 54.0 | DUP | 4064 | 2849 | | SP |
| 110 | SP | SP | 80.0 | DUP | 4831 | 3392 | | SP |
| 110 | SP | SP | 82.0 | DUP | 4831 | 3507 | | SP |
| 110 | SP | SP | 84.0 | DUP | 4831 | 3627 | | SP |
| 125 | SR | SPZ | 59.0 | DUP | 3031 | 3000 | | SR |
| 125 | SR | SPZ | 61.5 | DUP | 3031 | 3100 | | SR |
| 125 | SR | SPZ | 64.0 | DUP | 3031 | 3200 | | SR |
| 125 | SR | SPZ | 66.6 | DUP | 3031 | 3300 | | SR |
| 125 | SR | SPZ | 61.2 | DUP | 4064 | 3000 | | SR |
| 125 | SR | SPZ | 63.4 | DUP | 4064 | 3100 | | SR |
| 125 | SR | SPZ | 65.6 | DUP | 4064 | 3200 | | SR |
| 125 | SR | SPZ | 67.9 | DUP | 4064 | 3300 | | SR |
| 130 | SP | HP | 76.0 | DUP | 4831 | 3257 | | SP |
| 130 | SP | HP | 78.0 | DUP | 4831 | 3369 | | SP |
| 130 | SP | HP | 80.0 | DUP | 4831 | 3448 | | SP |
| 150 | SR | all | 54.3 | DUP | 3031 | 2800 | | SR |
| 150 | SR | all | 56.6 | DUP | 3031 | 2900 | | SR |
| 150 | SR | all | 58.9 | DUP | 3031 | 3000 | | SR |
| 150 | SR | all | 71.3 | DUP | 4831 | 2900 | | SR |
| 150 | SR | all | 73.4 | DUP | 4831 | 3000 | | SR |
| 150 | SR | all | 75.5 | DUP | 4831 | 3100 | | SR |
| 150 | SR | all | 77.5 | DUP | 4831 | 3200 | | SR |
| 150 | SP | all | 72.0 | DUP | 4831 | 2971 | | SP |
| 150 | SP | all | 74.0 | DUP | 4831 | 3082 | | SP |
| 150 | SP | all | 76.0 | DUP | 4831 | 3194 | | SP |
| 150 | NO | SPP | 72.0 | DUP | 4831 | 3223 | | NO |
| 150 | NO | SPP | 74.0 | DUP | 4831 | 3298 | | NO |
| 150 | NO | SPP | 76.0 | DUP | 4831 | 3389 | | NO |
| 150 | NO | SPZ | 72.0 | DUP | 4831 | 3246 | | NO |
| 150 | NO | SPZ | 74.0 | DUP | 4831 | 3321 | | NO |
| 150 | NO | SPZ | 76.0 | DUP | 4831 | 3394 | | NO |
| 150 | NO | HPBT | 70.0 | DUP | 4831 | 3111 | | NO |
| 150 | NO | HPBT | 72.0 | DUP | 4831 | 3193 | | NO |
| 150 | NO | HPBT | 74.0 | DUP | 4831 | 3287 | | NO |
| 165 | SR | all | 53.9 | DUP | 3031 | 2700 | | SR |
| 165 | SR | all | 56.6 | DUP | 3031 | 2800 | | SR |
| 165 | SR | all | 59.3 | DUP | 3031 | 2900 | | SR |
| 165 | SR | all | 68.6 | DUP | 4831 | 2800 | | SR |
| 165 | SR | all | 70.8 | DUP | 4831 | 2900 | | SR |

LOADS FOR .300 WINCHESTER MAGNUM .308

| BULLET | | | POWDER | | | | | |
|---|---|---|---|---|---|---|---|---|
| Wgt | Mfg | Type | GRN | Mfg | Type | VEL(fps) | OAL | Source |
| 165 | SR | all | 73.0 | DUP | 4831 | 3000 | | SR |
| 165 | SR | all | 75.2 | DUP | 4831 | 3100 | | SR |
| 165 | SR | all | 77.4 | DUP | 4831 | 3200 | | SR |
| 165 | SP | all | 72.0 | DUP | 4831 | 3043 | | SP |
| 165 | SP | all | 74.0 | DUP | 4831 | 3139 | | SP |
| 165 | SP | all | 76.0 | DUP | 4831 | 3226 | | SP |
| 165 | NO | SPP | 70.0 | DUP | 4831 | 3109 | | NO |
| 165 | NO | SPP | 72.0 | DUP | 4831 | 3176 | | NO |
| 165 | NO | SPP | 74.0 | DUP | 4831 | 3253 | | NO |
| 165 | NO | SPZ | 70.0 | DUP | 4831 | 3116 | | NO |
| 165 | NO | SPZ | 72.0 | DUP | 4831 | 3180 | | NO |
| 165 | NO | SPZ | 74.0 | DUP | 4831 | 3265 | | NO |
| 165 | NO | SPZ | 77.5 | DUP | 7828 | 3210 | | DUP |
| 168 | NO | HPBT | 68.0 | DUP | 4831 | 3020 | | NO |
| 168 | NO | HPBT | 70.0 | DUP | 4831 | 3081 | | NO |
| 168 | NO | HPBT | 72.0 | DUP | 4831 | 3173 | | NO |
| 180 | SR | all | 56.3 | DUP | 4064 | 2600 | | SR |
| 180 | SR | all | 58.9 | DUP | 4064 | 2700 | | SR |
| 180 | SR | all | 61.4 | DUP | 4064 | 2800 | | SR |
| 180 | SR | all | 67.3 | DUP | 4831 | 2700 | | SR |
| 180 | SR | all | 69.8 | DUP | 4831 | 2800 | | SR |
| 180 | SR | all | 72.2 | DUP | 4831 | 2900 | | SR |
| 180 | SR | all | 74.6 | DUP | 4831 | 3000 | | SR |
| 180 | SP | all | 69.0 | DUP | 4831 | 2771 | | SP |
| 180 | SP | all | 71.0 | DUP | 4831 | 2887 | | SP |
| 180 | SP | all | 73.0 | DUP | 4831 | 2996 | | SP |
| 180 | NO | SPP | 69.0 | DUP | 4831 | 2969 | | NO |
| 180 | NO | SPP | 71.0 | DUP | 4831 | 3058 | | NO |
| 180 | NO | SPP | 73.0 | DUP | 4831 | 3144 | | NO |
| 180 | NO | PPP | 69.0 | DUP | 4831 | 2973 | | NO |
| 180 | NO | PPP | 71.0 | DUP | 4831 | 3061 | | NO |
| 180 | NO | PPP | 73.0 | DUP | 4831 | 3148 | | NO |
| 180 | NO | SPZ | 69.0 | DUP | 4831 | 2981 | | NO |
| 180 | NO | SPZ | 71.0 | DUP | 4831 | 3074 | | NO |
| 180 | NO | SPZ | 73.0 | DUP | 4831 | 3162 | | NO |
| 180 | SP | SPZ | 74.0 | DUP | 7828 | 3050 | | DUP |
| 190 | SR | HP | 53.8 | DUP | 4064 | 2500 | | SR |
| 190 | SR | HP | 56.7 | DUP | 4064 | 2600 | | SR |
| 190 | SR | HP | 59.6 | DUP | 4064 | 2700 | | SR |
| 190 | SR | HP | 64.3 | DUP | 4831 | 2600 | | SR |
| 190 | SR | HP | 67.0 | DUP | 4831 | 2700 | | SR |
| 190 | SR | HP | 69.7 | DUP | 4831 | 2800 | | SR |
| 190 | SR | HP | 72.4 | DUP | 4831 | 2900 | | SR |
| 200 | SR | all | 51.8 | DUP | 4064 | 2400 | | SR |
| 200 | SR | all | 55.0 | DUP | 4064 | 2500 | | SR |
| 200 | SR | all | 58.2 | DUP | 4064 | 2600 | | SR |
| 200 | SR | all | 61.9 | DUP | 4831 | 2500 | | SR |
| 200 | SR | all | 64.8 | DUP | 4831 | 2600 | | SR |
| 200 | SR | all | 67.7 | DUP | 4831 | 2700 | | SR |
| 200 | SR | all | 70.6 | DUP | 4831 | 2800 | | SR |
| 200 | SP | SPZ | 67.0 | DUP | 4831 | 2673 | | SP |
| 200 | SP | SPZ | 69.0 | DUP | 4831 | 2765 | | SP |
| 200 | SP | SPZ | 71.0 | DUP | 4831 | 2841 | | SP |

LOADS FOR .300 WINCHESTER MAGNUM .308

| BULLET | | | POWDER | | | | | |
|---|---|---|---|---|---|---|---|---|
| Wgt | Mfg | Type | GRN | Mfg | Type | VEL(fps) | OAL | Source |
| 200 | NO | SPP | 67.0 | DUP | 4831 | 2860 | | NO |
| 200 | NO | SPP | 69.0 | DUP | 4831 | 2926 | | NO |
| 200 | NO | SPP | 71.0 | DUP | 4831 | 2972 | | NO |
| 200 | SR | SPBT | 71.0 | DUP | 7828 | 2900 | | DUP |
| 220 | SR | all | 50.1 | DUP | 4064 | 2300 | | SR |
| 220 | SR | all | 53.5 | DUP | 4064 | 2400 | | SR |
| 220 | SR | all | 60.9 | DUP | 4831 | 2400 | | SR |
| 220 | SR | all | 63.7 | DUP | 4831 | 2500 | | SR |
| 220 | SR | all | 66.5 | DUP | 4831 | 2600 | | SR |
| 220 | SR | all | 69.3 | DUP | 4831 | 2700 | | SR |
| 220 | SR | RN | 70.0 | DUP | 7828 | 2750 | | DUP |

LOADS FOR .308 NORMA MAGNUM .308

| BULLET | | | POWDER | | | | | |
|---|---|---|---|---|---|---|---|---|
| Wgt | Mfg | Type | GRN | Mfg | Type | VEL(fps) | OAL | Source |
| 100 | SP | PL | 62.0 | DUP | 4064 | 3179 | | SP |
| 100 | SP | PL | 64.0 | DUP | 4064 | 3296 | | SP |
| 100 | SP | PL | 66.0 | DUP | 4064 | 3388 | | SP |
| 110 | SR | HP | 60.5 | DUP | 4064 | 3000 | 3.250 | SR |
| 110 | SR | HP | 61.8 | DUP | 4064 | 3100 | | SR |
| 110 | SR | HP | 63.1 | DUP | 4064 | 3200 | | SR |
| 110 | SR | HP | 64.4 | DUP | 4064 | 3300 | | SR |
| 110 | SR | HP | 65.8 | DUP | 4064 | 3400 | | SR |
| 110 | SR | HP | 69.8 | DUP | 4350 | 3100 | | SR |
| 110 | SR | HP | 71.7 | DUP | 4350 | 3200 | | SR |
| 110 | SR | HP | 73.6 | DUP | 4350 | 3300 | | SR |
| 110 | SR | HP | 75.5 | DUP | 4350 | 3400 | | SR |
| 110 | SR | HP | 77.4 | DUP | 4350 | 3500 | | SR |
| 110 | SP | VAR | 50.0 | DUP | 4064 | 2679 | | SP |
| 110 | SP | PL | 52.0 | DUP | 4064 | 2792 | | SP |
| 110 | SP | PL | 54.0 | DUP | 4064 | 2882 | | SP |
| 110 | SP | RN&SP | 61.0 | DUP | 4064 | 3212 | | SP |
| 110 | SP | RN&SP | 63.0 | DUP | 4064 | 3349 | | SP |
| 110 | SP | RN&SP | 65.0 | DUP | 4064 | 3482 | | SP |
| 125 | SR | SPZ | 59.4 | DUP | 4064 | 2900 | | SR |
| 125 | SR | SPZ | 60.9 | DUP | 4064 | 3000 | | SR |
| 125 | SR | SPZ | 62.4 | DUP | 4064 | 3100 | | SR |
| 125 | SR | SPZ | 63.9 | DUP | 4064 | 3200 | | SR |
| 125 | SR | SPZ | 65.4 | DUP | 4064 | 3300 | | SR |
| 125 | SR | SPZ | 68.0 | DUP | 4350 | 3000 | | SR |
| 125 | SR | SPZ | 70.0 | DUP | 4350 | 3100 | | SR |
| 125 | SR | SPZ | 72.1 | DUP | 4350 | 3200 | | SR |
| 125 | SR | SPZ | 74.1 | DUP | 4350 | 3300 | | SR |
| 125 | SR | SPZ | 76.2 | DUP | 4350 | 3400 | | SR |
| 130 | SP | HP | 60.0 | DUP | 4064 | 3062 | | SP |
| 130 | SP | HP | 62.0 | DUP | 4064 | 3188 | | SP |
| 130 | SP | HP | 64.0 | DUP | 4064 | 3318 | | SP |
| 150 | SR | all | 59.7 | DUP | 4064 | 2900 | | SR |
| 150 | SR | all | 61.4 | DUP | 4064 | 3000 | | SR |
| 150 | SR | all | 63.0 | DUP | 4064 | 3100 | | SR |
| 150 | SR | all | 64.6 | DUP | 4064 | 3200 | | SR |
| 150 | SR | all | 66.8 | DUP | 4350 | 2900 | | SR |
| 150 | SR | all | 68.7 | DUP | 4350 | 3000 | | SR |
| 150 | SR | all | 70.5 | DUP | 4350 | 3100 | | SR |
| 150 | SR | all | 72.4 | DUP | 4350 | 3200 | | SR |
| 150 | SR | all | 74.2 | DUP | 4350 | 3300 | | SR |
| 150 | SP | all | 58.0 | DUP | 4064 | 2920 | | SP |
| 150 | SP | all | 60.0 | DUP | 4064 | 3031 | | SP |
| 150 | SP | all | 62.0 | DUP | 4064 | 3145 | | SP |
| 150 | NO | SPP | 58.0 | DUP | 4064 | 2963 | | NO |
| 150 | NO | SPP | 60.0 | DUP | 4064 | 3052 | | NO |
| 150 | NO | SPP | 62.0 | DUP | 4064 | 3195 | | NO |
| 150 | NO | SPZ | 58.0 | DUP | 4064 | 2971 | | NO |
| 150 | NO | SPZ | 60.0 | DUP | 4064 | 3064 | | NO |
| 150 | NO | SPZ | 62.0 | DUP | 4064 | 3212 | | NO |
| 150 | NO | HPBT | 52.0 | DUP | 4895 | 2691 | | NO |
| 150 | NO | HPBT | 54.0 | DUP | 4895 | 2811 | | NO |
| 150 | NO | HPBT | 56.0 | DUP | 4895 | 2985 | | NO |

LOADS FOR .308 NORMA MAGNUM .308

| BULLET | | | POWDER | | | | | |
|---|---|---|---|---|---|---|---|---|
| Wgt | Mfg | Type | GRN | Mfg | Type | VEL(fps) | OAL | Source |
| 165 | SR | all | 56.9 | DUP | 4064 | 2700 | | SR |
| 165 | SR | all | 58.7 | DUP | 4064 | 2800 | | SR |
| 165 | SR | all | 60.5 | DUP | 4064 | 2900 | | SR |
| 165 | SR | all | 62.8 | DUP | 4350 | 2700 | | SR |
| 165 | SR | all | 64.8 | DUP | 4350 | 2800 | | SR |
| 165 | SR | all | 66.9 | DUP | 4350 | 2900 | | SR |
| 165 | SR | all | 69.0 | DUP | 4350 | 3000 | | SR |
| 165 | SR | all | 71.0 | DUP | 4350 | 3100 | | SR |
| 165 | SP | all | 57.0 | DUP | 4064 | 2764 | | SP |
| 165 | SP | all | 59.0 | DUP | 4064 | 2899 | | SP |
| 165 | SP | all | 61.0 | DUP | 4064 | 3027 | | SP |
| 165 | NO | SPP | 57.0 | DUP | 4064 | 2884 | | NO |
| 165 | NO | SPP | 59.0 | DUP | 4064 | 2964 | | NO |
| 165 | NO | SPP | 61.0 | DUP | 4064 | 3073 | | NO |
| 165 | NO | SPZ | 57.0 | DUP | 4064 | 2892 | | NO |
| 165 | NO | SPZ | 59.0 | DUP | 4064 | 2978 | | NO |
| 165 | NO | SPZ | 61.0 | DUP | 4064 | 3083 | | NO |
| 168 | NO | HPBT | 51.0 | DUP | 4895 | 2553 | | NO |
| 168 | NO | HPBT | 53.0 | DUP | 4895 | 2694 | | NO |
| 168 | NO | HPBT | 55.0 | DUP | 4895 | 2835 | | NO |
| 180 | SR | all | 57.2 | DUP | 4064 | 2700 | | SR |
| 180 | SR | all | 59.1 | DUP | 4064 | 2800 | | SR |
| 180 | SR | all | 64.0 | DUP | 4350 | 2700 | | SR |
| 180 | SR | all | 65.9 | DUP | 4350 | 2800 | | SR |
| 180 | SR | all | 67.8 | DUP | 4350 | 2900 | | SR |
| 180 | SR | all | 69.7 | DUP | 4350 | 3000 | | SR |
| 180 | SP | all | 54.0 | DUP | 4064 | 2567 | | SP |
| 180 | SP | all | 56.0 | DUP | 4064 | 2692 | | SP |
| 180 | SP | all | 58.0 | DUP | 4064 | 2814 | | SP |
| 180 | NO | SPP | 54.0 | DUP | 4064 | 2651 | | NO |
| 180 | NO | SPP | 56.0 | DUP | 4064 | 2746 | | NO |
| 180 | NO | SPP | 58.0 | DUP | 4064 | 2886 | | NO |
| 180 | NO | PPP | 54.0 | DUP | 4064 | 2657 | | NO |
| 180 | NO | PPP | 56.0 | DUP | 4064 | 2755 | | NO |
| 180 | NO | PPP | 58.0 | DUP | 4064 | 2893 | | NO |
| 180 | NO | SPZ | 54.0 | DUP | 4064 | 2670 | | NO |
| 180 | NO | SPZ | 56.0 | DUP | 4064 | 2755 | | NO |
| 180 | NO | SPZ | 58.0 | DUP | 4064 | 2892 | | NO |
| 200 | SR | all | 54.0 | DUP | 4064 | 2500 | | SR |
| 200 | SR | all | 55.9 | DUP | 4064 | 2600 | | SR |
| 200 | SR | all | 57.8 | DUP | 4064 | 2700 | | SR |
| 200 | SR | all | 64.5 | DUP | 4350 | 2600 | | SR |
| 200 | SR | all | 66.0 | DUP | 4350 | 2700 | | SR |
| 200 | SR | all | 67.5 | DUP | 4350 | 2800 | | SR |
| 200 | SR | all | 69.0 | DUP | 4350 | 2900 | | SR |
| 200 | SP | SPZ | 66.0 | DUP | 4831 | 2780 | | SP |
| 200 | SP | SPZ | 68.0 | DUP | 4831 | 2868 | | SP |
| 200 | SP | SPZ | 70.0 | DUP | 4831 | 2941 | | SP |
| 200 | NO | SPP | 52.0 | DUP | 4064 | 2597 | | NO |
| 200 | NO | SPP | 54.0 | DUP | 4064 | 2627 | | NO |
| 200 | NO | SPP | 56.0 | DUP | 4064 | 2674 | | NO |
| 220 | SR | all | 52.6 | DUP | 4064 | 2400 | | SR |

## LOADS FOR .308 NORMA MAGNUM .308

| BULLET | | | POWDER | | | | | |
|---|---|---|---|---|---|---|---|---|
| Wgt | Mfg | Type | GRN | Mfg | Type | VEL(fps) | OAL | Source |
| 220 | SR | all | 54.6 | DUP | 4064 | 2500 | | SR |
| 220 | SR | all | 60.4 | DUP | 4350 | 2500 | | SR |
| 220 | SR | all | 62.9 | DUP | 4350 | 2600 | | SR |
| 220 | SR | all | 65.4 | DUP | 4350 | 2700 | | SR |
| 220 | SR | all | 67.8 | DUP | 4350 | 2800 | | SR |

CAUTION : BARREL THROAT VARIATIONS ARE COMMON WITH RIFLES CHAMBERED FOR THE .308 NORMA MAGNUM CARTRIDGE. WORK UP MAXIMUM LOADS WITH CAUTION.

LOADS FOR .300 WEATHERBY MAGNUM .308

| BULLET | | | POWDER | | | | | |
|---|---|---|---|---|---|---|---|---|
| Wgt | Mfg | Type | GRN | Mfg | Type | VEL(fps) | OAL | Source |
| 100 | SP | PL | 75.0 | DUP | 4350 | 3201 | | SP |
| 100 | SP | PL | 77.0 | DUP | 4350 | 3314 | | SP |
| 100 | SP | PL | 79.0 | DUP | 4350 | 3431 | | SP |
| 110 | SR | HP | 73.5 | DUP | 4064 | 3500 | 3.562 | SR |
| 110 | SR | HP | 75.3 | DUP | 4064 | 3600 | | SR |
| 110 | SR | HP | 77.0 | DUP | 4064 | 3700 | | SR |
| 110 | SR | HP | 78.7 | DUP | 4064 | 3800 | | SR |
| 110 | SR | HP | 82.2 | DUP | 4350 | 3500 | | SR |
| 110 | SR | HP | 84.2 | DUP | 4350 | 3600 | | SR |
| 110 | SR | HP | 86.1 | DUP | 4350 | 3700 | | SR |
| 110 | SR | HP | 88.0 | DUP | 4350 | 3800 | | SR |
| 110 | SR | HP | 89.9 | DUP | 4350 | 3900 | | SR |
| 110 | SP | VAR | 61.0 | DUP | 4350 | 2642 | | SP |
| 110 | SP | VAR | 63.0 | DUP | 4350 | 2736 | | SP |
| 110 | SP | VAR | 65.0 | DUP | 4350 | 2829 | | SP |
| 110 | SP | RN&SP | 83.0 | DUP | 4350 | 3685 | | SP |
| 110 | SP | RN&SP | 85.0 | DUP | 4350 | 3793 | | SP |
| 110 | SP | RN&SP | 87.0 | DUP | 4350 | 3876 | | SP |
| 125 | SR | SPZ | 69.4 | DUP | 4064 | 3200 | | SR |
| 125 | SR | SPZ | 71.2 | DUP | 4064 | 3300 | | SR |
| 125 | SR | SPZ | 72.9 | DUP | 4064 | 3400 | | SR |
| 125 | SR | SPZ | 77.9 | DUP | 4350 | 3200 | | SR |
| 125 | SR | SPZ | 79.8 | DUP | 4350 | 3300 | | SR |
| 125 | SR | SPZ | 81.6 | DUP | 4350 | 3400 | | SR |
| 125 | SR | SPZ | 83.4 | DUP | 4350 | 3500 | | SR |
| 125 | SR | SPZ | 85.2 | DUP | 4350 | 3600 | | SR |
| 130 | SP | all | 79.0 | DUP | 4350 | 3448 | | SP |
| 130 | SP | all | 81.0 | DUP | 4350 | 3552 | | SP |
| 130 | SP | all | 83.0 | DUP | 4350 | 3630 | | SP |
| 150 | SR | all | 66.6 | DUP | 4064 | 3000 | | SR |
| 150 | SR | all | 68.5 | DUP | 4064 | 3100 | | SR |
| 150 | SR | all | 70.4 | DUP | 4064 | 3200 | | SR |
| 150 | SR | all | 73.8 | DUP | 4350 | 3000 | | SR |
| 150 | SR | all | 75.9 | DUP | 4350 | 3100 | | SR |
| 150 | SR | all | 77.9 | DUP | 4350 | 3200 | | SR |
| 150 | SR | all | 80.0 | DUP | 4350 | 3300 | | SR |
| 150 | SR | all | 82.0 | DUP | 4350 | 3400 | | SR |
| 150 | SP | all | 77.0 | DUP | 4350 | 3249 | | SP |
| 150 | SP | all | 79.0 | DUP | 4350 | 3341 | | SP |
| 150 | SP | all | 81.0 | DUP | 4350 | 3430 | | SP |
| 150 | NO | SPP | 76.0 | DUP | 4350 | 3136 | | NO |
| 150 | NO | SPP | 78.0 | DUP | 4350 | 3261 | | NO |
| 150 | NO | SPP | 80.0 | DUP | 4350 | 3387 | | NO |
| 150 | NO | SPZ | 76.0 | DUP | 4350 | 3142 | | NO |
| 150 | NO | SPZ | 78.0 | DUP | 4350 | 3279 | | NO |
| 150 | NO | SPZ | 80.0 | DUP | 4350 | 3408 | | NO |
| 150 | NO | HPBT | 71.0 | DUP | 4350 | 3084 | | NO |
| 150 | NO | HPBT | 73.0 | DUP | 4350 | 3119 | | NO |
| 150 | NO | HPBT | 75.0 | DUP | 4350 | 3162 | | NO |
| 165 | SR | all | 63.9 | DUP | 4064 | 2800 | | SR |
| 165 | SR | all | 66.4 | DUP | 4064 | 2900 | | SR |
| 165 | SR | all | 68.9 | DUP | 4064 | 3000 | | SR |
| 165 | SR | all | 72.0 | DUP | 4350 | 2900 | | SR |

LOADS FOR .300 WEATHERBY MAGNUM .308

| BULLET | | | POWDER | | | | | |
|---|---|---|---|---|---|---|---|---|
| Wgt | Mfg | Type | GRN | Mfg | Type | VEL(fps) | OAL | Source |
| 165 | SR | all | 74.1 | DUP | 4350 | 3000 | | SR |
| 165 | SR | all | 76.2 | DUP | 4350 | 3100 | | SR |
| 165 | SR | all | 78.2 | DUP | 4350 | 3200 | | SR |
| 165 | SP | all | 76.0 | DUP | 4350 | 3046 | | SP |
| 165 | SP | all | 78.0 | DUP | 4350 | 3153 | | SP |
| 165 | SP | all | 80.0 | DUP | 4350 | 3268 | | SP |
| 165 | NO | SPP | 75.0 | DUP | 4350 | 3060 | | NO |
| 165 | NO | SPP | 77.0 | DUP | 4350 | 3165 | | NO |
| 165 | NO | SPP | 79.0 | DUP | 4350 | 3220 | | NO |
| 165 | NO | SPZ | 75.0 | DUP | 4350 | 3072 | | NO |
| 165 | NO | SPZ | 77.0 | DUP | 4350 | 3172 | | NO |
| 165 | NO | SPZ | 79.0 | DUP | 4350 | 3241 | | NO |
| 168 | NO | HPBT | 67.0 | DUP | 4350 | 2915 | | NO |
| 168 | NO | HPBT | 69.0 | DUP | 4350 | 2994 | | NO |
| 168 | NO | HPBT | 71.0 | DUP | 4350 | 3087 | | NO |
| 180 | SR | all | 63.0 | DUP | 4064 | 2700 | | SR |
| 180 | SR | all | 65.0 | DUP | 4064 | 2800 | | SR |
| 180 | SR | all | 67.0 | DUP | 4064 | 2900 | | SR |
| 180 | SR | all | 71.1 | DUP | 4350 | 2800 | | SR |
| 180 | SR | all | 73.1 | DUP | 4350 | 2900 | | SR |
| 180 | SR | all | 75.1 | DUP | 4350 | 3000 | | SR |
| 180 | SR | all | 77.1 | DUP | 4350 | 3100 | | SR |
| 180 | SP | all | 73.0 | DUP | 4350 | 2827 | | SP |
| 180 | SP | all | 75.0 | DUP | 4350 | 2937 | | SP |
| 180 | SP | all | 77.0 | DUP | 4350 | 3041 | | SP |
| 180 | NO | SPP | 72.0 | DUP | 4350 | 2906 | | NO |
| 180 | NO | SPP | 74.0 | DUP | 4350 | 2993 | | NO |
| 180 | NO | SPP | 76.0 | DUP | 4350 | 3077 | | NO |
| 180 | NO | PPP | 72.0 | DUP | 4350 | 2909 | | NO |
| 180 | NO | PPP | 74.0 | DUP | 4350 | 3001 | | NO |
| 180 | NO | PPP | 76.0 | DUP | 4350 | 3074 | | NO |
| 180 | NO | SPZ | 72.0 | DUP | 4350 | 2919 | | NO |
| 180 | NO | SPZ | 74.0 | DUP | 4350 | 3005 | | NO |
| 180 | NO | SPZ | 76.0 | DUP | 4350 | 3091 | | NO |
| 200 | SR | all | 60.8 | DUP | 4064 | 2600 | | SR |
| 200 | SR | all | 63.5 | DUP | 4064 | 2700 | | SR |
| 200 | SR | all | 69.5 | DUP | 4350 | 2700 | | SR |
| 200 | SR | all | 71.6 | DUP | 4350 | 2800 | | SR |
| 200 | SR | all | 73.7 | DUP | 4350 | 2900 | | SR |
| 200 | SR | all | 75.8 | DUP | 4350 | 3000 | | SR |
| 200 | SP | SPZ | 71.0 | DUP | 4350 | 2699 | | SP |
| 200 | SP | SPZ | 73.0 | DUP | 4350 | 2809 | | SP |
| 200 | SP | SPZ | 75.0 | DUP | 4350 | 2917 | | SP |
| 200 | NO | SPP | 70.0 | DUP | 4350 | 2795 | | NO |
| 200 | NO | SPP | 72.0 | DUP | 4350 | 2881 | | NO |
| 200 | NO | SPP | 74.0 | DUP | 4350 | 2962 | | NO |
| 220 | SR | all | 56.8 | DUP | 4064 | 2400 | | SR |
| 220 | SR | all | 59.6 | DUP | 4064 | 2500 | | SR |
| 220 | SR | all | 62.3 | DUP | 4064 | 2600 | | SR |
| 220 | SR | all | 65.9 | DUP | 4350 | 2500 | | SR |
| 220 | SR | all | 68.2 | DUP | 4350 | 2600 | | SR |
| 220 | SR | all | 70.5 | DUP | 4350 | 2700 | | SR |
| 220 | SR | all | 72.8 | DUP | 4350 | 2800 | | SR |

LOADS FOR .303 BRITISH .311

| BULLET | | | POWDER | | | | | |
|---|---|---|---|---|---|---|---|---|
| Wgt | Mfg | Type | GRN | Mfg | Type | VEL(fps) | OAL | Source |
| 100 | SP | PL | 42.0 | DUP | 4064 | 2637 | | SP |
| 100 | SP | PL | 44.0 | DUP | 4064 | 2768 | | SP |
| 100 | SP | PL | 46.0 | DUP | 4064 | 2882 | | SP |
| 150 | SR | SPZ | 36.9 | DUP | 3031 | 2200 | 3.075 | SR |
| 150 | SR | SPZ | 38.1 | DUP | 3031 | 2300 | | SR |
| 150 | SR | SPZ | 39.4 | DUP | 3031 | 2400 | | SR |
| 150 | SR | SPZ | 40.7 | DUP | 3031 | 2500 | | SR |
| 150 | SR | SPZ | 42.0 | DUP | 3031 | 2600 | | SR |
| 150 | SR | SPZ | 38.5 | DUP | 4064 | 2200 | | SR |
| 150 | SR | SPZ | 39.9 | DUP | 4064 | 2300 | | SR |
| 150 | SR | SPZ | 41.4 | DUP | 4064 | 2400 | | SR |
| 150 | SR | SPZ | 42.9 | DUP | 4064 | 2500 | | SR |
| 150 | SR | SPZ | 44.4 | DUP | 4064 | 2600 | | SR |
| 150 | SP | SPZ | 39.0 | DUP | 4064 | 2319 | | SP |
| 150 | SP | SPZ | 41.0 | DUP | 4064 | 2437 | | SP |
| 150 | SP | SPZ | 43.0 | DUP | 4064 | 2554 | | SP |
| 180 | SR | SPZ | 34.4 | DUP | 3031 | 2000 | | SR |
| 180 | SR | SPZ | 35.8 | DUP | 3031 | 2100 | | SR |
| 180 | SR | SPZ | 37.2 | DUP | 3031 | 2200 | | SR |
| 180 | SR | SPZ | 38.7 | DUP | 3031 | 2300 | | SR |
| 180 | SR | SPZ | 36.1 | DUP | 4064 | 2000 | | SR |
| 180 | SR | SPZ | 37.6 | DUP | 4064 | 2100 | | SR |
| 180 | SR | SPZ | 39.1 | DUP | 4064 | 2200 | | SR |

LOADS FOR 7.65 MAUSER .311

| BULLET | | | POWDER | | | | | |
|---|---|---|---|---|---|---|---|---|
| Wgt | Mfg | Type | GRN | Mfg | Type | VEL(fps) | OAL | Source |
| 150 | SR | SPZ | 36.3 | DUP | 3031 | 2300 | 2.970 | SR |
| 150 | SR | SPZ | 38.0 | DUP | 3031 | 2400 | | SR |
| 150 | SR | SPZ | 39.8 | DUP | 3031 | 2500 | | SR |
| 150 | SR | SPZ | 41.5 | DUP | 3031 | 2600 | | SR |
| 150 | SR | SPZ | 43.3 | DUP | 3031 | 2700 | | SR |
| 150 | SR | SPZ | 40.6 | DUP | 4064 | 2300 | | SR |
| 150 | SR | SPZ | 41.9 | DUP | 4064 | 2400 | | SR |
| 150 | SR | SPZ | 43.2 | DUP | 4064 | 2500 | | SR |
| 150 | SR | SPZ | 44.5 | DUP | 4064 | 2600 | | SR |
| 150 | SR | SPZ | 45.9 | DUP | 4064 | 2700 | | SR |
| 180 | SR | SPZ | 35.1 | DUP | 3031 | 2100 | | SR |
| 180 | SR | SPZ | 36.7 | DUP | 3031 | 2200 | | SR |
| 180 | SR | SPZ | 38.3 | DUP | 3031 | 2300 | | SR |
| 180 | SR | SPZ | 39.9 | DUP | 3031 | 2400 | | SR |
| 180 | SR | SPZ | 41.5 | DUP | 3031 | 2500 | | SR |
| 180 | SR | SPZ | 36.8 | DUP | 4064 | 2100 | | SR |
| 180 | SR | SPZ | 38.4 | DUP | 4064 | 2200 | | SR |
| 180 | SR | SPZ | 40.0 | DUP | 4064 | 2300 | | SR |
| 180 | SR | SPZ | 41.6 | DUP | 4064 | 2400 | | SR |
| 180 | SR | SPZ | 43.2 | DUP | 4064 | 2500 | | SR |

## LOADS FOR 7.7 JAPANESE .311

| BULLET | | | POWDER | | | | | |
|---|---|---|---|---|---|---|---|---|
| Wgt | Mfg | Type | GRN | Mfg | Type | VEL (fps) | OAL | Source |
| 150 | SR | SPZ | 39.9 | DUP | 3031 | 2300 | 3.150 | SR |
| 150 | SR | SPZ | 41.3 | DUP | 3031 | 2400 | | SR |
| 150 | SR | SPZ | 42.7 | DUP | 3031 | 2500 | | SR |
| 150 | SR | SPZ | 44.1 | DUP | 3031 | 2600 | | SR |
| 150 | SR | SPZ | 45.6 | DUP | 3031 | 2700 | | SR |
| 150 | SR | SPZ | 41.4 | DUP | 4895 | 2300 | | SR |
| 150 | SR | SPZ | 42.8 | DUP | 4895 | 2400 | | SR |
| 150 | SR | SPZ | 44.2 | DUP | 4895 | 2500 | | SR |
| 150 | SR | SPZ | 45.6 | DUP | 4895 | 2600 | | SR |
| 150 | SR | SPZ | 47.0 | DUP | 4895 | 2700 | | SR |
| 180 | SR | SPZ | 37.3 | DUP | 3031 | 2100 | | SR |
| 180 | SR | SPZ | 38.9 | DUP | 3031 | 2200 | | SR |
| 180 | SR | SPZ | 40.6 | DUP | 3031 | 2300 | | SR |
| 180 | SR | SPZ | 42.3 | DUP | 3031 | 2400 | | SR |
| 180 | SR | SPZ | 40.2 | DUP | 4064 | 2100 | | SR |
| 180 | SR | SPZ | 41.6 | DUP | 4064 | 2200 | | SR |
| 180 | SR | SPZ | 43.0 | DUP | 4064 | 2300 | | SR |
| 180 | SR | SPZ | 44.4 | DUP | 4064 | 2400 | | SR |
| 180 | SR | SPZ | 45.8 | DUP | 4064 | 2500 | | SR |

LOADS FOR 8MM MAUSER .323

| BULLET | | | POWDER | | | | | |
|---|---|---|---|---|---|---|---|---|
| Wgt | Mfg | Type | GRN | Mfg | Type | VEL (fps) | OAL | Source |
| 150 | SR | SPZ | 42.1 | DUP | 3031 | 2500 | 3.250 | SR |
| 150 | SR | SPZ | 43.8 | DUP | 3031 | 2600 | | SR |
| 150 | SR | SPZ | 45.5 | DUP | 3031 | 2700 | | SR |
| 150 | SR | SPZ | 47.2 | DUP | 3031 | 2800 | | SR |
| 150 | SR | SPZ | 49.0 | DUP | 3031 | 2900 | | SR |
| 150 | SR | SPZ | 45.7 | DUP | 4064 | 2500 | | SR |
| 150 | SR | SPZ | 47.4 | DUP | 4064 | 2600 | | SR |
| 150 | SR | SPZ | 49.1 | DUP | 4064 | 2700 | | SR |
| 150 | SR | SPZ | 50.9 | DUP | 4064 | 2800 | | SR |
| 150 | SR | SPZ | 52.7 | DUP | 4064 | 2900 | | SR |
| 150 | SP | SPZ | 47.0 | DUP | 4064 | 2695 | | SP |
| 150 | SP | SPZ | 49.0 | DUP | 4064 | 2807 | | SP |
| 150 | SP | SPZ | 51.0 | DUP | 4064 | 2915 | | SP |
| 170 | SP | SSP | 45.0 | DUP | 4064 | 2509 | | SP |
| 170 | SP | SSP | 47.0 | DUP | 4064 | 2631 | | SP |
| 170 | SP | SSP | 49.0 | DUP | 4064 | 2723 | | SP |
| 175 | SR | SPZ | 41.3 | DUP | 3031 | 2300 | | SR |
| 175 | SR | SPZ | 42.8 | DUP | 3031 | 2400 | | SR |
| 175 | SR | SPZ | 44.3 | DUP | 3031 | 2500 | | SR |
| 175 | SR | SPZ | 45.9 | DUP | 3031 | 2600 | | SR |
| 175 | SR | SPZ | 44.0 | DUP | 4064 | 2300 | | SR |
| 175 | SR | SPZ | 45.7 | DUP | 4064 | 2400 | | SR |
| 175 | SR | SPZ | 47.4 | DUP | 4064 | 2500 | | SR |
| 175 | SR | SPZ | 49.1 | DUP | 4064 | 2600 | | SR |
| 200 | SP | SPZ | 42.0 | DUP | 4064 | 2196 | | SP |
| 200 | SP | SPZ | 44.0 | DUP | 4064 | 2314 | | SP |
| 200 | SP | SPZ | 46.0 | DUP | 4064 | 2434 | | SP |
| 200 | NO | SPP | 42.0 | DUP | 4064 | 2446 | | NO |
| 200 | NO | SPP | 44.0 | DUP | 4064 | 2545 | | NO |
| 200 | NO | SPP | 46.0 | DUP | 4064 | 2644 | | NO |

LOADS FOR 8MM/06 .323

| BULLET | | | POWDER | | | | | |
|---|---|---|---|---|---|---|---|---|
| Wgt | Mfg | Type | GRN | Mfg | Type | VEL(fps) | OAL | Source |
| 150 | SR | SPZ | 45.4 | DUP | 3031 | 2600 | 3.250 | SR |
| 150 | SR | SPZ | 47.3 | DUP | 3031 | 2700 | | SR |
| 150 | SR | SPZ | 49.2 | DUP | 3031 | 2800 | | SR |
| 150 | SR | SPZ | 51.0 | DUP | 3031 | 2900 | | SR |
| 150 | SR | SPZ | 52.8 | DUP | 3031 | 3000 | | SR |
| 150 | SR | SPZ | 50.4 | DUP | 4064 | 2600 | | SR |
| 150 | SR | SPZ | 52.1 | DUP | 4064 | 2700 | | SR |
| 150 | SR | SPZ | 53.8 | DUP | 4064 | 2800 | | SR |
| 150 | SR | SPZ | 55.4 | DUP | 4064 | 2900 | | SR |
| 150 | SR | SPZ | 57.0 | DUP | 4064 | 3000 | | SR |
| 150 | SP | SPZ | 51.0 | DUP | 4064 | 2585 | | SP |
| 150 | SP | SPZ | 53.0 | DUP | 4064 | 2740 | | SP |
| 150 | SP | SPZ | 55.0 | DUP | 4064 | 2881 | | SP |
| 170 | SP | SSP | 50.0 | DUP | 4064 | 2437 | | SP |
| 170 | SP | SSP | 52.0 | DUP | 4064 | 2558 | | SP |
| 170 | SP | SSP | 54.0 | DUP | 4064 | 2675 | | SP |
| 175 | SR | SPZ | 43.6 | DUP | 3031 | 2400 | | SR |
| 175 | SR | SPZ | 45.5 | DUP | 3031 | 2500 | | SR |
| 175 | SR | SPZ | 47.4 | DUP | 3031 | 2600 | | SR |
| 175 | SR | SPZ | 49.3 | DUP | 3031 | 2700 | | SR |
| 175 | SR | SPZ | 51.1 | DUP | 3031 | 2800 | | SR |
| 175 | SR | SPZ | 47.7 | DUP | 4064 | 2400 | | SR |
| 175 | SR | SPZ | 49.4 | DUP | 4064 | 2500 | | SR |
| 175 | SR | SPZ | 51.1 | DUP | 4064 | 2600 | | SR |
| 175 | SR | SPZ | 52.7 | DUP | 4064 | 2700 | | SR |
| 175 | SR | SPZ | 54.3 | DUP | 4064 | 2800 | | SR |
| 200 | SP | SPZ | 49.0 | DUP | 4064 | 2411 | | SP |
| 200 | SP | SPZ | 51.0 | DUP | 4064 | 2519 | | SP |
| 200 | SP | SPZ | 53.0 | DUP | 4064 | 2647 | | SP |

LOADS FOR 8MM REMINGTON MAGNUM .323

| BULLET | | | POWDER | | | | | |
|---|---|---|---|---|---|---|---|---|
| Wgt | Mfg | Type | GRN | Mfg | Type | VEL(fps) | OAL | Source |
| 150 | SR | SPZ | 75.8 | DUP | 4350 | 3000 | 3.600 | SR |
| 150 | SR | SPZ | 77.8 | DUP | 4350 | 3100 | | SR |
| 150 | SR | SPZ | 79.8 | DUP | 4350 | 3200 | | SR |
| 150 | SR | SPZ | 81.8 | DUP | 4350 | 3300 | | SR |
| 150 | SR | SPZ | 83.8 | DUP | 4350 | 3400 | | SR |
| 150 | SR | SPZ | 84.8 | DUP | 4350 | 3450 | | SR |
| 150 | SR | SPZ | 78.9 | DUP | 4831 | 3000 | | SR |
| 150 | SR | SPZ | 80.7 | DUP | 4831 | 3100 | | SR |
| 150 | SR | SPZ | 82.5 | DUP | 4831 | 3200 | | SR |
| 150 | SR | SPZ | 84.3 | DUP | 4831 | 3300 | | SR |
| 150 | SR | SPZ | 86.0 | DUP | 4831 | 3400 | | SR |
| 150 | SP | SPZ | 80.0 | DUP | 4831 | 3194 | | SP |
| 150 | SP | SPZ | 83.0 | DUP | 4831 | 3320 | | SP |
| 150 | SP | SPZ | 86.0 | DUP | 4831 | 3436 | | SP |
| 170 | SP | SSP | 74.0 | DUP | 4831 | 2896 | | SP |
| 170 | SP | SSP | 77.0 | DUP | 4831 | 3001 | | SP |
| 170 | SP | SSP | 80.0 | DUP | 4831 | 3114 | | SP |
| 175 | SR | SPZ | 69.5 | DUP | 4350 | 2700 | | SR |
| 175 | SR | SPZ | 71.7 | DUP | 4350 | 2800 | | SR |
| 175 | SR | SPZ | 73.9 | DUP | 4350 | 2900 | | SR |
| 175 | SR | SPZ | 76.1 | DUP | 4350 | 3000 | | SR |
| 175 | SR | SPZ | 78.2 | DUP | 4350 | 3100 | | SR |
| 175 | SR | SPZ | 80.3 | DUP | 4350 | 3200 | | SR |
| 175 | SR | SPZ | 71.8 | DUP | 4831 | 2700 | | SR |
| 175 | SR | SPZ | 74.0 | DUP | 4831 | 2800 | | SR |
| 175 | SR | SPZ | 76.2 | DUP | 4831 | 2900 | | SR |
| 175 | SR | SPZ | 78.4 | DUP | 4831 | 3000 | | SR |
| 175 | SR | SPZ | 80.6 | DUP | 4831 | 3100 | | SR |
| 175 | SR | SPZ | 82.7 | DUP | 4831 | 3200 | | SR |
| 200 | SP | SPZ | 72.0 | DUP | 4831 | 2763 | | SP |
| 200 | SP | SPZ | 75.0 | DUP | 4831 | 2880 | | SP |
| 200 | SP | SPZ | 78.0 | DUP | 4831 | 2996 | | SP |
| 200 | NO | SPP | 72.0 | DUP | 4831 | 2836 | | NO |
| 200 | NO | SPP | 75.0 | DUP | 4831 | 2949 | | NO |
| 200 | NO | SPP | 78.0 | DUP | 4831 | 3047 | | NO |

LOADS FOR .338 WINCHESTER MAGNUM .338

| BULLET | | | POWDER | | | | | |
|---|---|---|---|---|---|---|---|---|
| Wgt | Mfg | Type | GRN | Mfg | Type | VEL (fps) | OAL | Source |
| 200 | SP | SPZ | 71.5 | DUP | 4831 | 2850 | | SP |
| 200 | SP | SPZ | 73.5 | DUP | 4831 | 2932 | | SP |
| 200 | SP | SPZ | 75.5 | DUP | 4831 | 3006 | | SP |
| 200 | NO | SSP | 71.0 | DUP | 4831 | 2684 | | NO |
| 200 | NO | SSP | 73.0 | DUP | 4831 | 2775 | | NO |
| 200 | NO | SSP | 75.0 | DUP | 4831 | 2870 | | NO |
| 250 | SR | SBT | 49.9 | DUP | 3031 | 2300 | 3.340 | SR |
| 250 | SR | SBT | 52.6 | DUP | 3031 | 2400 | | SR |
| 250 | SR | SBT | 55.3 | DUP | 3031 | 2500 | | SR |
| 250 | SR | SBT | 64.1 | DUP | 4831 | 2400 | | SR |
| 250 | SR | SBT | 66.8 | DUP | 4831 | 2500 | | SR |
| 250 | SR | SBT | 69.5 | DUP | 4831 | 2600 | | SR |
| 250 | SR | SBT | 72.1 | DUP | 4831 | 2700 | | SR |
| 250 | SP | GS | 67.0 | DUP | 4831 | 2521 | | SP |
| 250 | SP | GS | 69.0 | DUP | 4831 | 2616 | | SP |
| 250 | SP | GS | 71.0 | DUP | 4831 | 2710 | | SP |
| 250 | NO | RNP | 67.0 | DUP | 4831 | 2456 | | NO |
| 250 | NO | RNP | 69.0 | DUP | 4831 | 2526 | | NO |
| 250 | NO | RNP | 71.0 | DUP | 4831 | 2608 | | NO |
| 250 | SR | SPBT | 74.0C | DUP | 7828 | 2565 | | DUP |
| 275 | SP | SSP | 64.0 | DUP | 4831 | 2348 | | NO |
| 275 | SP | SSP | 66.0 | DUP | 4831 | 2422 | | SP |
| 275 | SP | SSP | 68.0 | DUP | 4831 | 2502 | | SP |
| 275 | SP | SPZ | 71.0C | DUP | 7828 | 2430 | | DUP |

Letter C indicates compressed powder load.

## LOADS FOR .340 WEATHERBY MAGNUM .338

| BULLET | | | POWDER | | | | | |
|---|---|---|---|---|---|---|---|---|
| Wgt | Mfg | Type | GRN | Mfg | Type | VEL (fps) | OAL | Source |
| 200 | SP | SPZ | 79.0 | DUP | 4831 | 2803 | | SP |
| 200 | SP | SPZ | 81.0 | DUP | 4831 | 2896 | | SP |
| 200 | SP | SPZ | 83.0 | DUP | 4831 | 2976 | | SP |
| 210 | NO | SPP | 81.5 | DUP | 4831 | 2961 | | NO |
| 210 | NO | SPP | 83.5 | DUP | 4831 | 3055 | | NO |
| 210 | NO | SPP | 85.5 | DUP | 4831 | 3140 | | NO |
| 250 | SR | SBT | 63.5 | DUP | 4064 | 2500 | 3.562 | SR |
| 250 | SR | SBT | 66.0 | DUP | 4064 | 2600 | | SR |
| 250 | SR | SBT | 70.7 | DUP | 4350 | 2500 | | SR |
| 250 | SR | SBT | 72.9 | DUP | 4350 | 2600 | | SR |
| 250 | SR | SBT | 75.1 | DUP | 4350 | 2700 | | SR |
| 250 | SR | SBT | 77.2 | DUP | 4350 | 2800 | | SR |
| 250 | SP | GS | 76.0 | DUP | 4831 | 2581 | | SP |
| 250 | SP | GS | 78.0 | DUP | 4831 | 2677 | | SP |
| 250 | SP | GS | 80.0 | DUP | 4831 | 2766 | | SP |
| 250 | NO | RNP | 73.5 | DUP | 4831 | 2542 | | NO |
| 250 | NO | RNP | 75.5 | DUP | 4831 | 2618 | | NO |
| 250 | NO | RNP | 77.5 | DUP | 4831 | 2697 | | NO |
| 275 | SP | SSP | 74.0 | DUP | 4831 | 2521 | | SP |
| 275 | SP | SSP | 76.0 | DUP | 4831 | 2610 | | SP |
| 275 | SP | SSP | 78.0 | DUP | 4831 | 2674 | | SP |

LOADS FOR .35 REMINGTON .358

| BULLET | | | POWDER | | | | | |
|---|---|---|---|---|---|---|---|---|
| Wgt | Mfg | Type | GRN | Mfg | Type | VEL(fps) | OAL | Source |
| 125 | SP | all | 39.0 | HER | RE7 | 2451 | | SP |
| 125 | SP | all | 41.0 | HER | RE7 | 2598 | | SP |
| 125 | SP | all | 43.0 | HER | RE7 | 2734 | | SP |
| 140 | SP | HP | 36.0 | HER | RE7 | 2247 | | SP |
| 140 | SP | HP | 38.0 | HER | RE7 | 2392 | | SP |
| 140 | SP | HP | 40.0 | HER | RE7 | 2539 | | SP |
| 158 | SP | all | 33.0 | HER | RE7 | 2091 | | SP |
| 158 | SP | all | 35.0 | HER | RE7 | 2212 | | SP |
| 158 | SP | all | 37.0 | HER | RE7 | 2357 | | SP |
| 180 | SP | FN | 33.0 | DUP | 4895 | 1811 | | SP |
| 180 | SP | FN | 35.0 | DUP | 4895 | 1929 | | SP |
| 180 | SP | FN | 37.0 | DUP | 4895 | 2055 | | SP |
| 200 | SR | RN | 31.4 | DUP | 3031 | 1700 | 2.525 | SR |
| 200 | SR | RN | 32.7 | DUP | 3031 | 1800 | | SR |
| 200 | SR | RN | 34.0 | DUP | 3031 | 1900 | | SR |
| 200 | SR | RN | 35.3 | DUP | 3031 | 2000 | | SR |
| 200 | SR | RN | 36.0 | DUP | 3031 | 2050 | | SR |
| 200 | SR | RN | 34.9 | DUP | 4064 | 1800 | | SR |
| 200 | SR | RN | 36.6 | DUP | 4064 | 1900 | | SR |
| 200 | SR | RN | 38.3 | DUP | 4064 | 2000 | | SR |
| 200 | SR | RN | 39.1 | DUP | 4064 | 2050 | | SR |

LOADS FOR .358 WINCHESTER .358

| BULLET | | | POWDER | | | | | |
|---|---|---|---|---|---|---|---|---|
| Wgt | Mfg | Type | GRN | Mfg | Type | VEL(fps) | OAL | Source |
| 125 | SP | all | 47.0 | HER | RE7 | 2766 | | SP |
| 125 | SP | all | 49.0 | HER | RE7 | 2891 | | SP |
| 125 | SP | all | 51.0 | HER | RE7 | 3026 | | SP |
| 140 | SP | HP | 46.0 | HER | RE7 | 2729 | | SP |
| 140 | SP | HP | 48.0 | HER | RE7 | 2856 | | SP |
| 140 | SP | HP | 50.0 | HER | RE7 | 2994 | | SP |
| 158 | SP | all | 44.0 | HER | RE7 | 2512 | | SP |
| 158 | SP | all | 46.0 | HER | RE7 | 2733 | | SP |
| 158 | SP | all | 48.0 | HER | RE7 | 2850 | | SP |
| 180 | SP | fn | 41.0 | HER | RE7 | 2456 | | SP |
| 180 | SP | FN | 43.0 | HER | RE7 | 2587 | | SP |
| 180 | SP | FN | 45.0 | HER | RE7 | 2728 | | SP |
| 200 | SR | RN | 35.2 | DUP | 4198 | 2200 | 2.780 | SR |
| 200 | SR | RN | 37.0 | DUP | 4198 | 2300 | | SR |
| 200 | SR | RN | 38.8 | DUP | 4198 | 2400 | | SR |
| 200 | SR | RN | 40.6 | DUP | 4198 | 2500 | | SR |
| 200 | SR | RN | 42.2 | DUP | 4064 | 2200 | | SR |
| 200 | SR | RN | 44.5 | DUP | 4064 | 2300 | | SR |
| 200 | SR | RN | 46.8 | DUP | 4064 | 2400 | | SR |
| 200 | SR | RN | 49.1 | DUP | 4064 | 2500 | | SR |
| 200 | SR | RN | 50.3 | DUP | 4064 | 2550 | | SR |
| 250 | SP | SPZ | 40.0 | DUP | 4064 | 2039 | | SP |
| 250 | SP | SPZ | 42.0 | DUP | 4064 | 2148 | | SP |
| 250 | SP | SPZ | 44.0 | DUP | 4064 | 2264 | | SP |

LOADS FOR .44 MAGNUM RIFLE ONLY .4295

| BULLET | | | POWDER | | | | | |
|---|---|---|---|---|---|---|---|---|
| Wgt | Mfg | Type | GRN | Mfg | Type | VEL(fps) | OAL | Source |
| 180 | SR | JHP | 10.1 | HER | Uniq | 1400 | 1.610 | SR |
| 180 | SR | JHP | 11.0 | HER | Uniq | 1500 | | SR |
| 180 | SR | JHP | 11.9 | HER | Uniq | 1600 | | SR |
| 180 | SR | JHP | 12.8 | HER | Uniq | 1700 | | SR |
| 180 | SR | JHP | 13.7 | HER | Uniq | 1800 | | SR |
| 180 | SR | JHP | 14.6 | HER | Uniq | 1900 | | SR |
| 180 | SR | JHP | 19.2 | HER | 2400 | 1500 | | SR |
| 180 | SR | JHP | 20.6 | HER | 2400 | 1600 | | SR |
| 180 | SR | JHP | 22.0 | HER | 2400 | 1700 | | SR |
| 180 | SR | JHP | 23.5 | HER | 2400 | 1800 | | SR |
| 180 | SR | JHP | 25.0 | HER | 2400 | 1900 | | SR |
| 180 | SR | JHP | 26.5 | HER | 2400 | 2000 | | SR |
| 200 | SP | MHP | 22.8 | HER | 2400 | 1843 | | SP |
| 200 | SP | MHP | 23.8 | HER | 2400 | 1891 | | SP |
| 200 | SP | MHP | 24.8 | HER | 2400 | 1965 | | SP |
| 225 | SP | HP | 19.0 | HER | 2400 | 1529 | | SP |
| 225 | SP | HP | 20.0 | HER | 2400 | 1615 | | SP |
| 225 | SP | HP | 21.0 | HER | 2400 | 1698 | | SP |
| 240 | SR | JHP | 10.2 | HER | Uniq | 1300 | | SR |
| 240 | SR | JHP | 11.4 | HER | Uniq | 1400 | | SR |
| 240 | SR | JHP | 12.6 | HER | Uniq | 1500 | | SR |
| 240 | SR | JHP | 13.8 | HER | Uniq | 1600 | | SR |
| 240 | SR | JHP | 15.0 | HER | Uniq | 1700 | | SR |
| 240 | SR | JHP | 19.7 | HER | 2400 | 1500 | | SR |
| 240 | SR | JHP | 21.1 | HER | 2400 | 1600 | | SR |
| 240 | SR | JHP | 22.6 | HER | 2400 | 1700 | | SR |
| 240 | SP | MHP | 20.2 | HER | 2400 | 1603 | | SP |
| 240 | SP | MHP | 21.2 | HER | 2400 | 1681 | | SP |
| 240 | SP | MHP | 22.2 | HER | 2400 | 1756 | | SP |

When reloading for a Ruger Carbine, use only the data in the velocity ranges of 1600 to 1900 fps for the 180 grain bullet, and 1500 to 1800 fps for the 240 grain bullet. This is to assure functioning in the semiautomatic action of the Ruger.

Under no circumstances should these loads be used for pistol ammunition since dangerous overloads would be result.

LOADS FOR .444 MARLIN .4295

| BULLET | | | POWDER | | | | | |
|---|---|---|---|---|---|---|---|---|
| Wgt | Mfg | Type | GRN | Mfg | Type | VEL(fps) | OAL | Source |
| 180 | SR | JHP | 46.4 | DUP | 4198 | 2300 | 2.570 | SR |
| 180 | SR | JHP | 48.7 | DUP | 4198 | 2400 | | SR |
| 180 | SR | JHP | 51.0 | DUP | 4198 | 2500 | | SR |
| 180 | SR | JHP | 58.8 | DUP | 4895 | 2400 | | SR |
| 180 | SR | JHP | 61.5 | DUP | 4895 | 2500 | | SR |
| 240 | SR | JHP | 42.5 | DUP | 4198 | 2000 | | SR |
| 240 | SR | JHP | 44.4 | DUP | 4198 | 2100 | | SR |
| 240 | SR | JHP | 46.2 | DUP | 4198 | 2200 | | SR |
| 240 | SR | JHP | 48.0 | DUP | 4198 | 2300 | | SR |
| 240 | SR | JHP | 50.5 | DUP | 4895 | 2000 | | SR |
| 240 | SR | JHP | 53.3 | DUP | 4895 | 2100 | | SR |
| 240 | SR | JHP | 56.0 | DUP | 4895 | 2200 | | SR |
| 240 | SP | MHP | 42.0 | DUP | 4198 | 2121 | | SP |
| 240 | SP | MHP | 44.0 | DUP | 4198 | 2218 | | SP |
| 240 | SP | MHP | 46.0 | DUP | 4198 | 2302 | | SP |

LOADS FOR .45-70 GOVERNMENT .458

| BULLET | | | POWDER | | | | | |
|---|---|---|---|---|---|---|---|---|
| Wgt | Mfg | Type | GRN | Mfg | Type | VEL(fps) | UAL | Source |
| 1. Loads for the Model 1873 Springfield, Remington rolling block, other old blackpowder rifles, and replicas. | | | | | | | | |
| 300 | SR | FN | 24.5 | HER | 2400 | 1450 | 2.525 | SR |
| 300 | SR | FN | 25.6 | HER | 2400 | 1500 | | SR |
| 300 | SR | FN | 26.7 | HER | 2400 | 1550 | | SR |
| 300 | SR | FN | 27.7 | HER | 2400 | 1600 | | SR |
| 300 | SR | FN | 28.7 | HER | 2400 | 1650 | | SR |
| 300 | SR | FN | 29.7 | HER | 2400 | 1700 | | SR |
| 300 | SR | FN | 31.1 | DUP | 4198 | 1450 | | SR |
| 300 | SR | FN | 32.4 | DUP | 4198 | 1500 | | SR |
| 300 | SR | FN | 33.6 | DUP | 4198 | 1550 | | SR |
| 300 | SR | FN | 34.8 | DUP | 4198 | 1600 | | SR |
| 300 | SR | FN | 36.0 | DUP | 4198 | 1650 | | SR |
| 2. Loads for Model 1886 Winchester | | | | | | | | |
| 300 | SR | FN | 29.8 | HER | 2400 | 1650 | 2.625 | SR |
| 300 | SR | FN | 30.8 | HER | 2400 | 1700 | | SR |
| 300 | SR | FN | 31.8 | HER | 2400 | 1750 | | SR |
| 300 | SR | FN | 32.7 | HER | 2400 | 1800 | | SR |
| 300 | SR | FN | 33.6 | HER | 2400 | 1850 | | SR |
| 300 | SR | FN | 36.3 | DUP | 4198 | 1550 | | SR |
| 300 | SR | FN | 37.3 | DUP | 4198 | 1600 | | SR |
| 300 | SR | FN | 38.3 | DUP | 4198 | 1650 | | SR |
| 300 | SR | FN | 39.2 | DUP | 4198 | 1700 | | SR |
| 300 | SR | FN | 40.1 | DUP | 4198 | 1750 | | SR |
| 300 | SR | FN | 41.0 | DUP | 4198 | 1800 | | SR |
| 400 | SP | FN | 52.0 | DUP | 4064 | 1636 | | SP |
| 400 | SP | FN | 54.0 | DUP | 4064 | 1712 | | SP |
| 400 | SP | FN | 56.0 | DUP | 4064 | 1754 | | SP |
| 3. Loads for the Model 1895 Marlin and Ruger No. 1 and No. 3 rifles. | | | | | | | | |
| 300 | SR | FN | 35.5 | HER | 2400 | 1950 | 2.525 | SR |
| 300 | SR | FN | 37.0 | HER | 2400 | 2000 | | SR |
| 300 | SR | FN | 38.5 | HER | 2400 | 2050 | | SR |
| 300 | SR | FN | 40.0 | HER | 2400 | 2100 | | SR |
| 300 | SR | FN | 41.4 | HER | 2400 | 2150 | | SR |
| 300 | SR | FN | 47.4 | DUP | 4198 | 1950 | | SR |
| 300 | SR | FN | 48.5 | DUP | 4198 | 2000 | | SR |
| 300 | SR | FN | 49.6 | DUP | 4198 | 2050 | | SR |
| 300 | SR | FN | 50.7 | DUP | 4198 | 2100 | | SR |
| 300 | SR | FN | 51.8 | DUP | 4198 | 2150 | | SR |
| 400 | SP | FN | 60.0 | DUP | 4064 | 1848 | | SP |
| 400 | SP | FN | 62.0 | DUP | 4064 | 1911 | | SP |
| 400 | SP | FN | 64.0 | DUP | 4064 | 1964 | | SP |

## LOADS FOR .458 WINCHESTER MAGNUM .458

| BULLET | | | POWDER | | | | | |
|---|---|---|---|---|---|---|---|---|
| Wgt | Mfg | Type | GRN | Mfg | Type | VEL(fps) | OAL | Source |
| 300 | SR | FN | 50.7 | DUP | 4198 | 2000 | 2.930 | SR |
| 300 | SR | FN | 53.3 | DUP | 4198 | 2100 | | SR |
| 300 | SR | FN | 55.9 | DUP | 4198 | 2200 | | SR |
| 300 | SR | FN | 58.5 | DUP | 4198 | 2300 | | SR |
| 300 | SR | FN | 61.1 | DUP | 4198 | 2400 | | SR |
| 300 | SR | FN | 63.7 | DUP | 4198 | 2500 | | SR |
| 300 | SR | FN | 59.9 | DUP | 3031 | 2000 | | SR |
| 300 | SR | FN | 62.5 | DUP | 3031 | 2100 | | SR |
| 300 | SR | FN | 65.1 | DUP | 3031 | 2200 | | SR |
| 300 | SR | FN | 67.7 | DUP | 3031 | 2300 | | SR |
| 300 | SR | FN | 70.3 | DUP | 3031 | 2400 | | SR |
| 300 | SR | FN | 72.9 | DUP | 3031 | 2500 | | SR |
| 400 | SP | FN | 71.0 | DUP | 4064 | 2062 | | SP |
| 400 | SP | FN | 73.0 | DUP | 4064 | 2113 | | SP |
| 400 | SP | FN | 75.0 | DUP | 4064 | 2165 | | SP |

# Shotshell Data

12 GAUGE 2 3/4" PAPER CASE

(FEDERAL CHAMPION & FEDERAL SPECIAL TARGET LOADS)

| Shot Weight Oz. | Powder Mfg. | | Grains | Velocity f.p.s. | Wad Mfg/Type | | Primer Mfg/No. | | Source |
|---|---|---|---|---|---|---|---|---|---|
| 1 | HER | RDT | 20.0 | 1290 | REM | R12L | FED | 209 | HER |
| 1 | HER | RDT | 20.5 | 1290 | FED | 12S3 | FED | 209 | HER |
| 1 | HER | RDT | 20.5 | 1290 | FED | 12S0 | FED | 209 | HER |
| 1 | HER | RDT | 21.0 | 1290 | FED | 12S3 | CCI | 209M | HER |
| 1 | HER | GDT | 21.5 | 1290 | REM | R12L | FED | 209 | HER |
| 1 | HER | GDT | 23.5 | 1290 | FED | 12S3 | FED | 209 | HER |
| 1 | HER | GDT | 22.5 | 1290 | FED | 12S0 | FED | 209 | HER |
| 1 | HER | GDT | 23.0 | 1290 | FED | 12S3 | CCI | 209M | HER |
| 1-1/8 | HER | RDT | 18.0 | 1145 | FED | 12C1 | FED | 209 | HER |
| 1-1/8 | HER | RDT | 18.5 | 1145 | FED | 12S1 | FED | 209 | HER |
| 1-1/8 | HER | RDT | 18.0 | 1145 | FED | 12S3 | FED | 209 | HER |
| 1-1/8 | HER | RDT | 18.5 | 1145 | REM | R12L | FED | 209 | HER |
| 1-1/8 | HER | RDT | 18.0 | 1145 | REM | RXP12 | FED | 209 | HER |
| 1-1/8 | HER | RDT | 18.5 | 1145 | FED | 12C1 | CCI | 109 | HER |
| 1-1/8 | HER | RDT | 18.5 | 1145 | FED | 12S1 | CCI | 109 | HER |
| 1-1/8 | HER | RDT | 18.5 | 1145 | FED | 12S3 | CCI | 109 | HER |
| 1-1/8 | HER | RDT | 18.5 | 1145 | REM | R12L | CCI | 109 | HER |
| 1-1/8 | HER | RDT | 18.0 | 1145 | REM | RXP12 | CCI | 109 | HER |
| 1-1/8 | HER | GDT | 19.0 | 1145 | FED | 12C1 | FED | 209 | HER |
| 1-1/8 | HER | GDT | 19.0 | 1145 | FED | 12S1 | FED | 209 | HER |
| 1-1/8 | HER | GDT | 19.5 | 1145 | FED | 12S3 | FED | 209 | HER |
| 1-1/8 | HER | GDT | 19.0 | 1145 | REM | R12L | FED | 209 | HER |
| 1-1/8 | HER | GDT | 18.5 | 1145 | REM | RXP12 | FED | 209 | HER |
| 1-1/8 | HER | GDT | 19.0 | 1145 | FED | 12C1 | CCI | 109 | HER |
| 1-1/8 | HER | GDT | 19.0 | 1145 | FED | 12S1 | CCI | 109 | HER |
| 1-1/8 | HER | GDT | 20.0 | 1145 | FED | 12S3 | CCI | 109 | HER |
| 1-1/8 | HER | GDT | 19.0 | 1145 | REM | R12L | CCI | 109 | HER |
| 1-1/8 | HER | GDT | 19.0 | 1145 | REM | RXP12 | CCI | 109 | HER |
| 1-1/8 | HER | RDT | 19.0 | 1200 | FED | 12C1 | FED | 209 | HER |
| 1-1/8 | HER | RDT | 19.0 | 1200 | FED | 12S1 | FED | 209 | HER |
| 1-1/8 | HER | RDT | 19.0 | 1200 | FED | 12S3 | FED | 209 | HER |
| 1-1/8 | HER | RDT | 19.5 | 1200 | REM | R12L | FED | 209 | HER |
| 1-1/8 | HER | RDT | 19.0 | 1200 | REM | R12H | FED | 209 | HER |
| 1-1/8 | HER | RDT | 19.0 | 1200 | REM | RXP12 | FED | 209 | HER |
| 1-1/8 | HER | RDT | 19.0 | 1200 | FED | 12C1 | CCI | 109 | HER |
| 1-1/8 | HER | RDT | 19.5 | 1200 | FED | 12S1 | CCI | 109 | HER |
| 1-1/8 | HER | RDT | 20.0 | 1200 | FED | 12S3 | CCI | 109 | HER |
| 1-1/8 | HER | RDT | 19.5 | 1200 | REM | R12L | CCI | 109 | HER |
| 1-1/8 | HER | RDT | 19.0 | 1200 | REM | R12H | CCI | 109 | HER |
| 1-1/8 | HER | RDT | 19.0 | 1200 | REM | RXP12 | CCI | 109 | HER |
| 1-1/8 | HER | GDT | 20.0 | 1200 | FED | 12C1 | FED | 209 | HER |
| 1-1/8 | HER | GDT | 20.0 | 1200 | FED | 12S1 | FED | 209 | HER |
| 1-1/8 | HER | GDT | 21.0 | 1200 | FED | 12S3 | FED | 209 | HER |

12 GAUGE 2 3/4" PAPER CASE

(FEDERAL CHAMPION & FEDERAL SPECIAL TARGET LOADS)

| Shot Weight Oz. | Powder Mfg. | Grains | Velocity f.p.s. | Wad Mfg/Type | | Primer Mfg/No. | Source |
|---|---|---|---|---|---|---|---|
| 1-1/8 HER | GDT | 20.0 | 1200 | REM | R12L | FED 209 | HER |
| 1-1/8 HER | GDT | 19.5 | 1200 | REM | R12H | FED 209 | HER |
| 1-1/8 HER | GDT | 20.0 | 1200 | REM | RXP12 | FED 209 | HER |
| 1-1/8 HER | GDT | 20.5 | 1200 | FED | 12C1 | CCI 109 | HER |
| 1-1/8 HER | GDT | 20.0 | 1200 | FED | 12S1 | CCI 109 | HER |
| 1-1/8 HER | GDT | 22.0 | 1200 | FED | 12S3 | CCI 109 | HER |
| 1-1/8 HER | GDT | 20.0 | 1200 | REM | R12L | CCI 109 | HER |
| 1-1/8 HER | GDT | 20.0 | 1200 | REM | R12H | CCI 109 | HER |
| 1-1/8 HER | GDT | 20.0 | 1200 | REM | RXP12 | CCI 109 | HER |
| 1-1/8 HER | GDT | 21.5 | 1255 | FED | 12C1 | FED 209 | HER |
| 1-1/8 HER | GDT | 21.5 | 1255 | FED | 12S1 | FED 209 | HER |
| 1-1/8 HER | GDT | 23.0 | 1255 | FED | 12S3 | FED 209 | HER |
| 1-1/8 HER | GDT | 21.5 | 1255 | REM | R12H | FED 209 | HER |
| 1-1/8 HER | GDT | 21.5 | 1255 | REM | RXP12 | FED 209 | HER |
| 1-1/8 HER | UNIQ | 22.0 | 1200 | FED | 12C1 | FED 209 | HER |
| 1-1/8 HER | UNIQ | 22.0 | 1200 | FED | 12S1 | FED 209 | HER |
| 1-1/8 HER | UNIQ | 22.0 | 1200 | FED | 12s3 | FED 209 | HER |
| 1-1/8 HER | UNIQ | 22.0 | 1200 | REM | R12L | FED 209 | HER |
| 1-1/8 HER | UNIQ | 21.0 | 1200 | REM | RXP12 | FED 209 | HER |
| 1-1/8 HER | UNIQ | 22.0 | 1200 | FED | 12C1 | CCI 109 | HER |
| 1-1/8 HER | UNIQ | 22.0 | 1200 | FED | 12S1 | CCI 109 | HER |
| 1-1/8 HER | UNIQ | 23.0 | 1200 | FED | 12S3 | CCI 109 | HER |
| 1-1/8 HER | UNIQ | 22.0 | 1200 | REM | R12L | CCI 109 | HER |
| 1-1/8 HER | UNIQ | 22.0 | 1200 | REM | RXP12 | CCI 109 | HER |
| 1-1/8 HER | UNIQ | 22.5 | 1255 | FED | 12C1 | FED 209 | HER |
| 1-1/8 HER | UNIQ | 23.0 | 1255 | FED | 12S1 | FED 209 | HER |
| 1-1/8 HER | UNIQ | 23.0 | 1255 | FED | 12S3 | FED 209 | HER |
| 1-1/8 HER | UNIQ | 22.5 | 1255 | REM | R12H | FED 209 | HER |
| 1-1/8 HER | UNIQ | 22.0 | 1255 | REM | RXP12 | FED 209 | HER |
| 1-1/8 DUP | 800X | 22.5 | 1140 | FED | 12S3 | FED 209 | DUP |
| 1-1/8 DUP | 800X | 23.0 | 1140 | REM | RXP12 | FED 209 | DUP |
| 1-1/8 DUP | 800X | 22.5 | 1140 | FED | 12S3 | CCI 209M | DUP |
| 1-1/8 DUP | 800X | 23.0 | 1135 | REM | RXP12 | CCI 209M | DUP |
| 1-1/8 DUP | 800X | 24.0 | 1205 | FED | 12S3 | FED 209 | DUP |
| 1-1/8 DUP | 800X | 24.5 | 1200 | REM | RXP12 | FED 209 | DUP |
| 1-1/8 DUP | 800X | 24.0 | 1200 | FED | 12S3 | CCI 209M | DUP |
| 1-1/8 DUP | 800X | 24.5 | 1190 | REM | RXP12 | CCI 209M | DUP |
| 1-1/8 DUP | 800X | 25.5 | 1255 | FED | 12S3 | FED 209 | DUP |
| 1-1/8 DUP | 800X | 26.0 | 1250 | REM | R12L | FED 209 | DUP |
| 1-1/8 DUP | 800X | 25.5 | 1255 | FED | 12S3 | CCI 209M | DUP |
| 1-1/8 DUP | 800X | 26.0 | 1250 | REM | R12L | CCI 209M | DUP |
| 1-1/8 DUP | 700X | 17.5 | 1145 | FED | 12S3 | FED 209 | FED |
| 1-1/8 DUP | 700X | 18.5 | 1200 | FED | 12S3 | FED 209 | FED |
| 1-1/8 DUP | 700X | 18 | 1145 | FED | 12C1 | FED 209 | FED |
| 1-1/8 DUP | 7625 | 23 | 1145 | FED | 12C1 | FED 209 | FED |
| 1-1/8 DUP | PB | 22 | 1145 | FED | 12C1 | FED 209 | FED |
| 1-1/8 DUP | 700X | 19.5 | 1200 | FED | 12C1 | FED 209 | FED |
| 1-1/8 DUP | 7625 | 24 | 1200 | FED | 12C1 | FED 209 | FED |
| 1-1/8 DUP | PB | 23.5 | 1200 | FED | 12C1 | FED 209 | FED |
| 1-1/8 DUP | 700X | 21.5 | 1255 | FED | 12C1 | FED 209 | FED |

12 GAUGE 2 3/4" PAPER CASE

(FEDERAL CHAMPION & FEDERAL SPECIAL TARGET LOADS)

| Shot Weight Oz. | Powder Mfg. | | Grains | Velocity f.p.s. | Wad Mfg/Type | | Primer Mfg/No. | | Source |
|---|---|---|---|---|---|---|---|---|---|
| 1-1/8 | DUP | PB | 24.5 | 1255 | FED | 12C1 | FED | 209 | FED |
| 1-1/4 | HER | GDT | 21.0 | 1220 | FED | 12C1 | FED | 209 | HER |
| 1-1/4 | HER | GDT | 23.0 | 1220 | FED | 12S4 | FED | 209 | HER |
| 1-1/4 | HER | GDT | 21.0 | 1220 | REM | SP12 | FED | 209 | HER |
| 1-1/4 | HER | GDT | 21.0 | 1220 | REM | RXP12 | FED | 209 | HER |
| 1-1/4 | HER | UNIQ | 22.5 | 1220 | FED | 12C1 | FED | 209 | HER |
| 1-1/4 | HER | UNIQ | 23.0 | 1220 | FED | 12S1 | FED | 209 | HER |
| 1-1/4 | HER | UNIQ | 24.0 | 1220 | FED | 12S4 | FED | 209 | HER |
| 1-1/4 | HER | UNIQ | 22.0 | 1200 | REM | SP12 | FED | 209 | HER |
| 1-1/4 | HER | UNIQ | 22.0 | 1200 | REM | RXP12 | FED | 209 | HER |
| 1-1/4 | HER | BDT | 37.0 | 1330 | FED | 12S4 | FED | 209 | HER |
| 1-1/4 | HER | HERC | 29.0 | 1330 | REM | RP12 | FED | 209 | HER |
| 1-1/4 | HER | HERC | 29.5 | 1330 | REM | SP12 | FED | 209 | HER |
| 1-1/4 | DUP | 7625 | 29.0 | 1330 | FED | 12S4 | FED | 209 | FED |
| 1-1/4 | DUP | 4756 | 33.5 | 1330 | FED | 12S4 | FED | 209 | FED |
| 1-1/4 | HER | BDT | 36.5 | 1330 | FED | 12S4 | FED | 209 | FED |
| 1-1/4 | DUP | 800X | 25.0 | 1235 | FED | 12S4 | FED | 209 | DUP |
| 1-1/4 | DUP | 800X | 25.5 | 1215 | REM | R12H | FED | 209 | DUP |
| 1-1/4 | DUP | 800X | 28.0 | 1325 | FED | 12S4 | FED | 209 | DUP |
| 1-1/4 | DUP | 800X | 29.0 | 1325 | REM | SP12 | FED | 209 | DUP |
| 1-3/8 | HER | BDT | 34.0 | 1240 | REM | SP12 | FED | 209 | HER |
| 1-3/8 | HER | BDT | 35.5 | 1295 | REM | SP12 | FED | 209 | HER |
| 1-3/8 | HER | BDT | 37.5 | 1350 | REM | RP12 | FED | 209 | HER |
| 1-3/8 | DUP | 7625 | 30.0 | 1290 | REM | RP12 | FED | 209 | DUP |
| 1-3/8 | DUP | 4756 | 36.0 | 1370 | REM | RP12 | FED | 209 | DUP |
| 1-3/8 | DUP | 800X | 26.5 | 1255 | FED | 12S4 | FED | 209 | DUP |
| 1-3/8 | DUP | 800X | 28.0 | 1270 | REM | SP12 | FED | 209 | DUP |
| 1-3/8 | DUP | 800X | 26.5 | 1260 | FED | 12S4 | CCI | 209M | DUP |
| 1-3/8 | DUP | 800X | 27.5 | 1265 | REM | SP12 | CCI | 209M | DUP |
| 1-1/2 | DUP | 800X | 27.0 | 1230 | REM | RP12 | FED | 209 | DUP |
| 1-1/2 | DUP | 800X | 27.0 | 1240 | REM | RP12 | CCI | 209M | DUP |

## 12 GAUGE 2 3/4" FEDERAL GOLD MEDAL PLASTIC TARGET

| Shot Weight Oz. | Powder Mfg. | Grains | Velocity f.p.s. | Wad Mfg/Type | | Primer Mfg/No. | | Source |
|---|---|---|---|---|---|---|---|---|
| 1-1/8 | HER | RDT | 17.5 | 1145 | REM | 12S3 | FED 209 | HER |
| 1-1/8 | HER | RDT | 18.0 | 1145 | REM | RXP12 | FED 209 | HER |
| 1-1/8 | HER | RDT | 18.0 | 1145 | REM | 12S3 | CCI 209 | HER |
| 1-1/8 | HER | RDT | 18.5 | 1145 | REM | RXP12 | CCI 209 | HER |
| 1-1/8 | HER | RDT | 18.0 | 1145 | FED | 12S3 | CCI 209M | HER |
| 1-1/8 | HER | RDT | 18.5 | 1200 | FED | 12S3 | FED 209 | HER |
| 1-1/8 | HER | RDT | 19.0 | 1200 | REM | RXP12 | FED 209 | HER |
| 1-1/8 | HER | RDT | 20.0 | 1200 | FED | 12S3 | CCI 209 | HER |
| 1-1/8 | HER | RDT | 20.0 | 1200 | REM | RXP12 | CCI 209 | HER |
| 1-1/8 | HER | RDT | 19.0 | 1200 | FED | 12S3 | CCI 209M | HER |
| 1-1/8 | HER | RDT | 21.0 | 1255 | REM | RXP12 | FED 209 | HER |
| 1-1/8 | HER | GDT | 19.5 | 1145 | FED | 12S3 | FED 209 | HER |
| 1-1/8 | HER | GDT | 19.5 | 1145 | REM | RXP12 | FED 209 | HER |
| 1-1/8 | HER | GDT | 19.0 | 1145 | FED | 12S3 | CCI 209 | HER |
| 1-1/8 | HER | GDT | 21.0 | 1145 | REM | RXP12 | CCI 209 | HER |
| 1-1/8 | HER | GDT | 19.5 | 1145 | FED | 12S3 | CCI 209M | HER |
| 1-1/8 | HER | GDT | 20.5 | 1200 | FED | 12S3 | FED 209 | HER |
| 1-1/8 | HER | GDT | 21.5 | 1200 | REM | RXP12 | FED 209 | HER |
| 1-1/8 | HER | GDT | 22.0 | 1200 | FED | 12S3 | CCI 209 | HER |
| 1-1/8 | HER | GDT | 22.0 | 1200 | REM | RXP12 | CCI 209 | HER |
| 1-1/8 | HER | GDT | 21.0 | 1200 | FED | 12S3 | CCI 209M | HER |
| 1-1/8 | HER | GDT | 22.5 | 1255 | FED | 12S3 | FED 209 | HER |
| 1-1/8 | HER | GDT | 21.0 | 1255 | REM | RXP12 | FED 209 | HER |
| 1-1/8 | HER | GDT | 22.5 | 1255 | FED | 12S3 | CCI 209M | HER |
| 1-1/8 | HER | GDT | 24.0 | 1310 | REM | RXP12 | FED 209 | FED |
| 1-1/8 | DUP | 700X | 17.5 | 1145 | FED | 12S3 | FED 209 | FED |
| 1-1/8 | DUP | 700X | 18.5 | 1200 | FED | 12S3 | FED 209 | HER |
| 1-1/8 | DUP | PB | 22.5 | 1200 | FED | 12S3 | FED 209 | FED |
| 1-1/8 | DUP | 700X | 20 | 1255 | FED | 12S3 | FED 209 | FED |
| 1-1/8 | DUP | 7625 | 26 | 1255 | FED | 12S3 | FED 209 | FED |
| 1-1/8 | DUP | PB | 24.5 | 1255 | FED | 12S3 | FED 209 | FED |
| 1-1/8 | DUP | 800X | 23.5 | 1195 | FED | 12S3 | CCI 209M | DUP |
| 1-1/8 | DUP | 800X | 24.5 | 1215 | REM | R12L | CCI 209M | DUP |
| 1-1/8 | DUP | 800X | 25.0 | 1250 | FED | 12S3 | FED 209 | DUP |
| 1-1/8 | DUP | 800X | 25.5 | 1240 | REM | R12L | FED 209 | DUP |
| 1-1/8 | DUP | 800X | 25.0 | 1255 | FED | 12S3 | CCI 209M | DUP |
| 1-1/8 | DUP | 800X | 25.5 | 1250 | REM | R12L | CCI 209M | DUP |
| 1-1/4 | HER | UNIQ | 24.0 | 1220 | FED | 12S4 | FED 209 | HER |
| 1-1/4 | HER | UNIQ | 24.0 | 1220 | REM | SP12 | FED 209 | HER |
| 1-1/4 | HER | UNIQ | 24.5 | 1220 | FED | 12S4 | CCI 209M | HER |
| 1-1/4 | HER | HERC | 25.0 | 1220 | FED | 12S4 | FED 209 | HER |
| 1-1/4 | HER | HERC | 26.0 | 1220 | REM | SP12 | FED 209 | HER |
| 1-1/4 | HER | HERC | 25.5 | 1220 | FED | 12S4 | CCI 209M | HER |
| 1-1/4 | HER | HERC | 27.0 | 1275 | REM | SP12 | FED 209 | HER |
| 1-1/4 | HER | BDT | 34.0 | 1275 | FED | 12S4 | FED 209 | HER |
| 1-1/4 | HER | BDT | 35.0 | 1275 | FED | 12S4 | CCI 209M | HER |
| 1-1/4 | HER | BDT | 35.0 | 1330 | REM | SP12 | FED 209 | HER |
| 1-1/4 | HER | BDT | 37.5 | 1330 | REM | SP12 | CCI 209M | HER |
| 1-1/4 | DUP | 7625 | 26 | 1220 | FED | 12C1 | FED 209 | FED |
| 1-1/4 | DUP | PB | 24.5 | 1220 | FED | 12C1 | FED 209 | FED |
| 1-1/4 | DUP | 7625 | 26.0 | 1220 | FED | 12S4 | FED 209 | FED |

12 GAUGE 2 3/4" FEDERAL GOLD MEDAL PLASTIC TARGET

| Shot Weight Oz. | Powder Mfg. | Grains | Velocity f.p.s. | Wad Mfg/Type | | Primer Mfg/No. | Source |
|---|---|---|---|---|---|---|---|
| 1-1/4 | DUP 4756 | 29.5 | 1220 | FED | 12S4 | FED 209 | HER |
| 1-1/4 | DUP PB | 25 | 1220 | FED | 12S4 | FED 209 | FED |
| 1-1/4 | DUP 7625 | 28.0 | 1330 | FED | 12S4 | FED 209 | FED |
| 1-1/4 | DUP 4756 | 32 | 1330 | FED | 12S4 | FED 209 | FED |
| 1-1/4 | DUP 800X | 25.0 | 1235 | FED | 12S3 | FED 209 | DUP |
| 1-1/4 | DUP 800X | 25.0 | 1220 | REM | R12H | FED 209 | DUP |
| 1-1/4 | DUP 800X | 27.5 | 1325 | FED | 12S4 | FED 209 | DUP |
| 1-1/4 | DUP 800X | 28.0 | 1325 | REM | R12H | FED 209 | DUP |
| 1-3/8 | HER BDT | 34.0 | 1240 | REM | RP12 | FED 209 | HER |
| 1-3/8 | HER BDT | 35.0 | 1240 | REM | RP12 | CCI 209M | HER |
| 1-3/8 | HER BDT | 35.5 | 1295 | REM | RP12 | FED 209 | HER |
| 1-3/8 | HER BDT | 36.5 | 1295 | REM | RP12 | CCI 209M | HER |
| 1-3/8 | DUP 800X | 26.5 | 1280 | FED | 12S4 | FED 209 | DUP |
| 1-3/8 | DUP 800X | 28.0 | 1310 | REM | SP12 | FED 209 | DUP |
| 1-3/8 | DUP 800X | 26.5 | 1290 | FED | 12S4 | CCI 209M | DUP |
| 1-3/8 | DUP 800X | 28.0 | 1305 | REM | SP12 | CCI 209M | DUP |
| 1-1/2 | DUP 800X | 27.5 | 1255 | REM | RP12 | FED 209 | DUP |
| 1-1/2 | DUP 800X | 27.5 | 1245 | REM | RP12 | CCI 209M | DUP |

12 GAUGE 2 3/4" ACTIV PLASTIC SHELLS

| Shot Weight Oz. | Powder Mfg. | Grains | Velocity f.p.s. | Wad Mfg/Type | Primer Mfg/No. | Source |
|---|---|---|---|---|---|---|
| 1-1/8 | HER RDT | 19.0 | 1145 | ACTIV L-29 | CCI 209 | HER |
| 1-1/8 | HER RDT | 18.5 | 1145 | REM 12S0 | CCI 209 | HER |
| 1-1/8 | HER RDT | 18.0 | 1145 | ACTIV L-29 | CCI 209M | HER |
| 1-1/8 | HER RDT | 17.5 | 1145 | REM 12S0 | CCI 209M | HER |
| 1-1/8 | HER RDT | 18.0 | 1145 | ACTIV L-29 | FED 209 | HER |
| 1-1/8 | HER RDT | 18.0 | 1145 | REM 12S0 | FED 209 | HER |
| 1-1/8 | HER RDT | 18.0 | 1145 | ACTIV L-29 | WIN 209 | HER |
| 1-1/8 | HER RDT | 18.5 | 1145 | REM 12S0 | WIN 209 | HER |
| 1-1/8 | HER RDT | 18.0 | 1145 | REM 12S3 | WIN 209 | HER |
| 1-1/8 | HER RDT | 17.5 | 1145 | REM PT12 | WIN 209 | HER |
| 1-1/8 | HER RDT | 20.5 | 1200 | ACTIV L-29 | CCI 209 | HER |
| 1-1/8 | HER RDT | 20.0 | 1200 | REM 12S0 | CCI 209 | HER |
| 1-1/8 | HER RDT | 19.5 | 1200 | ACTIV L-29 | CCI 209M | HER |
| 1-1/8 | HER GDT | 21.0 | 1200 | ACTIV L-29 | CCI 209M | HER |
| 1-1/8 | HER RDT | 19.5 | 1200 | REM 12S0 | CCI 209M | HER |
| 1-1/8 | HER GDT | 21.5 | 1200 | REM 12S0 | CCI 209M | HER |
| 1-1/8 | HER RDT | 19.5 | 1200 | ACTIV L-29 | FED 209 | HER |
| 1-1/8 | HER RDT | 19.5 | 1200 | REM12S0 | FED 209 | HER |
| 1-1/8 | HER GDT | 21.5 | 1200 | REM 12S0 | FED 209 | HER |
| 1-1/8 | HER RDT | 20.0 | 1200 | ACTIV L-29 | WIN 209 | HER |
| 1-1/8 | HER GDT | 21.0 | 1200 | ACTIV L-29 | WIN 209 | HER |
| 1-1/8 | HER RDT | 19.5 | 1200 | REM 12S0 | WIN 209 | HER |
| 1-1/8 | HER GDT | 21.0 | 1200 | REM 12S0 | WIN 209 | HER |
| 1-1/8 | HER RDT | 19.5 | 1200 | REM 12S3 | WIN 209 | HER |
| 1-1/8 | HER GDT | 21.5 | 1200 | REM 12S3 | WIN 209 | HER |
| 1-1/8 | HER RDT | 19.5 | 1200 | REM PT12 | WIN 209 | HER |
| 1-1/8 | HER GDT | 22.0 | 1200 | REM PT12 | WIN 209 | HER |
| 1-1/8 | HER GDT | 23.0 | 1255 | ACTIV L-29 | CCI 209M | HER |
| 1-1/8 | HER GDT | 22.0 | 1255 | REM 12S3 | CCI 209M | HER |
| 1-1/8 | HER RDT | 20.5 | 1255 | ACTIV L-29 | WIN 209 | HER |
| 1-1/8 | HER GDT | 23.5 | 1255 | ACTIV L-29 | WIN 209 | HER |
| 1-1/4 | HER GDT | 23.0 | 1220 | ACTIV T-32 | CCI 209 | HER |
| 1-1/4 | HER UNIQ | 25.5 | 1220 | ACTIV T-32 | CCI 209 | HER |
| 1-1/4 | HER GDT | 22.0 | 1220 | ACTIV T-32 | CCI 209M | HER |
| 1-1/4 | HER UNIQ | 24.5 | 1220 | ACTIV T-32 | CCI 209M | HER |
| 1-1/4 | HER GDT | 22.5 | 1220 | ACTIV T-32 | FED 209 | HER |
| 1-1/4 | HER UNIQ | 24.5 | 1220 | ACTIV T-32 | FED 209 | HER |
| 1-1/4 | HER HERC | 30.5 | 1330 | REM 12S4 | CCI 209 | HER |
| 1-1/4 | HER BLDT | 39.5 | 1330 | REM 12S4 | CCI 209 | HER |
| 1-1/4 | HER HERC | 29.0 | 1330 | ACTIV T-32 | CCI 209M | HER |
| 1-1/4 | HER UNIQ | 27.5 | 1330 | ACTIV T-32 | FED 209 | HER |
| 1-1/4 | HER HERC | 29.5 | 1330 | ACTIV T-32 | FED 209 | HER |
| 1-1/4 | HER BLDT | 37.0 | 1330 | REM 12S4 | FED 209 | HER |
| 1-3/8 | HER HERC | 30.5 | 1295 | ACTIV T-35 | CCI 209 | HER |
| 1-3/8 | HER BLDT | 40.0 | 1295 | REM RP12 | CCI 209 | HER |
| 1-3/8 | HER HERC | 29.5 | 1295 | ACTIV T-35 | CCI 209M | HER |
| 1-3/8 | HER BLDT | 38.5 | 1295 | ACTIV T-35 | CCI 209M | HER |
| 1-3/8 | HER BLDT | 38.0 | 1295 | REM RP12 | CCI 209M | HER |
| 1-3/8 | HER BLDT | 38.0 | 1295 | ACTIV T-35 | FED 209 | HER |
| 1-3/8 | HER BLDT | 37.0 | 1295 | REM RP12 | FED 209 | HER |

12 GAUGE 2 3/4" ACTIV PLASTIC SHELLS

| Shot Weight Oz. | Powder Mfg. | Grains | Velocity f.p.s. | Wad Mfg/Type | Primer Mfg/No. | Source |
|---|---|---|---|---|---|---|
| 1-3/8 | HER BLDT | 40.0 | 1350 | ACTIV T-35 | FED 209 | HER |
| 1-1/2 | HER BLDT | 38.5 | 1260 | ACTIV T-42 | CCI 209 | HER |
| 1-1/2 | HER BLDT | 36.5 | 1260 | ACTIV T-42 | CCI 209M | HER |
| 1-1/2 | HER BLDT | 35.5 | 1260 | REM RP12 | CCI 209M | HER |

12 GAUGE 2 3/4" FIOCCHI PLASTIC SHELLS

| Shot Weight Oz. | Powder Mfg. | Grains | Velocity f.p.s. | Wad Mfg/Type | Primer Mfg/No. | Source |
|---|---|---|---|---|---|---|
| 1-1/8 | HER RDT | 18.0 | 1145 | REM 12S3 | FIO 615 | HER |
| 1-1/8 | HER RDT | 18.0 | 1145 | REM 12S3 | FED 209 | HER |
| 1-1/8 | HER GDT | 19.5 | 1145 | REM 12S3 | FED 209 | HER |
| 1-1/8 | HER GDT | 19.5 | 1145 | REM RXP12 | CCI 209M | HER |
| 1-1/8 | HER RDT | 19.5 | 1200 | REM 12S3 | FIO 615 | HER |
| 1-1/8 | HER GDT | 22.5 | 1200 | REM 12S3 | FIO 615 | HER |
| 1-1/8 | HER UNIQ | 25.5 | 1200 | REM 12S3 | FIO 615 | HER |
| 1-1/8 | HER RDT | 20.0 | 1200 | REM 12S3 | FIO 616 | HER |
| 1-1/8 | HER GDT | 22.5 | 1200 | REM 12S3 | FIO 616 | HER |
| 1-1/8 | HER UNIQ | 23.5 | 1200 | REM 12S3 | FIO 616 | HER |
| 1-1/8 | HER GDT | 21.5 | 1200 | REM 12S3 | FED 209 | HER |
| 1-1/8 | HER RDT | 18.5 | 1200 | REM RXP12 | CCI 209M | HER |
| 1-1/8 | HER GDT | 21.0 | 1200 | REM RXP12 | CCI 209M | HER |
| 1-1/8 | HER GDT | 24.0 | 1255 | REM 12S3 | FIO 616 | HER |
| 1-1/8 | HER UNIQ | 25.5 | 1255 | REM 12S3 | FIO 616 | HER |
| 1-1/8 | HER GDT | 23.5 | 1255 | REM 12S3 | FED 209 | HER |
| 1-1/8 | HER UNIQ | 25.0 | 1255 | REM 12S3 | FED 209 | HER |
| 1-1/8 | HER GDT | 23.5 | 1255 | REM RXP12 | CCI 209M | HER |
| 1-1/8 | HER GDT | 25.0 | 1310 | REM 12S3 | FIO 616 | HER |
| 1-1/8 | HER UNIQ | 27.0 | 1310 | REM 12S3 | FIO 616 | HER |
| 1-1/8 | HER GDT | 24.5 | 1310 | REM 12S3 | FED 209 | HER |
| 1-1/8 | HER UNIQ | 27.0 | 1310 | REM 12S3 | FED 209 | HER |
| 1-1/8 | HER GDT | 24.0 | 1310 | REM RXP12 | CCI 209M | HER |
| 1-1/8 | HER UNIQ | 26.5 | 1310 | REM RXP12 | CCI 209M | HER |
| 1-1/4 | HER GDT | 23.0 | 1220 | REM 12S4 | FIO 616 | HER |
| 1-1/4 | HER UNIQ | 25.0 | 1220 | REM 12S4 | FIO 616 | HER |
| 1-1/4 | HER GDT | 23.0 | 1220 | REM 12S4 | FED 209 | HER |
| 1-1/4 | HER UNIQ | 24.5 | 1220 | REM 12S4 | FED 209 | HER |
| 1-1/4 | HER GDT | 24.5 | 1220 | REM R12H | CCI 209M | HER |
| 1-1/4 | HER UNIQ | 27.0 | 1275 | REM 12S4 | FIO 616 | HER |
| 1-1/4 | HER HERC | 28.0 | 1275 | REM 12S4 | FIO 616 | HER |
| 1-1/4 | HER UNIQ | 26.0 | 1275 | REM 12S4 | FED 209 | HER |
| 1-1/4 | HER HERC | 27.5 | 1275 | REM 12S4 | FED 209 | HER |
| 1-1/4 | HER HERC | 28.0 | 1275 | REM SP12 | CCI 209M | HER |
| 1-1/4 | HER HERC | 30.0 | 1300 | REM 12S4 | FIO 616 | HER |
| 1-1/4 | HER BLDT | 40.0 | 1300 | REM 12S4 | FIO 616 | HER |
| 1-1/4 | HER HERC | 30.5 | 1300 | REM SP12 | FIO 616 | HER |
| 1-1/4 | HER BLDT | 41.0 | 1300 | REM SP12 | FIO 616 | HER |
| 1-1/4 | HER HERC | 30.0 | 1300 | REM 12S4 | FED 209 | HER |
| 1-1/4 | HER BLDT | 37.0 | 1300 | REM 12S4 | FED 209 | HER |
| 1-1/4 | HER HERC | 30.0 | 1300 | REM SP12 | CCI 209M | HER |
| 1-1/4 | HER BLDT | 41.0 | 1300 | REM SP12 | CCI 209M | HER |
| 1-3/8 | HER BLDT | 38.0 | 1295 | REM RP12 | FIO 616 | HER |
| 1-3/8 | HER BLDT | 36.0 | 1295 | REM RP12 | FED 209 | HER |
| 1-3/8 | HER BLDT | 37.0 | 1295 | REM RP12 | CCI 209M | HER |
| 1-3/8 | HER BLDT | 41.5 | 1350 | REM RP12 | FIO 616 | HER |
| 1-3/8 | HER BLDT | 39.0 | 1350 | REM RP12 | FED 209 | HER |
| 1-3/8 | HER BLDT | 40.0 | 1350 | REM RP12 | CCI 209M | HER |
| 1-1/2 | HER BLDT | 36.5 | 1205 | REM RP12 | FIO 616 | HER |
| 1-1/2 | HER BLDT | 34.5 | 1205 | REM RP12 | FED 209 | HER |
| 1-1/2 | HER BLDT | 33.0 | 1205 | REM RP12 | CCI 209M | HER |
| 1-1/2 | HER BLDT | 37.5 | 1275 | REM RP12 | FIO 616 | HER |
| 1-1/2 | HER BLDT | 36.5 | 1275 | REM RP12 | CCI 209M | HER |

## 12 GAUGE 2 3/4" PETERS 'BLUE MAGIC' PLASTIC TARGET SHELLS

| Shot Weight Oz. | Powder Mfg. | Grains | Velocity f.p.s. | Wad Mfg/Type | Primer Mfg/No. | Source |
|---|---|---|---|---|---|---|
| 1 | HER RDT | 20.5 | 1290 | FED 12S3 | REM 97* | HER |
| 1 | HER RDT | 19.5 | 1290 | REM R12L | REM 97* | HER |
| 1 | HER RDT | 20.5 | 1290 | REM RXP12 | REM 97* | HER |
| 1 | HER RDT | 19.5 | 1290 | REM R12L | FED 209 | HER |
| 1 | HER RDT | 20.5 | 1290 | REM R12L | CCI 209M | HER |
| 1 | HER GDT | 21.5 | 1290 | FED 12S3 | REM 97* | HER |
| 1 | HER GDT | 21.5 | 1290 | REM R12L | REM 97* | HER |
| 1 | HER GDT | 21.5 | 1290 | REM RXP12 | REM 97* | HER |
| 1 | HER GDT | 21.5 | 1290 | REM R12L | FED 209 | HER |
| 1 | HER GDT | 22.0 | 1290 | REM R12L | CCI 209M | HER |
| 1 | DUP 700X | 17.0 | 1150 | REM RXP12 | REM 97* | DUP |
| 1 | DUP 700X | 16.5 | 1155 | FED 12S0 | REM 97* | DUP |
| 1 | DUP 700X | 18.0 | 1195 | REM. RXP12 | REM 97* | DUP |
| 1 | DUP 700X | 17.5 | 1205 | FED 12S0 | REM 97* | DUP |
| 1 | DUP 700X | 16.5 | 1145 | REM RXP12 | FED 209 | DUP |
| 1 | DUP 700X | 16.5 | 1160 | FED 12S0 | FED 209 | DUP |
| 1 | DUP 700X | 17.5 | 1195 | REM RXP12 | FED 209 | DUP |
| 1 | DUP 700X | 17.5 | 1210 | FED 12S0 | FED 209 | DUP |
| 1 | DUP PB | 19.5 | 1135 | REM RXP12 | FED 209 | DUP |
| 1 | DUP PB | 19.0 | 1140 | FED 12S3 | FED 209 | DUP |
| 1 | DUP PB | 21.0 | 1205 | REM RXP12 | FED 209 | DUP |
| 1 | DUP PB | 20.5 | 1200 | FED 12S3 | FED 209 | DUP |
| 1 | DUP PB | 20.0 | 1155 | REM RXP12 | CCI 209 | DUP |
| 1 | DUP PB | 20.0 | 1155 | FED 12S3 | CCI 209 | DUP |
| 1 | DUP PB | 21.0 | 1200 | REM RXP12 | CCI 209 | DUP |
| 1 | DUP PB | 21.0 | 1195 | FED 12S3 | CCI 209 | DUP |
| 1 | DUP 7625 | 22.0 | 1140 | REM RXP12 | FED 209 | DUP |
| 1 | DUP 7625 | 21.0 | 1150 | FED 12S3 | FED 209 | DUP |
| 1 | DUP 7625 | 23.5 | 1205 | REM RXP12 | FED 209 | DUP |
| 1 | DUP 7625 | 22.5 | 1210 | FED 12S3 | FED 209 | DUP |
| 1 | DUP 7625 | 22.5 | 1135 | REM RXP12 | CCI 209 | DUP |
| 1 | DUP 7625 | 21.0 | 1135 | FED 12S3 | CCI 209 | DUP |
| 1 | DUP 7625 | 24.0 | 1210 | REM RXP12 | CCI 209 | DUP |
| 1 | DUP 7625 | 22.5 | 1190 | FED 12S3 | CCI 209 | DUP |
| 1 | DUP 800X | 22.0 | 1150 | REM RXP12 | REM 97* | DUP |
| 1 | DUP 800X | 21.0 | 1145 | FED 12S4 | REM 97* | DUP |
| 1 | DUP 800X | 21.5 | 1140 | REM RXP12 | CCI 209M* | DUP |
| 1 | DUP 800X | 21.0 | 1145 | FED 12S4 | CCI 209M* | DUP |
| 1 | DUP 800X | 22.5 | 1205 | REM RXP12 | REM 97* | DUP |
| 1 | DUP 800X | 22.5 | 1200 | FED 12S4 | REM 97* | DUP |
| 1 | DUP 800X | 22.5 | 1205 | REM RXP12 | CCI 209M* | DUP |
| 1 | DUP 800X | 22.5 | 1215 | FED 12S4 | CCI 209M* | DUP |
| 1-1/8 | HER RDT | 18.0 | 1145 | FED 12C1 | REM 97* | HER |
| 1-1/8 | HER RDT | 17.5 | 1145 | FED 12S1 | REM 97* | HER |
| 1-1/8 | HER RDT | 17.5 | 1145 | REM R12H | REM 97* | HER |
| 1-1/8 | HER RDT | 17.5 | 1145 | REM RXP12 | REM 97* | HER |
| 1-1/8 | HER RDT | 17.0 | 1145 | REM RXP12 | FED 209 | HER |
| 1-1/8 | HER RDT | 17.5 | 1145 | FED 12C1 | CCI 109 | HER |
| 1-1/8 | HER RDT | 17.5 | 1145 | FED 12S1 | CCI 109 | HER |
| 1-1/8 | HER RDT | 17.5 | 1145 | REM R12H | CCI 109 | HER |
| 1-1/8 | HER RDT | 17.5 | 1145 | REM RXP12 | CCI 109 | HER |
| 1-1/8 | HER RDT | 17.5 | 1145 | REM RXP12 | CCI 209M | HER |

12 GAUGE 2 3/4" PETERS 'BLUE MAGIC' PLASTIC TARGET SHELLS

| Shot Weight Oz. | Powder Mfg. | Grains | Velocity f.p.s. | Wad Mfg/Type | | Primer Mfg/No. | Source |
|---|---|---|---|---|---|---|---|
| 1-1/8 | HER RDT | 18.5 | 1200 | FED | 12C1 | REM 97* | HER |
| 1-1/8 | HER RDT | 18.5 | 1200 | FED | 12S1 | REM 97* | HER |
| 1-1/8 | HER RDT | 18.5 | 1200 | REM | RXP12 | REM 97* | HER |
| 1-1/8 | HER RDT | 18.0 | 1200 | REM | RXP12 | FED 209 | HER |
| 1-1/8 | HER RDT | 18.5 | 1200 | FED | 12C1 | CCI 109 | HER |
| 1-1/8 | HER RDT | 19.0 | 1200 | FED | 12S1 | CCI 109 | HER |
| 1-1/8 | HER RDT | 19.0 | 1200 | REM | R12H | CCI 109 | HER |
| 1-1/8 | HER RDT | 19.0 | 1200 | REM | RXP12 | CCI 109 | HER |
| 1-1/8 | HER RDT | 18.5 | 1200 | REM | RXP12 | CCI 209M | HER |
| 1-1/8 | HER GDT | 18.5 | 1145 | FED | 12C1 | REM 97* | HER |
| 1-1/8 | HER GDT | 18.5 | 1145 | FED | 12S1 | REM 97* | HER |
| 1-1/8 | HER GDT | 18.5 | 1145 | REM | R12H | REM 97* | HER |
| 1-1/8 | HER GDT | 18.5 | 1145 | REM | RXP12 | REM 97* | HER |
| 1-1/8 | HER GDT | 19.0 | 1145 | REM | RXP12 | FED 209 | HER |
| 1-1/8 | HER GDT | 18.5 | 1145 | FED | 12C1 | CCI 109 | HER |
| 1-1/8 | HER GDT | 19.5 | 1145 | FED | 12S1 | CCI 109 | HER |
| 1-1/8 | HER GDT | 19.5 | 1145 | REM | R12H | CCI 109 | HER |
| 1-1/8 | HER GDT | 19.5 | 1145 | REM | RXP12 | CCI 109 | HER |
| 1-1/8 | HER GDT | 19.0 | 1145 | REM | RXP12 | CCI 209M | HER |
| 1-1/8 | HER GDT | 20.5 | 1200 | FED | 12C1 | REM 97* | HER |
| 1-1/8 | HER GDT | 20.5 | 1200 | FED | 12S1 | REM 97* | HER |
| 1-1/8 | HER GDT | 20.5 | 1200 | REM | R12H | REM 97* | HER |
| 1-1/8 | HER GDT | 20.0 | 1200 | REM | RXP12 | REM 97* | HER |
| 1-1/8 | HER GDT | 20.5 | 1200 | REM | RXP12 | FED 209 | HER |
| 1-1/8 | HER GDT | 20.5 | 1200 | FED | 12C1 | CCI 109 | HER |
| 1-1/8 | HER GDT | 21.0 | 1200 | FED | 12S1 | CCI 109 | HER |
| 1-1/8 | HER GDT | 20.5 | 1200 | REM | R12H | CCI 109 | HER |
| 1-1/8 | HER GDT | 20.5 | 1200 | REM | RXP12 | CCI 109 | HER |
| 1-1/8 | HER GDT | 20.0 | 1200 | REM | RXP12 | CCI 209M | HER |
| 1-1/8 | HER GDT | 22.0 | 1255 | FED | 12C1 | REM 97* | HER |
| 1-1/8 | HER GDT | 21.5 | 1255 | FED | 12S1 | REM 97* | HER |
| 1-1/8 | HER GDT | 21.5 | 1255 | REM | R12H | REM 97* | HER |
| 1-1/8 | HER GDT | 22.0 | 1255 | REM | RXP12 | REM 97* | HER |
| 1-1/8 | HER GDT | 21.5 | 1255 | REM | RXP12 | FED 209 | HER |
| 1-1/8 | HER GDT | 23.0 | 1310 | REM | RXP12 | REM 97* | HER |
| 1-1/8 | HER GDT | 23.0 | 1310 | FED | 12S3 | REM 97* | HER |
| 1-1/8 | HER UNIQ | 22.0 | 1200 | FED | 12C1 | REM 97* | HER |
| 1-1/8 | HER UNIQ | 21.5 | 1200 | FED | 12S1 | REM 97* | HER |
| 1-1/8 | HER UNIQ | 22.0 | 1200 | REM | RXP12 | REM 97* | HER |
| 1-1/8 | HER UNIQ | 22.0 | 1200 | REM | RXP12 | FED 209 | HER |
| 1-1/8 | HER UNIQ | 22.5 | 1200 | FED | 12C1 | CCI 109 | HER |
| 1-1/8 | HER UNIQ | 22.0 | 1200 | FED | 12S1 | CCI 109 | HER |
| 1-1/8 | HER UNIQ | 22.5 | 1200 | REM | R12H | CCI 109 | HER |
| 1-1/8 | HER UNIQ | 22.5 | 1200 | REM | RXP12 | CCI 109 | HER |
| 1-1/8 | HER UNIQ | 22.0 | 1200 | REM | RXP12 | CCI 209M | HER |
| 1-1/8 | HER UNIQ | 23.0 | 1255 | FED | 12C1 | REM 97* | HER |
| 1-1/8 | HER UNIQ | 23.5 | 1255 | FED | 12S1 | REM 97* | HER |
| 1-1/8 | HER UNIQ | 23.0 | 1255 | REM | R12H | REM 97* | HER |
| 1-1/8 | HER UNIQ | 23.0 | 1255 | REM | RXP12 | REM 97* | HER |
| 1-1/8 | HER UNIQ | 23.5 | 1255 | REM | RXP12 | FED 209 | HER |
| 1-1/8 | HER UNIQ | 23.5 | 1255 | REM | RXP12 | CCI 209M | HER |

12 GAUGE 2 3/4" PETERS 'BLUE MAGIC' PLASTIC TARGET SHELLS

| Shot Weight Oz. | Powder Mfg. | Grains | Velocity f.p.s. | Wad Mfg/Type | | Primer Mfg/No. | Source |
|---|---|---|---|---|---|---|---|
| 1-1/8 HER | UNIQ | 26.0 | 1310 | REM | RXP12 | REM 97* | HER |
| 1-1/8 HER | UNIQ | 26.5 | 1310 | FED | 12S3 | REM 97* | HER |
| 1-1/8 HER | UNIQ | 25.0 | 1310 | REM | RXP12 | CCI 209M | HER |
| 1-1/8 DUP | 700X | 17.5 | 1135 | REM | RXP12 | REM 97* | DUP |
| 1-1/8 DUP | 700X | 18.5 | 1195 | REM | RXP12 | REM 97* | DUP |
| 1-1/8 DUP | 700X | 17.0 | 1135 | REM | RXP12 | FED 209 | DUP |
| 1-1/8 DUP | 700X | 18.5 | 1200 | REM | RXP12 | FED 209 | DUP |
| 1-1/8 DUP | 700X | 17.5 | 1155 | REM | RXP12 | CCI 209 | DUP |
| 1-1/8 DUP | 700X | 18.5 | 1195 | REM | RXP12 | CCI 209 | DUP |
| 1-1/8 DUP | PB | 20.5 | 1145 | REM | R12H | CCI 209 | DUP |
| 1-1/8 DUP | PB | 20.0 | 1155 | FED | 12S4 | CCI 209 | DUP |
| 1-1/8 DUP | PB | 22.0 | 1195 | REM | R12H | CCI 209 | DUP |
| 1-1/8 DUP | PB | 21.5 | 1210 | FED | 12S4 | CCI 209 | DUP |
| 1-1/8 DUP | PB | 20.0 | 1135 | REM | R12H | FED 209 | DUP |
| 1-1/8 DUP | PB | 19.5 | 1150 | FED | 12S4 | FED 209 | DUP |
| 1-1/8 DUP | PB | 22.0 | 1200 | REM | R12H | FED 209 | DUP |
| 1-1/8 DUP | PB | 21.0 | 1200 | FED | 12S4 | FED 209 | DUP |
| 1-1/8 DUP | 7625 | 22.5 | 1135 | REM | RXP12 | CCI 209 | DUP |
| 1-1/8 DUP | 7625 | 21.5 | 1145 | FED | 12S4 | CCI 209 | DUP |
| 1-1/8 DUP | 7625 | 23.5 | 1210 | REM | RXP12 | CCI 209 | DUP |
| 1-1/8 DUP | 7625 | 23.0 | 1210 | FED | 12S4 | CCI 209 | DUP |
| 1-1/8 DUP | 7625 | 22.5 | 1145 | REM | RXP12 | FED 209 | DUP |
| 1-1/8 DUP | 7625 | 21.0 | 1150 | FED | 12S4 | FED 209 | DUP |
| 1-1/8 DUP | 7625 | 23.5 | 1200 | REM | RXP12 | FED 209 | DUP |
| 1-1/8 DUP | 7625 | 22.5 | 1205 | FED | 12S4 | FED 209 | DUP |
| 1-1/8 DUP | 800X | 21.5 | 1145 | REM | RXP12 | REM 97* | DUP |
| 1-1/8 DUP | 800X | 21.5 | 1140 | FED | 12S4 | REM 97* | DUP |
| 1-1/8 DUP | 800X | 21.5 | 1145 | REM | RXP12 | CCI 209M | DUP |
| 1-1/8 DUP | 800X | 21.0 | 1150 | FED | 12S4 | CCI 209M | DUP |
| 1-1/8 DUP | 800X | 23.0 | 1205 | REM | RXP12 | REM 97* | DUP |
| 1-1/8 DUP | 800X | 23.0 | 1205 | FED | 12S4 | REM 97* | DUP |
| 1-1/8 DUP | 800X | 23.0 | 1215 | REM | RXP12 | CCI 209M | DUP |
| 1-1/8 DUP | 800X | 22.5 | 1200 | FED | 12S4 | CCI 209M | DUP |
| 1-1/4 HER | UNIQ | 22.5 | 1220 | FED | 12S4 | REM 97* | HER |
| 1-1/4 HER | UNIQ | 23.0 | 1220 | REM | SP12 | REM 97* | HER |
| 1-1/4 HER | HERC | 24.0 | 1220 | FED | 12S4 | REM 97* | HER |
| 1-1/4 HER | HERC | 24.0 | 1220 | REM | SP12 | REM 97* | HER |
| 1-1/4 HER | HERC | 24.5 | 1220 | REM | SP12 | CCI 209M | HER |
| 1-1/4 HER | HERC | 26.0 | 1275 | REM | SP12 | REM 97* | HER |
| 1-1/4 HER | BDT | 35.0 | 1275 | REM | SP12 | REM 97* | HER |
| 1-1/4 HER | BDT | 34.5 | 1275 | REM | SP12 | FED 209 | HER |
| 1-1/4 HER | BDT | 34.5 | 1275 | REM | SP12 | CCI 209M | HER |
| 1-1/4 HER | BDT | 36.5 | 1330 | REM | SP12 | REM 97* | HER |
| 1-1/4 HER | BDT | 37.0 | 1330 | REM | RP12 | REM 97* | HER |
| 1-1/4 HER | BDT | 35.5 | 1330 | REM | SP12 | FED 209 | HER |
| 1-1/4 HER | BDT | 35.5 | 1330 | REM | SP12 | CCI 209M | HER |
| 1-1/4 DUP | 800X | 24.5 | 1225 | REM | SP12 | REM 97* | DUP |
| 1-1/4 DUP | 800X | 24.0 | 1230 | FED | 12S4 | REM 97* | DUP |
| 1-1/4 DUP | 800X | 24.0 | 1225 | REM | SP12 | CCI 209M | DUP |
| 1-1/4 DUP | 800X | 24.0 | 1235 | FED | 12S4 | CCI 209M | DUP |

## 12 GAUGE 2 3/4" PETERS 'BLUE MAGIC' PLASTIC TARGET SHELLS

| Shot Weight Oz. | Powder Mfg. | | Grains | Velocity f.p.s. | Wad Mfg/Type | | Primer Mfg/No. | Source |
|---|---|---|---|---|---|---|---|---|
| 1-1/4 | DUP | 800X | 27.0 | 1335 | REM | SP12 | REM 97* | DUP |
| 1-3/8 | HER | BDT | 34.5 | 1240 | REM | SP12 | REM 97* | HER |
| 1-3/8 | HER | BDT | 33.5 | 1240 | FED | 12S4 | REM 97* | HER |
| 1-3/8 | HER | BDT | 35.5 | 1295 | REM | RP12 | REM 97* | HER |
| 1-3/8 | HER | BDT | 35.5 | 1295 | REM | SP12 | REM 97* | HER |
| 1-3/8 | DUP | 800X | 25.0 | 1240 | REM | SP12 | REM 97* | DUP |
| 1-3/8 | DUP | 800X | 24.5 | 1220 | REM | SP12 | CCI 209 | DUP |

NOTE * Insert one (1) 20 gauge .135 inch card inside bottom of shot cup.

# 12 GAUGE 2 3/4" WINCHESTER WESTERN AA

## UPLAND & SUPER-X

| Shot Weight Oz. | Powder Mfg. | | Grains | Velocity f.p.s. | Wad Mfg/Type | | Primer Mfg/No. | | Source |
|---|---|---|---|---|---|---|---|---|---|
| 1 | HER | RDT | 17.0 | 1145 | FED | 12S0 | WIN | 209 | HER |
| 1 | HER | RDT | 17.0 | 1145 | FED | 12S1 | WIN | 209 | HER |
| 1 | HER | RDT | 17.0 | 1145 | REM | R12L | WIN | 209* | HER |
| 1 | HER | RDT | 17.0 | 1145 | REM | RXP12 | WIN | 209* | HER |
| 1 | HER | RDT | 18.0 | 1200 | FED | 12S0 | WIN | 209 | HER |
| 1 | HER | RDT | 18.0 | 1200 | FED | 12S1 | WIN | 209 | HER |
| 1 | HER | RDT | 18.0 | 1200 | REM | RXP12 | WIN | 209 | HER |
| 1 | HER | RDT | 18.0 | 1200 | REM | RXP12 | WIN | 209 | HER |
| 1 | HER | RDT | 18.5 | 1200 | FED | 12S0 | CCI | 209 | HER |
| 1 | HER | RDT | 18.0 | 1200 | FED | 12S1 | CCI | 209 | HER |
| 1 | HER | RDT | 18.0 | 1200 | REM | RXP12 | CCI | 209 | HER |
| 1 | HER | RDT | 20.0 | 1290 | FED | 12C1 | WIN | 209 | HER |
| 1 | HER | RDT | 20.0 | 1290 | FED | 12S3 | WIN | 209 | HER |
| 1 | HER | RDT | 20.0 | 1290 | REM | RXP12 | WIN | 209 | HER |
| 1 | HER | BEYE | 16.5 | 1145 | FED | 12S0 | WIN | 209 | HER |
| 1 | HER | BEYE | 16.0 | 1145 | FED | 12S1 | WIN | 209* | HER |
| 1 | HER | BEYE | 16.5 | 1145 | REM | RXP12 | WIN | 209* | HER |
| 1 | HER | BEYE | 18.0 | 1200 | FED | 12S0 | WIN | 209 | HER |
| 1 | HER | BEYE | 17.5 | 1200 | FED | 12S1 | WIN | 209* | HER |
| 1 | HER | BEYE | 18.0 | 1200 | REM | R12L | WIN | 209* | HER |
| 1 | HER | BEYE | 17.5 | 1200 | REM | RXP12 | WIN | 209 | HER |
| 1 | HER | BEYE | 18.0 | 1200 | FED | 12S0 | CCI | 209 | HER |
| 1 | HER | BEYE | 17.5 | 1200 | FED | 12S1 | CCI | 209 | HER |
| 1 | HER | BEYE | 17.5 | 1200 | REM | R12L | CCI | 209 | HER |
| 1 | HER | BEYE | 17.5 | 1200 | REM | RXP12 | CCI | 209* | HER |
| 1 | HER | GDT | 19.5 | 1200 | FED | 12S0 | WIN | 209 | HER |
| 1 | HER | GDT | 20.0 | 1200 | REM | R12L | WIN | 209 | HER |
| 1 | HER | GDT | 20.0 | 1200 | REM | RXP12 | WIN | 209 | HER |
| 1 | HER | GDT | 20.5 | 1200 | FED | 12S0 | CCI | 209 | HER |
| 1 | HER | GDT | 21.0 | 1290 | FED | 12C1 | WIN | 209 | HER |
| 1 | HER | GDT | 22.5 | 1290 | FED | 12S3 | WIN | 209 | HER |
| 1 | HER | GDT | 21.0 | 1290 | REM | RXP12 | WIN | 209 | HER |
| 1 | DUP | 700X | 16.5 | 1145 | FED | 12S0 | FED | 209 | DUP |
| 1 | DUP | 700X | 17.5 | 1195 | FED | 12S0 | FED | 209 | DUP |
| 1 | DUP | 7625 | 24.0 | 1140 | REM | RXP12 | FED | 209 | DUP |
| 1 | DUP | 7625 | 25.5 | 1215 | REM | RXP12 | FED | 209 | DUP |
| 1 | DUP | 800X | 21.0 | 1130 | REM | RXP12 | FED | 209* | DUP |
| 1 | DUP | 800X | 21.0 | 1145 | FED | 12S4 | FED | 209* | DUP |
| 1 | DUP | 800X | 22.5 | 1195 | REM | RXP12 | FED | 209* | DUP |
| 1 | DUP | 800X | 22.5 | 1215 | FED | 12S4 | FED | 209* | DUP |
| 1-1/8 | HER | RDT | 17.5 | 1145 | FED | 12C1 | WIN | 209 | HER |
| 1-1/8 | HER | RDT | 17.5 | 1145 | FED | 12C2 | WIN | 209 | HER |
| 1-1/8 | HER | RDT | 17.5 | 1145 | FED | 12S1 | WIN | 209 | HER |
| 1-1/8 | HER | RDT | 17.0 | 1145 | REM | R12H | WIN | 209 | HER |
| 1-1/8 | HER | RDT | 17.0 | 1145 | REM | RXP12 | WIN | 209 | HER |
| 1-1/8 | HER | RDT | 17.5 | 1145 | FED | 12C1 | CCI | 109 | HER |
| 1-1/8 | HER | RDT | 17.5 | 1145 | FED | 12C2 | CCI | 109 | HER |
| 1-1/8 | HER | RDT | 17.5 | 1145 | FED | 12S1 | CCI | 109 | HER |
| 1-1/8 | HER | RDT | 17.0 | 1145 | REM | R12H | CCI | 109 | HER |
| 1-1/8 | HER | RDT | 17.0 | 1145 | REM | RXP12 | CCI | 109 | HER |

12 GAUGE 2 3/4" WINCHESTER WESTERN AA

UPLAND & SUPER-X

| Shot Weight Oz. | Powder Mfg. | | Grains | Velocity f.p.s. | Wad Mfg/Type | | Primer Mfg/No. | | Source |
|---|---|---|---|---|---|---|---|---|---|
| 1-1/8 | HER | RDT | 18.5 | 1200 | FED | 12C1 | WIN | 209 | HER |
| 1-1/8 | HER | RDT | 19.0 | 1200 | FED | 12C2 | WIN | 209 | HER |
| 1-1/8 | HER | RDT | 19.0 | 1200 | FED | 12S1 | WIN | 209 | HER |
| 1-1/8 | HER | RDT | 18.5 | 1200 | REM | R12H | WIN | 209 | HER |
| 1-1/8 | HER | RDT | 18.5 | 1200 | REM | RXP12 | WIN | 209 | HER |
| 1-1/8 | HER | RDT | 18.5 | 1200 | FED | 12C1 | CCI | 109 | HER |
| 1-1/8 | HER | RDT | 19.0 | 1200 | FED | 12C2 | CCI | 109 | HER |
| 1-1/8 | HER | RDT | 19.0 | 1200 | FED | 12S1 | CCI | 109 | HER |
| 1-1/8 | HER | RDT | 18.5 | 1200 | REM | R12H | CCI | 109 | HER |
| 1-1/8 | HER | RDT | 18.5 | 1200 | REM | RXP12 | CCI | 109 | HER |
| 1-1/8 | HER | RDT | 20.0 | 1145 | FED | 12S1 | WIN | 209 | HER |
| 1-1/8 | HER | GDT | 18.5 | 1145 | FED | 12C1 | WIN | 209 | HER |
| 1-1/8 | HER | GDT | 19.0 | 1145 | FED | 12C2 | WIN | 209 | HER |
| 1-1/8 | HER | GDT | 18.0 | 1145 | FED | 12S1 | WIN | 209 | HER |
| 1-1/8 | HER | GDT | 18.0 | 1145 | REM | R12H | WIN | 209 | HER |
| 1-1/8 | HER | GDT | 18.0 | 1145 | REM | RXP12 | WIN | 209 | HER |
| 1-1/8 | HER | GDT | 18.5 | 1145 | FED | 12C1 | CCI | 109 | HER |
| 1-1/8 | HER | GDT | 19.0 | 1145 | FED | 12C2 | CCI | 109 | HER |
| 1-1/8 | HER | GDT | 19.0 | 1145 | FED | 12S1 | CCI | 109 | HER |
| 1-1/8 | HER | GDT | 18.0 | 1145 | REM | R12H | CCI | 109 | HER |
| 1-1/8 | HER | GDT | 18.5 | 1145 | REM | RXP12 | CCI | 109 | HER |
| 1-1/8 | HER | GDT | 19.5 | 1200 | FED | 12C1 | WIN | 209 | HER |
| 1-1/8 | HER | GDT | 20.0 | 1200 | FED | 12C2 | WIN | 209 | HER |
| 1-1/8 | HER | GDT | 20.0 | 1200 | FED | 12S1 | WIN | 209 | HER |
| 1-1/8 | HER | GDT | 19.5 | 1200 | REM | R12H | WIN | 209 | HER |
| 1-1/8 | HER | GDT | 19.5 | 1200 | REM | RXP12 | WIN | 209 | HER |
| 1-1/8 | HER | GDT | 20.0 | 1200 | FED | 12C1 | CCI | 109 | HER |
| 1-1/8 | HER | GDT | 20.0 | 1200 | FED | 12C2 | CCI | 109 | HER |
| 1-1/8 | HER | GDT | 20.0 | 1200 | FED | 12S1 | CCI | 109 | HER |
| 1-1/8 | HER | GDT | 20.0 | 1200 | REM | R12H | CCI | 109 | HER |
| 1-1/8 | HER | GDT | 19.5 | 1200 | REM | RXP12 | CCI | 109 | HER |
| 1-1/8 | HER | GDT | 21.0 | 1255 | FED | 12C1 | WIN | 209 | HER |
| 1-1/8 | HER | GDT | 23.0 | 1255 | FED | 12C2 | WIN | 209 | HER |
| 1-1/8 | HER | GDT | 22.0 | 1255 | FED | 12S1 | WIN | 209 | HER |
| 1-1/8 | HER | GDT | 22.0 | 1255 | FED | 12S3 | WIN | 209 | HER |
| 1-1/8 | HER | GDT | 21.0 | 1255 | REM | R12H | WIN | 209 | HER |
| 1-1/8 | HER | GDT | 21.0 | 1255 | REM | RXP12 | WIN | 209 | HER |
| 1-1/8 | HER | UNIQ | 22.0 | 1200 | FED | 12C1 | WIN | 209 | HER |
| 1-1/8 | HER | UNIQ | 22.5 | 1200 | FED | 12C2 | WIN | 209 | HER |
| 1-1/8 | HER | UNIQ | 22.5 | 1200 | FED | 12S1 | WIN | 209 | HER |
| 1-1/8 | HER | UNIQ | 22.0 | 1200 | REM | RXP12 | WIN | 209 | HER |
| 1-1/8 | HER | UNIQ | 22.0 | 1200 | FED | 12C1 | CCI | 109 | HER |
| 1-1/8 | HER | UNIQ | 22.5 | 1200 | FED | 12C2 | CCI | 109 | HER |
| 1-1/8 | HER | UNIQ | 22.5 | 1200 | FED | 12S1 | CCI | 109 | HER |
| 1-1/8 | HER | UNIQ | 22.0 | 1200 | REM | RXP12 | CCI | 109 | HER |
| 1-1/8 | HER | UNIQ | 23.0 | 1255 | FED | 12C1 | WIN | 209 | HER |
| 1-1/8 | HER | UNIQ | 24.0 | 1255 | FED | 12S3 | WIN | 209 | HER |
| 1-1/8 | HER | UNIQ | 22.0 | 1255 | REM | R12H | WIN | 209 | HER |
| 1-1/8 | HER | UNIQ | 22.5 | 1255 | REM | RXP12 | WIN | 209 | HER |

## 12 GAUGE 2 3/4" WINCHESTER WESTERN AA

### UPLAND & SUPER-X

| Shot Weight Oz. | Powder Mfg. | | Grains | Velocity f.p.s. | Wad Mfg/Type | | Primer Mfg/No. | | Source |
|---|---|---|---|---|---|---|---|---|---|
| 1-1/8 | DUP | 700X | 17.0 | 1135 | FED | 12S1 | FED | 209 | DUP |
| 1-1/8 | DUP | 700X | 18.0 | 1145 | REM | RXP12 | FED | 209 | DUP |
| 1-1/8 | DUP | 700X | 18.5 | 1195 | FED | 12S1 | FED | 209 | DUP |
| 1-1/8 | DUP | 700X | 19.0 | 1190 | REM | RXP12 | FED | 209 | DUP |
| 1-1/8 | DUP | 700X | 17.5 | 1155 | FED | 12S1 | CCI | 209 | DUP |
| 1-1/8 | DUP | 700X | 17.5 | 1150 | REM | RXP12 | CCI | 209 | DUP |
| 1-1/8 | DUP | 700X | 19.0 | 1185 | FED | 12S1 | CCI | 209 | DUP |
| 1-1/8 | DUP | 700X | 19.0 | 1195 | REM | RXP12 | CCI | 209 | DUP |
| 1-1/8 | DUP | 700X | 18.0 | 1155 | FED | 12S1 | REM | 97* | DUP |
| 1-1/8 | DUP | 700X | 18.0 | 1150 | REM | RXP12 | REM | 97* | DUP |
| 1-1/8 | DUP | 700X | 19.5 | 1205 | FED | 12S1 | REM | 97* | DUP |
| 1-1/8 | DUP | 700X | 19.5 | 1215 | REM | RXP12 | REM | 97* | DUP |
| 1-1/8 | DUP | PB | 21.0 | 1140 | DUP | 12S4 | FED | 209 | DUP |
| 1-1/8 | DUP | PB | 20.0 | 1135 | REM | RXP12 | FED | 209 | DUP |
| 1-1/8 | DUP | PB | 22.5 | 1200 | FED | 12S4 | FED | 209 | DUP |
| 1-1/8 | DUP | PB | 22.0 | 1205 | REM | RXP12 | FED | 209 | DUP |
| 1-1/8 | DUP | PB | 21.5 | 1155 | FED | 12S4 | CCI | 209 | DUP |
| 1-1/8 | DUP | PB | 21.5 | 1150 | REM | RXP12 | CCI | 209 | DUP |
| 1-1/8 | DUP | PB | 22.5 | 1190 | FED | 12S4 | CCI | 209 | DUP |
| 1-1/8 | DUP | PB | 22.5 | 1195 | REM | RXP12 | CCI | 209 | DUP |
| 1-1/8 | DUP | 7625 | 22.5 | 1160 | REM | RXP12 | FED | 209 | DUP |
| 1-1/8 | DUP | 7625 | 23.5 | 1195 | REM | RXP12 | FED | 209 | DUP |
| 1-1/8 | DUP | 7625 | 23.0 | 1160 | REM | RXP12 | CCI | 209 | DUP |
| 1-1/8 | DUP | 7625 | 24.5 | 1210 | REM | RXP12 | CCI | 209 | DUP |
| 1-1/8 | DUP | 800X | 21.5 | 1130 | REM | RXP12 | FED | 209 | DUP |
| 1-1/8 | DUP | 800X | 21.0 | 1135 | FED | 12S4 | FED | 209 | DUP |
| 1-1/8 | DUP | 800X | 23.0 | 1190 | REM | RXP12 | FED | 209 | DUP |
| 1-1/8 | DUP | 800X | 22.5 | 1190 | FED | 12S4 | FED | 209 | DUP |
| 1-1/8 | DUP | 800X | 25.0 | 1260 | REM | RXP12 | FED | 209 | DUP |
| 1-1/8 | DUP | 800X | 24.0 | 1245 | FED | 12S4 | FED | 209 | DUP |
| 1-1/4 | HER | UNIQ | 23.5 | 1220 | FED | 12S4 | WIN | 209 | HER |
| 1-1/4 | HER | UNIQ | 22.5 | 1220 | REM | RP12 | WIN | 209 | HER |
| 1-1/4 | HER | HERC | 25.0 | 1220 | FED | 12S4 | WIN | 209 | HER |
| 1-1/4 | HER | HERC | 26.0 | 1275 | FED | 12S4 | WIN | 209 | HER |
| 1-1/4 | HER | BLDT | 35.0 | 1275 | REM | SP12 | WIN | 209 | HER |
| 1-1/4 | HER | BLDT | 34.0 | 1275 | FED | 12S4 | WIN | 209 | HER |
| 1-1/4 | HER | BLDT | 38.0 | 1330 | REM | RP12 | WIN | 209 | HER |
| 1-1/4 | HER | BLDT | 37.0 | 1330 | REM | SP12 | WIN | 209 | HER |
| 1-1/4 | DUP | 800X | 24.5 | 1220 | REM | SP12 | FED | 209 | DUP |
| 1-1/4 | DUP | 800X | 24.5 | 1225 | FED | 12S4 | FED | 209 | DUP |
| 1-1/4 | DUP | 800X | 28.0 | 1335 | REM | SP12 | FED | 209 | DUP |
| 1-3/8 | HER | BLDT | 33.0 | 1240 | REM | SP12 | WIN | 209 | HER |
| 1-3/8 | HER | BLDT | 33.0 | 1240 | FED | 12S4 | WIN | 209 | HER |
| 1-3/8 | DUP | 800X | 25.5 | 1205 | REM | SP12 | FED | 209 | DUP |

* NOTE: for each star, add one 20-gauge, 0.135 inch thick card inside bottom of shot cup.

12 GAUGE 2 3/4" FEDERAL HI POWER PLASTIC SHELLS

| Shot Weight Oz. | Powder Mfg. | | Grains | Velocity f.p.s. | Wad Mfg/Type | | Primer Mfg/No. | | Source |
|---|---|---|---|---|---|---|---|---|---|
| 1 | HER | RDT | 21.0 | 1290 | FED | 12S3 | FED | 209 | HER |
| 1 | HER | RDT | 20.5 | 1290 | REM | R12L | FED | 209 | HER |
| 1 | HER | GDT | 23.0 | 1290 | FED | 12S3 | FED | 209 | HER |
| 1 | HER | GDT | 22.5 | 1290 | REM | R12L | FED | 209 | HER |
| 1-1/8 | HER | RDT | 18.5 | 1145 | FED | 12S3 | FED | 209 | HER |
| 1-1/8 | HER | RDT | 18.5 | 1145 | REM | RXP12 | FED | 209 | HER |
| 1-1/8 | HER | RDT | 18.5 | 1145 | FED | 12S3 | REM | 97* | HER |
| 1-1/8 | HER | RDT | 18.5 | 1145 | FED | 12S3 | CCI | 209M | HER |
| 1-1/8 | HER | RDT | 19.5 | 1200 | FED | 12S1 | FED | 209 | HER |
| 1-1/8 | HER | RDT | 19.0 | 1200 | FED | 12S3 | FED | 209 | HER |
| 1-1/8 | HER | RDT | 19.5 | 1200 | REM | RXP12 | FED | 209 | HER |
| 1-1/8 | HER | RDT | 19.5 | 1200 | FED | 12S3 | REM | 97* | HER |
| 1-1/8 | HER | RDT | 20.0 | 1200 | FED | 12S3 | CCI | 209M | HER |
| 1-1/8 | HER | RDT | 21.0 | 1255 | FED | 12C1 | FED | 209 | HER |
| 1-1/8 | HER | RDT | 21.0 | 1255 | FED | 12S1 | FED | 209 | HER |
| 1-1/8 | HER | RDT | 21.5 | 1255 | FED | 12S3 | FED | 209 | HER |
| 1-1/8 | HER | RDT | 21.0 | 1255 | REM | RXP12 | FED | 209 | HER |
| 1-1/8 | HER | RDT | 21.5 | 1255 | FED | 12S3 | REM | 97* | HER |
| 1-1/8 | HER | RDT | 21.5 | 1255 | FED | 12S3 | CCI | 209M | HER |
| 1-1/8 | HER | GDT | 20.0 | 1145 | FED | 12S3 | FED | 209 | HER |
| 1-1/8 | HER | GDT | 19.0 | 1145 | REM | RXP12 | FED | 209 | HER |
| 1-1/8 | HER | GDT | 20.0 | 1145 | FED | 12S3 | REM | 97* | HER |
| 1-1/8 | HER | GDT | 20.0 | 1145 | FED | 12S3 | CCI | 209M | HER |
| 1-1/8 | HER | GDT | 20.5 | 1200 | FED | 12C1 | FED | 209 | HER |
| 1-1/8 | HER | GDT | 20.5 | 1200 | FED | 12S1 | FED | 209 | HER |
| 1-1/8 | HER | GDT | 21.0 | 1200 | FED | 12S3 | FED | 209 | HER |
| 1-1/8 | HER | GDT | 20.5 | 1200 | REM | RXP12 | FED | 209 | HER |
| 1-1/8 | HER | GDT | 21.0 | 1200 | FED | 12S3 | REM | 97* | HER |
| 1-1/8 | HER | GDT | 21.5 | 1200 | FED | 12S3 | CCI | 209M | HER |
| 1-1/8 | HER | GDT | 22.0 | 1255 | FED | 12C1 | FED | 209 | HER |
| 1-1/8 | HER | GDT | 22.0 | 1255 | FED | 12S1 | FED | 209 | HER |
| 1-1/8 | HER | GDT | 22.0 | 1255 | FED | 12S3 | FED | 209 | HER |
| 1-1/8 | HER | GDT | 22.5 | 1255 | REM | RXP12 | FED | 209 | HER |
| 1-1/8 | HER | GDT | 22.0 | 1255 | FED | 12S3 | REM | 97* | HER |
| 1-1/8 | HER | GDT | 22.0 | 1255 | FED | 12S3 | CCI | 209M | HER |
| 1-1/8 | HER | UNIQ | 23.0 | 1200 | FED | 12S3 | FED | 209 | HER |
| 1-1/8 | HER | UNIQ | 22.0 | 1200 | REM | RXP12 | FED | 209 | HER |
| 1-1/8 | HER | UNIQ | 23.5 | 1200 | FED | 12S3 | REM | 97* | HER |
| 1-1/8 | HER | UNIQ | 24.0 | 1200 | FED | 12S3 | CCI | 209M | HER |
| 1-1/8 | HER | UNIQ | 23.0 | 1255 | FED | 12S1 | FED | 209 | HER |
| 1-1/8 | HER | UNIQ | 24.0 | 1255 | FED | 12S3 | FED | 209 | HER |
| 1-1/8 | HER | UNIQ | 23.0 | 1255 | REM | RXP12 | FED | 209 | HER |
| 1-1/8 | HER | UNIQ | 25.5 | 1255 | FED | 12S3 | REM | 97* | HER |
| 1-1/8 | HER | UNIQ | 25.5 | 1255 | FED | 12S3 | CCI | 209M | HER |
| 1-1/8 | DUP | 700X | 18.5 | 1200 | FED | 12S3 | FED | 209 | FED |
| 1-1/8 | DUP | PB | 22.5 | 1200 | FED | 12S3 | FED | 209 | FED |
| 1-1/8 | DUP | 700X | 20.0 | 1255 | FED | 12S3 | FED | 209 | FED |
| 1-1/8 | DUP | 7625 | 26.0 | 1255 | FED | 12S3 | FED | 209 | FED |
| 1-1/8 | DUP | PB | 24.5 | 1255 | FED | 12S3 | FED | 209 | FED |
| 1-1/4 | HER | GDT | 22.0 | 1220 | FED | 12S1 | FED | 209 | HER |
| 1-1/4 | HER | GDT | 23.0 | 1220 | FED | 12S4 | FED | 209 | HER |
| 1-1/4 | HER | GDT | 22.0 | 1220 | REM | R12H | FED | 209 | HER |
| 1-1/4 | HER | GDT | 22.0 | 1220 | REM | RXP12 | FED | 209 | HER |

12 GAUGE 2 3/4" FEDERAL HI POWER PLASTIC SHELLS

| Shot Weight Oz. | | Powder Mfg. | Grains | Velocity f.p.s. | Wad Mfg/Type | | Primer Mfg/No. | Source |
|---|---|---|---|---|---|---|---|---|
| 1-1/4 | HER | UNIQ | 23.0 | 1220 | FED | 12C1 | FED 209 | HER |
| 1-1/4 | HER | UNIQ | 23.0 | 1220 | FED | 12S1 | FED 209 | HER |
| 1-1/4 | HER | UNIQ | 23.0 | 1220 | FED | 12S4 | FED 209 | HER |
| 1-1/4 | HER | UNIQ | 23.0 | 1220 | REM | RXP12 | FED 209 | HER |
| 1-1/4 | HER | UNIQ | 24.5 | 1220 | FED | 12S4 | REM 97* | HER |
| 1-1/4 | HER | UNIQ | 25.0 | 1220 | FED | 12S4 | CCI 209M | HER |
| 1-1/4 | HER | UNIQ | 25.5 | 1330 | FED | 12C1 | FED 209 | HER |
| 1-1/4 | HER | UNIQ | 25.5 | 1330 | REM | SP12 | FED 209 | HER |
| 1-1/4 | HER | HERC | 28.5 | 1330 | FED | 12C1 | FED 209 | HER |
| 1-1/4 | HER | HERC | 29.0 | 1330 | FED | 12S4 | FED 209 | HER |
| 1-1/4 | HER | HERC | 28.5 | 1330 | REM | SP12 | FED 209 | HER |
| 1-1/4 | HER | HERC | 30.0 | 1330 | FED | 12S4 | REM 97* | HER |
| 1-1/4 | HER | HERC | 30.0 | 1330 | FED | 12S4 | CCI 209M | HER |
| 1-1/4 | HER | BLDT | 39.0 | 1330 | FED | 12S4 | REM 97* | HER |
| 1-1/4 | HER | BLDT | 38.0 | 1330 | FED | 12S4 | CCI 209M | HER |
| 1-1/4 | DUP | 7625 | 26.0 | 1220 | FED | 12C1 | FED 209 | FED |
| 1-1/4 | DUP | PB | 24.5 | 1220 | FED | 12C1 | FED 209 | FED |
| 1-1/4 | DUP | 7625 | 26.0 | 1220 | FED | 12S4 | FED 209 | FED |
| 1-1/4 | DUP | 4756 | 29.5 | 1220 | FED | 12S4 | FED 209 | FED |
| 1-1/4 | DUP | PB | 25.0 | 1220 | FED | 12S4 | FED 209 | FED |
| 1-1/4 | DUP | 7625 | 28.0 | 1330 | FED | 12S4 | FED 209 | FED |
| 1-1/4 | DUP | 4756 | 32.0 | 1330 | FED | 12S4 | FED 209 | FED |
| 1-3/8 | HER | BLDT | 38.5 | 1295 | REM | RP12 | FED 209 | HER |
| 1-3/8 | HER | BLDT | 38.0 | 1295 | REM | SP12 | FED 209 | HER |
| 1-3/8 | HER | BLDT | 38.5 | 1295 | REM | RP12 | REM 97* | HER |
| 1-3/8 | HER | BLDT | 39.0 | 1295 | REM | RP12 | CCI 209M | HER |
| 1-3/8 | HER | BLDT | 39.5 | 1350 | REM | RP12 | FED 209 | HER |
| 1-3/8 | HER | BLDT | 40.5 | 1350 | REM | RP12 | REM 97* | HER |
| 1-3/8 | HER | BLDT | 39.5 | 1350 | REM | RP12 | CCI 209M | HER |
| 1-1/2 | HER | BLDT | 36.0 | 1275 | REM | RP12 | FED 209 | HER |
| 1-1/2 | HER | BLDT | 37.0 | 1275 | REM | SP12 | FED 209 | HER |
| 1-1/2 | HER | BLDT | 36.5 | 1275 | REM | RP12 | REM 97* | HER |
| 1-1/2 | HER | BLDT | 37.0 | 1275 | REM | RP12 | CCI 209M | HER |

* NOTE: Insert one (1) 20-gauge, 0.135 inch thick card wad inside bottom of shot cup.

12 GAUGE 2 3/4" REMINGTON-PETERS PLASTIC

RXP TARGET SHELLS

| Shot Weight Oz. | Powder Mfg. | | Grains | Velocity f.p.s. | Wad Mfg/Type | | Primer Mfg/No. | Source |
|---|---|---|---|---|---|---|---|---|
| 1 | HER | RDT | 20.0 | 1290 | FED | 12S3 | REM 97* | HER |
| 1 | HER | RDT | 19.5 | 1290 | REM | R12L | REM 97* | HER |
| 1 | HER | RDT | 19.5 | 1290 | REM | RXP12 | REM 97* | HER |
| 1 | HER | GDT | 22.0 | 1290 | FED | 12S3 | REM 97* | HER |
| 1 | HER | GDT | 22.5 | 1290 | REM | R12L | REM 97* | HER |
| 1 | HER | GDT | 22.0 | 1290 | REM | RXP12 | REM 97* | HER |
| 1-1/8 | HER | RDT | 17.0 | 1145 | FED | 12C2 | REM 97* | HER |
| 1-1/8 | HER | RDT | 17.0 | 1145 | FED | 12S1 | REM 97* | HER |
| 1-1/8 | HER | RDT | 17.0 | 1145 | REM | R12H | REM 97* | HER |
| 1-1/8 | HER | RDT | 17.0 | 1145 | REM | RXP12 | REM 97* | HER |
| 1-1/8 | HER | RDT | 17.0 | 1145 | FED | 12C2 | CCI 109 | HER |
| 1-1/8 | HER | RDT | 17.0 | 1145 | FED | 12C2 | CCI 109 | HER |
| 1-1/8 | HER | RDT | 17.0 | 1145 | REM | R12H | CCI 109 | HER |
| 1-1/8 | HER | RDT | 17.0 | 1145 | REM | RXP12 | CCI 109 | HER |
| 1-1/8 | HER | RDT | 18.0 | 1200 | FED | 12C2 | REM 97* | HER |
| 1-1/8 | HER | RDT | 18.0 | 1200 | FED | 12S1 | REM 97* | HER |
| 1-1/8 | HER | RDT | 18.0 | 1200 | REM | R12H | REM 97* | HER |
| 1-1/8 | HER | RDT | 18.0 | 1200 | REM | RXP12 | REM 97* | HER |
| 1-1/8 | HER | RDT | 18.0 | 1200 | FED | 12C2 | CCI 109 | HER |
| 1-1/8 | HER | RDT | 18.0 | 1200 | FED | 12S1 | CCI 109 | HER |
| 1-1/8 | HER | RDT | 18.0 | 1200 | REM | R12H | CCI 109 | HER |
| 1-1/8 | HER | RDT | 18.0 | 1200 | REM | RXP12 | CCI 109 | HER |
| 1-1/8 | HER | RDT | 19.5 | 1255 | REM | RXP12 | REM 97* | HER |
| 1-1/8 | HER | GDT | 18.0 | 1145 | FED | 12C2 | REM 97* | HER |
| 1-1/8 | HER | GDT | 18.0 | 1145 | FED | 12S1 | REM 97* | HER |
| 1-1/8 | HER | GDT | 18.0 | 1145 | REM | R12H | REM 97* | HER |
| 1-1/8 | HER | GDT | 18.0 | 1145 | REM | RXP12 | REM 97* | HER |
| 1-1/8 | HER | GDT | 18.0 | 1145 | FED | 12C2 | CCI 109 | HER |
| 1-1/8 | HER | GDT | 18.0 | 1145 | FED | 12S1 | CCI 109 | HER |
| 1-1/8 | HER | GDT | 18.0 | 1145 | REM | R12H | CCI 109 | HER |
| 1-1/8 | HER | GDT | 18.0 | 1145 | REM | RXP12 | CCI 109 | HER |
| 1-1/8 | HER | GDT | 19.0 | 1200 | FED | 12C2 | REM 97* | HER |
| 1-1/8 | HER | GDT | 19.0 | 1200 | FED | 12S1 | REM 97* | HER |
| 1-1/8 | HER | GDT | 19.0 | 1200 | REM | R12H | REM 97* | HER |
| 1-1/8 | HER | GDT | 19.0 | 1200 | REM | RXP12 | REM 97* | HER |
| 1-1/8 | HER | GDT | 19.0 | 1200 | FED | 12C2 | CCI 109 | HER |
| 1-1/8 | HER | GDT | 19.0 | 1200 | FED | 12S1 | CCI 109 | HER |
| 1-1/8 | HER | GDT | 19.0 | 1200 | REM | R12H | CCI 109 | HER |
| 1-1/8 | HER | GDT | 19.0 | 1200 | REM | RXP12 | CCI 109 | HER |
| 1-1/8 | HER | GDT | 21.0 | 1255 | FED | 12C2 | REM 97* | HER |
| 1-1/8 | HER | GDT | 21.0 | 1255 | REM | RXP12 | REM 97* | HER |
| 1-1/8 | HER | UNIQ | 21.0 | 1200 | FED | 12C2 | REM 97* | HER |
| 1-1/8 | HER | UNIQ | 21.0 | 1200 | FED | 12S1 | REM 97* | HER |
| 1-1/8 | HER | UNIQ | 21.0 | 1200 | REM | R12H | REM 97* | HER |
| 1-1/8 | HER | UNIQ | 21.0 | 1200 | REM | RXP12 | REM 97* | HER |
| 1-1/8 | HER | UNIQ | 21.0 | 1200 | FED | 12C2 | CCI 109 | HER |
| 1-1/8 | HER | UNIQ | 21.0 | 1200 | FED | 12S1 | CCI 109 | HER |
| 1-1/8 | HER | UNIQ | 21.0 | 1200 | REM | R12H | CCI 109 | HER |
| 1-1/8 | HER | UNIQ | 21.0 | 1200 | REM | RXP12 | CCI 109 | HER |

## 12 GAUGE 2 3/4" REMINGTON-PETERS PLASTIC

### RXP TARGET SHELLS

| Shot Weight Oz. | Powder Mfg. | Grains | Velocity f.p.s. | Wad Mfg/Type | | Primer Mfg/No. | Source |
|---|---|---|---|---|---|---|---|
| 1-1/8 HER | UNIQ | 22.0 | 1255 | FED | 12C2 | REM 97* | HER |
| 1-1/8 HER | UNIQ | 23.0 | 1255 | FED | 12S1 | REM 97* | HER |
| 1-1/8 HER | UNIQ | 23.0 | 1255 | REM | RXP12 | REM 97* | HER |
| 1-1/4 HER | UNIQ | 22.0 | 1220 | REM | RP12 | REM 97* | HER |
| 1-1/4 HER | UNIQ | 22.5 | 1220 | REM | SP12 | REM 97* | HER |
| 1-1/4 HER | HERC | 25.0 | 1220 | REM | RP12 | REM 97* | HER |
| 1-1/4 HER | HERC | 25.0 | 1220 | REM | SP12 | REM 97* | HER |
| 1-1/4 HER | HERC | 27.0 | 1330 | REM | RP12 | REM 97* | HER |
| 1-1/4 HER | HERC | 27.0 | 1330 | REM | SP12 | REM 97* | HER |
| 1-3/8 HER | BLDT | 35.0 | 1295 | REM | RP12 | REM 97* | HER |
| 1-3/8 HER | BLDT | 35.0 | 1295 | REM | SP12 | REM 97* | HER |

* NOTE: Insert one (1) 20-gauge, 0.135 inch thick card wad inside bottom of shot cup.

12 GAUGE 3 INCH WINCHESTER-WESTERN
PLASTIC AA-TYPE SHELLS

| Shot Weight Oz. | Powder Mfg. | Grains | Velocity f.p.s. | Wad Mfg/Type | | Primer Mfg/No. | Source |
|---|---|---|---|---|---|---|---|
| 1-3/8 | HER BLDT | 37.5 | 1295 | FED | 12S3 | WIN 209 | HER |
| 1-3/8 | HER BLDT | 38.0 | 1295 | REM | RXP12 | WIN 209 | HER |
| 1-3/8 | HER BLDT | 40.0 | 1350 | FED | 12S4 | WIN 209 | HER |
| 1-3/8 | HER BLDT | 40.5 | 1350 | REM | RXP12 | WIN 209 | HER |
| 1-1/2 | HER BLDT | 38.5 | 1315 | REM | SP12 | WIN 209 | HER |
| 1-5/8 | HER 2400 | 50.0 | 1335 | REM | RP12 | WIN 209 | HER |
| 1-3/4 | HER 2400 | 45.0 | 1245 | REM | RP12 | WIN 209 | HER |

12 GAUGE 3 inch WINCHESTER-WESTERN

COMPRESSION-FORMED, SUPER-X

| Shot Weight Oz. | Powder Mfg. | Grains | Velocity f.p.s. | Wad Mfg/Type | | Primer Mfg/No. | Source |
|---|---|---|---|---|---|---|---|
| 1-1/4 | DUP PB | 28.0 | 1310 | REM | RXP12 | WIN 209 | DUP |
| 1-1/4 | DUP PB | 32.0 | 1375 | REM | RXP12 | FED 209 | DUP |
| 1-3/8 | DUP 7625 | 29.0 | 1260 | FED | 12S4 | FED 209 | DUP |
| 1-3/8 | DUP 7625 | 31.0 | 1310 | REM | RXP12 | FED 209 | DUP |
| 1-3/8 | DUP 4756 | 36.0 | 1365 | REM | SP12 | FED 209 | DUP |
| 1-1/2 | DUP 7625 | 29.0 | 1210 | REM | SP12 | WIN 209 | DUP |
| 1-1/2 | DUP 4756 | 34.0 | 1290 | REM | SP12 | FED 209 | DUP |
| 1-5/8 | DUP 4756 | 32.0 | 1200 | REM | RP12 | WIN 209 | DUP |
| 1-3/4 | DUP 4756 | 31.0 | 1135 | REM | RP12 | CCI 209 | DUP |

## 12 GAUGE 3 inch FEDERAL HI-POWER PLASTIC

| Shot Weight Oz. | | Powder Mfg. | Grains | Velocity f.p.s. | Wad Mfg/Type | | Primer Mfg/No. | Source |
|---|---|---|---|---|---|---|---|---|
| 1-1/4 | DUP | 800X | 33.0 | 1420 | FED | 12S3 | FED 209* | DUP |
| 1-1/4 | DUP | 700X | 24.0 | 1275 | FED | 12S3 | FED 209 | DUP |
| 1-1/4 | DUP | PB | 31.0 | 1365 | FED | 12S4 | FED 209 | DUP |
| 1-1/4 | DUP | 7625 | 37.0 | 1470 | FED | 12S4 | WIN 209 | DUP |
| 1-3/8 | HER | HERC | 30.5 | 1295 | FED | 12S3 | FED 209 | HER |
| 1-3/8 | HER | HERC | 30.5 | 1295 | REM | RXP12 | FED 209 | HER |
| 1-3/8 | HER | BLDT | 38.0 | 1295 | REM | RXP12 | FED 209 | HER |
| 1-3/8 | HER | BLDT | 40.0 | 1350 | FED | 12S4 | FED 209 | HER |
| 1-3/8 | HER | BLDT | 40.0 | 1350 | REM | SP12 | FED 209 | HER |
| 1-3/8 | DUP | PB | 29.0 | 1265 | FED | 12S4 | WIN 209 | DUP |
| 1-3/8 | DUP | 7625 | 35.0 | 1375 | FED | 12S4 | FED 209 | DUP |
| 1-3/8 | DUP | 7625 | 35.0 | 1370 | REM | R12H | FED 209 | DUP |
| 1-3/8 | DUP | 4756 | 42.0 | 1460 | REM | SP12 | FED 209 | DUP |
| 1-3/8 | DUP | 800X | 26.5 | 1200 | FED | 12S3 | FED 209* | DUP |
| 1-3/8 | DUP | 800X | 29.5 | 1295 | FED | 12S3 | FED 209 | DUP |
| 1-3/8 | DUP | 800X | 32.0 | 1355 | FED | 12S3 | FED 209 | DUP |
| 1-1/2 | DUP | 800X | 27.5 | 1215 | FED | 12S3 | FED 209 | DUP |
| 1-1/2 | DUP | 800X | 30.0 | 1285 | FED | 12S3 | FED 209 | DUP |
| 1-1/2 | DUP | 7625 | 31.0 | 1235 | FED | 12S4 | FED 209 | DUP |
| 1-1/2 | DUP | 4756 | 38.0 | 1350 | REM | SP12 | FED 209 | DUP |
| 1-5/8 | HER | BLDT | 39.0 | 1280 | REM | SP12 | FED 209 | HER |
| 1-5/8 | HER | 2400 | 56.0 | 1245 | REM | RP12 | FED 209 | HER |
| 1-5/8 | DUP | 7625 | 32.0 | 1215 | REM | RP12 | FED 209 | DUP |
| 1-5/8 | DUP | 4756 | 36.0 | 1280 | REM | RP12 | FED 209 | DUP |
| 1-5/8 | DUP | 800X | 28.0 | 1200 | FED | 12S4 | FED 209 | DUP |
| 1-3/4 | DUP | 800X | 25.5 | 1105 | FED | 12S4 | FED 209 | DUP |
| 1-3/4 | DUP | 4756 | 33.0 | 1180 | REM | RP12 | FED 209 | DUP |
| 1-3/4 | HER | BLDT | 39.0 | 1245 | REM | RP12 | FED 209 | HER |

* NOTE: Insert one (1) 20-gauge, 0.135 inch thick card wad inside bottom of shot cup.

## 20 GAUGE, 2 3/4 INCH FEDERAL

## PAPER TARGET SHELLS

| Shot Weight Oz. | Powder Mfg. | | Grains | Velocity f.p.s. | Wad Mfg/Type | | Primer Mfg/No. | Source |
|---|---|---|---|---|---|---|---|---|
| 3/4 | HER | RDT | 12.5 | 1200 | FED | 20S1 | FED 209* | HER |
| 3/4 | HER | RDT | 13.5 | 1200 | REM | RXP20 | FED 209* | HER |
| 3/4 | HER | RDT | 15.0 | 1200 | FED | 20S1 | CCI 109 | HER |
| 3/4 | HER | GDT | 14.0 | 1200 | FED | 20S1 | FED 209 | HER |
| 3/4 | HER | GDT | 14.0 | 1200 | REM | RXP20 | FED 209 | HER |
| 3/4 | HER | GDT | 17.0 | 1200 | REM | RXP20 | CCI 109 | HER |
| 3/4 | HER | UNIQ | 15.5 | 1200 | FED | 20S1 | FED 209 | HER |
| 3/4 | HER | UNIQ | 16.0 | 1200 | REM | RXP20 | FED 209 | HER |
| 7/8 | HER | RDT | 14.0 | 1155 | FED | 20S1 | FED 209 | HER |
| 7/8 | HER. | RDT | 14.0 | 1155 | REM | RXP20 | CCI 109 | HER |
| 7/8 | HER | GDT | 15.0 | 1155 | FED | 20S1 | FED 209 | HER |
| 7/8 | HER | GDT | 14.5 | 1155 | REM | RXP20 | FED 209 | HER |
| 7/8 | HER | GDT | 14.5 | 1155 | FED | 20S1 | CCI 109 | HER |
| 7/8 | HER | GDT | 15.0 | 1155 | REM | RXP20 | CCI 109 | HER |
| 7/8 | HER | GDT | 14.5 | 1155 | FED | 20S1 | CCI 209M | HER |
| 7/8 | HER | GDT | 15.0 | 1200 | FED | 20S1 | FED 209 | HER |
| 7/8 | HER | GDT | 15.5 | 1200 | REM | RXP20 | FED 209 | HER |
| 7/8 | HER | GDT | 15.0 | 1200 | FED | 20S1 | CCI 109 | HER |
| 7/8 | HER | GDT | 16.0 | 1200 | REM | RXP20 | CCI 109 | HER |
| 7/8 | HER | GDT | 15.0 | 1200 | FED | 20S1 | CCI 209M | HER |
| 7/8 | HER | UNIQ | 15.5 | 1155 | FED | 20S1 | FED 209 | HER |
| 7/8 | HER | UNIQ | 15.5 | 1155 | REM | RXP20 | FED 209 | HER |
| 7/8 | HER | UNIQ | 15.0 | 1155 | REM | RXP20 | CCI 109 | HER |
| 7/8 | HER | UNIQ | 16.0 | 1155 | FED | 20S1 | CCI 209M | HER |
| 7/8 | HER | UNIQ | 15.5 | 1200 | FED | 20S1 | FED 209 | HER |
| 7/8 | HER | UNIQ | 16.5 | 1200 | REM | RXP20 | FED 209 | HER |
| 7/8 | HER | UNIQ | 17.0 | 1200 | FED | 20S1 | CCI 109 | HER |
| 7/8 | HER | UNIQ | 17.0 | 1200 | REM | RXP20 | CCI 109 | HER |
| 7/8 | HER | UNIQ | 17.0 | 1200 | FED | 20S1 | CCI 209M | HER |
| 7/8 | HER | HERC | 17.0 | 1200 | FED | 20S1 | CCI 209M | HER |
| 7/8 | DUP | 800X | 17.0 | 1195 | FED | 20S1 | FED 209 | DUP |
| 7/8 | DUP | 800X | 17.0 | 1190 | REM | RXP20 | FED 209 | DUP |
| 7/8 | DUP | 800X | 17.0 | 1190 | FED | 20S1 | CCI 209M | DUP |
| 7/8 | DUP | 800X | 17.5 | 1200 | REM | RXP20 | CCI 209M | DUP |
| 1 | HER | HERC | 17.0 | 1165 | REM | RP20 | FED 209 | HER |
| 1 | HER | HERC | 17.0 | 1165 | REM | RXP20 | FED 209 | HER |
| 1 | HER | HERC | 16.5 | 1165 | REM | SP20 | FED 209 | HER |
| 1 | DUP | 800X | 17.5 | 1175 | FED | 20S1 | FED 209 | DUP |
| 1 | DUP | 800X | 17.0 | 1160 | REM | SP20 | FED 209 | DUP |
| 1 | DUP | 800X | 17.5 | 1175 | FED | 20S1 | CCI 209M | DUP |
| 1 | DUP | 800X | 17.0 | 1160 | REM | SP20 | CCI 209M | DUP |
| 1 | DUP | 800X | 19.0 | 1215 | FED | 20S1 | REM 97-4 | DUP |
| 1 | DUP | 800X | 18.0 | 1205 | REM | SP20 | FED 209 | DUP |
| 1 | DUP | PB | 17.0 | 1145 | REM | SP20 | WIN 209 | DUP |
| 1 | DUP | 7625 | 17.5 | 1155 | REM | SP20 | FED 209 | DUP |
| 1 | DUP | 7625 | 19.0 | 1200 | REM | SP20 | WIN 209 | DUP |
| 1 | DUP | 4756 | 20.5 | 1175 | REM | SP20 | FED 209 | DUP |
| 1 | DUP | 4756 | 21.5 | 1215 | REM | SP20 | FED 209 | DUP |

* NOTE: Insert two (2) 28-gauge, 0.135 inch thick card wads inside bottom of shot cup.

## 20 GAUGE, 2 3/4 INCH FEDERAL PLASTIC TARGET SHELLS

| Shot Weight Oz. | Powder Mfg. | | Grains | Velocity f.p.s. | Wad Mfg/Type | | Primer Mfg/No. | Source |
|---|---|---|---|---|---|---|---|---|
| 3/4 | HER | RDT | 13.0 | 1200 | FED | 20S1 | FED 209*** | HER |
| 3/4 | HER | RDT | 13.0 | 1200 | REM | RXP20 | FED 209*** | HER |
| 3/4 | HER | GDT | 15.0 | 1200 | FED | 20S1 | FED 209* | HER |
| 3/4 | HER | GDT | 15.0 | 1200 | REM | RXP20 | FED 209* | HER |
| 3/4 | HER | UNIQ | 15.0 | 1200 | FED | 20S1 | FED 209** | HER |
| 3/4 | HER | UNIQ | 16.0 | 1200 | REM | RXP20 | FED 209** | HER |
| 7/8 | HER | RDT | 13.5 | 1155 | FED | 20S1 | CCI 109 | HER |
| 7/8 | HER | RDT | 15.0 | 1155 | REM | RXP20 | CCI 109 | HER |
| 7/8 | HER | GDT | 14.5 | 1155 | FED | 20S1 | FED 209 | HER |
| 7/8 | HER | GDT | 15.0 | 1155 | REM | RXP20 | FED 209 | HER |
| 7/8 | HER | GDT | 14.5 | 1155 | FED | 20S1 | CCI 109 | HER |
| 7/8 | HER | GDT | 14.5 | 1155 | FED | 20S1 | CCI 209M | HER |
| 7/8 | HER | GDT | 15.5 | 1200 | FED | 20S1 | FED 209 | HER |
| 7/8 | HER | GDT | 16.0 | 1200 | REM | RXP20 | FED 209 | HER |
| 7/8 | HER | GDT | 15.5 | 1200 | FED | 20S1 | CCI 109 | HER |
| 7/8 | HER | GDT | 16.0 | 1200 | REM | RXP20 | CCI 109 | HER |
| 7/8 | HER | GDT | 16.5 | 1200 | FED | 20S1 | CCI 209M | HER |
| 7/8 | HER | UNIQ | 15.0 | 1155 | FED | 20S1 | FED 209 | HER |
| 7/8 | HER | UNIQ | 15.5 | 1155 | REM | RXP20 | FED 209 | HER |
| 7/8 | HER | UNIQ | 16.0 | 1155 | REM | RXP20 | CCI 109 | HER |
| 7/8 | HER | UNIQ | 16.0 | 1155 | FED | 20S1 | CCI 209M | HER |
| 7/8 | HER | UNIQ | 16.0 | 1200 | FED | 20S1 | FED 209 | HER |
| 7/8 | HER | UNIQ | 16.0 | 1200 | FED | 20S1 | FED 209 | HER |
| 7/8 | HER | UNIQ | 16.0 | 1200 | REM | RXP20 | FED 209 | HER |
| 7/8 | HER | UNIQ | 17.0 | 1200 | FED | 20S1 | CCI 109 | HER |
| 7/8 | HER | UNIQ | 17.0 | 1200 | REM | RXP20 | CCI 109 | HER |
| 7/8 | HER | UNIQ | 17.0 | 1200 | FED | 20S1 | CCI 209M | HER |
| 7/8 | HER | HERC | 17.0 | 1200 | FED | 20S1 | FED 209 | HER |
| 7/8 | HER | HERC | 17.0 | 1200 | REM | RXP20 | FED 209 | HER |
| 7/8 | HER | HERC | 17.0 | 1200 | FED | 20S1 | CCI 109 | HER |
| 7/8 | HER | HERC | 18.0 | 1200 | REM | RXP20 | CCI 109 | HER |
| 7/8 | HER | HERC | 17.5 | 1200 | FED | 20S1 | CCI 209M | HER |
| 7/8 | DUP | 800X | 18.0 | 1210 | FED | 20S1 | FED 209 | DUP |
| 7/8 | DUP | 800X | 18.0 | 1205 | REM | RXP20 | FED 209 | DUP |
| 7/8 | DUP | 800X | 18.5 | 1215 | FED | 20S1 | CCI 209M | DUP |
| 7/8 | DUP | 800X | 18.5 | 1210 | REM | RXP20 | CCI 209M | DUP |
| 1 | HER | UNIQ | 16.0 | 1165 | REM | SP20 | FED 209 | HER |
| 1 | HER | HERC | 17.0 | 1165 | REM | SP20 | FED 209 | HER |
| 1 | HER | HERC | 17.0 | 1165 | REM | RXP20 | FED 209 | HER |
| 1 | HER | HERC | 17.0 | 1220 | FED | 20S1 | CCI 209M | HER |
| 1 | HER | BLDT | 22.0 | 1220 | REM | SP20 | FED 209 | HER |
| 1 | DUP | 800X | 17.5 | 1160 | FED | 20S1 | FED 209 | DUP |
| 1 | DUP | 800X | 17.5 | 1160 | REM | SP20 | FED 209 | DUP |
| 1 | DUP | 800X | 18.0 | 1175 | FED | 20S1 | CCI 209M | DUP |
| 1 | DUP | 800X | 18.0 | 1165 | REM | SP20 | CCI 209M | DUP |
| 1 | DUP | 800X | 18.5 | 1215 | FED | 20S1 | FED 209 | DUP |
| 1 | DUP | 800X | 19.0 | 1220 | REM | SP20 | FED 209 | DUP |
| 1 | DUP | PB | 17.5 | 1155 | REM | SP20 | CCI 209 | DUP |
| 1 | DUP | 7625 | 18.0 | 1160 | REM | SP20 | WIN 209 | DUP |
| 1 | DUP | 4756 | 20.5 | 1180 | REM | SP20 | FED 209 | DUP |
| 1 | DUP | 4756 | 21.5 | 1210 | REM | SP20 | FED 209 | DUP |

*** NOTE: For each star (*) add one (1) 28-gauge, 0.135 inch thick card wad inside bottom of shot cup.

20 GAUGE 2 3/4", FEDERAL HI-POWER

PLASTIC SHELLS

| Shot Weight Oz. | Powder Mfg. | | Grains | Velocity f.p.s. | Wad Mfg/Type | | Primer Mfg/No. | | Source |
|---|---|---|---|---|---|---|---|---|---|
| 7/8 | HER | GDT | 15.5 | 1200 | FED | 20S1 | FED | 209 | HER |
| 7/8 | HER | GDT | 15.5 | 1200 | REM | RXP20 | FED | 209 | HER |
| 7/8 | HER | UNIQ | 16.0 | 1200 | FED | 20S1 | FED | 209 | HER |
| 7/8 | HER | UNIQ | 16.0 | 1200 | REM | RXP20 | FED | 209 | HER |
| 7/8 | HER | HERC | 17.0 | 1200 | FED | 20S1 | FED | 209 | HER |
| 7/8 | HER | HERC | 18.5 | 1200 | REM | RXP20 | FED | 209 | HER |
| 1 | HER | UNIQ | 16.5 | 1165 | FED | 20S1 | FED | 209 | HER |
| 1 | HER | UNIQ | 18.0 | 1165 | REM | RXP20 | FED | 209 | HER |
| 1 | HER | UNIQ | 17.5 | 1220 | FED | 20S1 | FED | 209 | HER |
| 1 | HER | HERC | 17.5 | 1165 | FED | 20S1 | FED | 209 | HER |
| 1 | HER | HERC | 18.5 | 1165 | REM | RXP20 | FED | 209 | HER |
| 1 | HER | HERC | 18.0 | 1220 | FED | 20S1 | FED | 209 | HER |
| 1 | HER | BLDT | 22.5 | 1220 | REM | RXP20 | FED | 209 | HER |
| 1 | DUP | PB | 17.5 | 1160 | REM | SP20 | FED | 410 | DUP |
| 1 | DUP | 7625 | 17.5 | 1160 | REM | SP20 | FED | 209 | DUP |
| 1 | DUP | 7625 | 19.0 | 1205 | REM | SP20 | WIN | 209 | DUP |
| 1 | DUP | 4756 | 20.5 | 1170 | REM | SP20 | FED | 209 | DUP |
| 1 | DUP | 4756 | 21.5 | 1220 | REM | SP20 | FED | 209 | DUP |

## 20 GAUGE 2 3/4", WINCHESTER-WESTERN AA PLASTIC SHELLS

| Shot Weight Oz. | | Powder Mfg. | Grains | Velocity f.p.s. | Wad Mfg/Type | | Primer Mfg/No. | Source |
|---|---|---|---|---|---|---|---|---|
| 3/4 | HER | RDT | 12.0 | 1200 | FED | 20S1 | WIN 209** | HER |
| 3/4 | HER | RDT | 12.5 | 1200 | REM | RXP20 | WIN 209** | HER |
| 3/4 | HER | GDT | 14.0 | 1200 | FED | 20S1 | WIN 209* | HER |
| 3/4 | HER | GDT | 14.0 | 1200 | REM | RXP20 | WIN 209* | HER |
| 3/4 | HER | UNIQ | 15.5 | 1200 | FED | 20S1 | WIN 209 | HER |
| 3/4 | HER | UNIQ | 15.5 | 1200 | REM | RXP20 | WIN 209* | HER |
| 7/8 | HER | GDT | 14.0 | 1155 | FED | 20S1 | WIN 209 | HER |
| 7/8 | HER | GDT | 14.5 | 1155 | REM | RXP20 | WIN 209 | HER |
| 7/8 | HER | GDT | 14.0 | 1155 | FED | 20S1 | CCI 109 | HER |
| 7/8 | HER | GDT | 14.5 | 1155 | REM | RXP20 | CCI 109 | HER |
| 7/8 | HER | GDT | 14.5 | 1200 | FED | 20S1 | WIN 209 | HER |
| 7/8 | HER | GDT | 15.0 | 1200 | REM | RXP20 | WIN 209 | HER |
| 7/8 | HER | GDT | 14.5 | 1200 | FED | 20S1 | CCI 109 | HER |
| 7/8 | HER | GDT | 15.0 | 1200 | REM | RXP20 | WIN 209 | HER |
| 7/8 | HER | UNIQ | 15.0 | 1155 | FED | 20S1 | WIN 209 | HER |
| 7/8 | HER | UNIQ | 15.0 | 1155 | REM | RXP20 | WIN 209 | HER |
| 7/8 | HER | UNIQ | 15.5 | 1155 | FED | 20S1 | CCI 109 | HER |
| 7/8 | HER | UNIQ | 15.5 | 1155 | REM | RXP20 | CCI 109 | HER |
| 7/8 | HER | UNIQ | 15.5 | 1200 | FED | 20S1 | WIN 209 | HER |
| 7/8 | HER | UNIQ | 16.0 | 1200 | REM | RXP20 | WIN 209 | HER |
| 7/8 | HER | UNIQ | 16.0 | 1200 | FED | 20S1 | CCI 109 | HER |
| 7/8 | HER | UNIQ | 16.0 | 1200 | REM | RXP20 | CCI 109 | HER |
| 7/8 | HER | HERC | 16.5 | 1200 | FED | 20S1 | WIN 209 | HER |
| 7/8 | HER | HERC | 16.5 | 1200 | REM | RXP20 | WIN 209 | HER |
| 7/8 | HER | HERC | 16.5 | 1200 | FED | 20S1 | CCI 109 | HER |
| 7/8 | HER | HERC | 16.5 | 1200 | REM | RXP20 | CCI 109 | HER |
| 7/8 | DUP | 800X | 17.0 | 1210 | REM | RXP20 | WIN 209 | DUP |
| 7/8 | DUP | 800X | 17.0 | 1200 | FED | 20S1 | WIN 209 | DUP |
| 7/8 | DUP | 800X | 16.5 | 1195 | REM | RXP20 | FED 209 | DUP |
| 7/8 | DUP | 800X | 16.5 | 1200 | FED | 20S1 | FED 209 | DUP |
| 1 | HER | HERC | 16.5 | 1165 | REM | RXP20 | WIN 209 | HER |
| 1 | HER | HERC | 16.5 | 1165 | REM | SP20 | WIN 209 | HER |
| 1 | HER | BLDT | 23.0 | 1220 | REM | RXP20 | WIN 209 | HER |
| 1 | HER | BLDT | 23.5 | 1220 | REM | SP20 | WIN 209 | HER |
| 1 | DUP | 800X | 17.0 | 1165 | REM | RXP20 | WIN 209 | DUP |
| 1 | DUP | 800X | 17.0 | 1175 | REM | RXP20 | FED 209 | DUP |
| 1 | DUP | 4756 | 20.5 | 1165 | REM | RP20 | WIN 209 | DUP |
| 1 | DUP | 4756 | 19.5 | 1165 | REM | RP20 | FED 209 | DUP |
| 1 | DUP | 4756 | 21.5 | 1215 | REM | RP20 | WIN 209 | DUP |
| 1 | DUP | 4756 | 22.0 | 1220 | REM | RP20 | CCI 209 | DUP |

*** NOTE: For each star (*) add one (1) 28-gauge, 0.135 inch thick card wad inside bottom of shot cup.

## 20 GAUGE 2 3/4", REMINGTON-PETERS RXP

### PLASTIC TARGET SHELLS

| Shot Weight Oz. | Powder Mfg. | | Grains | Velocity f.p.s. | Wad Mfg/Type | | Primer Mfg/No. | Source |
|---|---|---|---|---|---|---|---|---|
| 3/4 | HER | RDT | 12.5 | 1200 | REM | RXP20 | REM 97* | HER |
| 3/4 | HER | RDT | 12.5 | 1200 | FED | 20S1 | REM 97* | HER |
| 3/4 | HER | GDT | 13.5 | 1200 | REM | RXP20 | REM 97* | HER |
| 3/4 | HER | GDT | 13.5 | 1200 | FED | 20S1 | REM 97* | HER |
| 3/4 | HER | UNIQ | 15.5 | 1200 | REM | RXP20 | REM 97* | HER |
| 3/4 | HER | UNIQ | 15.5 | 1200 | FED | 20S1 | REM 97* | HER |
| 7/8 | HER | RDT | 12.0 | 1155 | REM | RXP20 | REM 97* | HER |
| 7/8 | HER | RDT | 13.0 | 1155 | REM | RXP20 | CCI 109 | HER |
| 7/8 | HER | GDT | 13.0 | 1155 | FED | 20S1 | REM 97* | HER |
| 7/8 | HER | GDT | 14.0 | 1155 | REM | RXP20 | REM 97* | HER |
| 7/8 | HER | GDT | 13.5 | 1155 | FED | 20S1 | CCI 109 | HER |
| 7/8 | HER | GDT | 14.0 | 1155 | REM | RXP20 | CCI 109 | HER |
| 7/8 | HER | GDT | 14.5 | 1200 | REM | RXP20 | CCI 109 | HER |
| 7/8 | HER | UNIQ | 16.0 | 1200 | FED | 20S1 | REM 97* | HER |
| 7/8 | HER | UNIQ | 16.0 | 1200 | REM | RXP20 | REM 97* | HER |
| 7/8 | HER | UNIQ | 16.0 | 1200 | FED | 20S1 | CCI 109 | HER |
| 7/8 | HER | UNIQ | 16.0 | 1200 | REM | RXP20 | CCI 209M | HER |
| 7/8 | HER | HERC | 17.0 | 1200 | REM | RXP20 | REM 97* | HER |
| 7/8 | HER | HERC | 17.0 | 1200 | FED | 20S1 | CCI 109 | HER |
| 7/8 | HER | HERC | 17.0 | 1200 | REM | RXP20 | CCI 109 | HER |
| 7/8 | HER | HERC | 16.5 | 1200 | REM | RXP20 | CCI 209M | HER |
| 7/8 | DUP | 800X | 17.0 | 1205 | REM | RXP20 | REM 97* | DUP |
| 7/8 | DUP | 800X | 16.5 | 1195 | FED | 20S1 | REM 97* | DUP |
| 7/8 | DUP | 800X | 17.0 | 1205 | REM | RXP20 | CCI 209M | DUP |
| 7/8 | DUP | 800X | 16.5 | 1195 | FED | 20S1 | CCI 209M | DUP |
| 7/8 | DUP | 700X | 13.2 | 1140 | REM | RXP20 | REM 97* | DUP |
| 7/8 | DUP | 700X | 12.1 | 1105 | REM | RXP20 | FED 209 | DUP |
| 1 | HER | UNIQ | 15.5 | 1165 | FED | 20S1 | REM 97* | HER |
| 1 | HER | UNIQ | 16.0 | 1165 | REM | RXP20 | REM 97* | HER |
| 1 | HER | HERC | 18.0 | 1220 | REM | RXP20 | REM 97* | HER |
| 1 | DUP | 800X | 16.5 | 1160 | REM | RXP20 | REM 97* | DUP |
| 1 | DUP | 800X | 17.0 | 1165 | REM | RXP20 | REM 97-4 | DUP |
| 1 | DUP | 4756 | 20.5 | 1165 | REM | RP20 | CCI 209 | DUP |
| 1 | DUP | 4756 | 20.0 | 1165 | REM | RP20 | WIN 209 | DUP |
| 1 | DUP | 4756 | 21.5 | 1205 | REM | RP20 | CCI 209 | DUP |
| 1 | DUP | 4756 | 21.5 | 1205 | REM | RP20 | FED 410 | DUP |

20 GAUGE, 3 inch, FEDERAL HI-POWER

PLASTIC SHELLS

| Shot Weight Oz. | Powder Mfg. | Grains | Velocity f.p.s. | Wad Mfg/Type | | Primer Mfg/No. | Source |
|---|---|---|---|---|---|---|---|
| 1-1/8 | HER BLDT | 26.0 | 1230 | REM | RXP20 | FED 209 | HER |
| 1-1/8 | HER BLDT | 27.0 | 1285 | REM | RXP20 | FED 209 | HER |
| 1-1/8 | HER 2400 | 36.0 | 1285 | REM | SP20 | FED 209 | HER |
| 1-1/8 | DUP 4756 | 24.0 | 1215 | REM | SP20 | FED 209 | DUP |
| 1-1/8 | DUP 4756 | 26.0 | 1270 | REM | SP20 | WIN 209 | DUP |
| 1-1/4 | HER BLDT | 25.0 | 1190 | REM | RXP20 | FED 209 | HER |
| 1-1/4 | DUP 4756 | 26.0 | 1195 | REM | RP20 | FED 410 | DUP |
| 1-1/4 | DUP 4756 | 24.0 | 1155 | REM | RP20 | WIN 209 | DUP |
| 1-1/4 | DUP 4227 | 40.0 | 1175 | REM | RP20 | FED 207 | DUP |
| 1-1/4 | DUP 4227 | 40.0 | 1150 | REM | RP20 | WIN 209 | DUP |

20 GAUGE 3 inch, WINCHESTER-WESTERN

COMPRESSION FORMED, SUPER-X

| Shot Weight Oz. | Powder Mfg. | Grains | Velocity f.p.s. | Wad Mfg/Type | | Primer Mfg/No. | Source |
|---|---|---|---|---|---|---|---|
| 1-1/8 | DUP 4756 | 23.5 | 1200 | REM | SP20 | WIN 209 | DUP |
| 1-1/8 | DUP 4756 | 25.0 | 1215 | REM | SP20 | FED 410 | DUP |
| 1-1/4 | DUP 4227 | 39.0 | 1135 | REM | RP20 | WIN 209 | DUP |
| 1-1/4 | DUP 4227 | 39.0 | 1185 | REM | SP20 | FED 209 | DUP |

## 20 GAUGE, 3 inch, REMINGTON SP EXPRESS

| Shot Weight Oz. | Powder Mfg. | | Grains | Velocity f.p.s. | Wad Mfg/Type | | Primer Mfg./No. | Source |
|---|---|---|---|---|---|---|---|---|
| 1 | DUP | 7625 | 21.5 | 1225 | REM | RXP20 | CCI 157 | DUP |
| 1 | DUP | 7625 | 21.5 | 1225 | FED | 2051 | CCI 157 | DUP |
| 1 | DUP | 4756 | 28.5 | 1375 | REM | RP20 | CCI 157 | DUP |
| 1-1/8 | DUP | 4756 | 24.5 | 1205 | REM | RP20 | CCI 157 | DUP |

## 20 GAUGE, 3 inch, WINCHESTER-WESTERN

### PLASTIC AA-TYPE SHELLS

| Shot Weight Oz. | Powder Mfg. | | Grains | Velocity f.p.s. | Wad Mfg/Type | | Primer Mfg./No. | Source |
|---|---|---|---|---|---|---|---|---|
| 1-1/8 | HER | BLDT | 25.5 | 1230 | REM | SP20 | WIN 209 | HER |
| 1-1/4 | HER | BLDT | 24.0 | 1135 | REM | SP20 | WIN 209 | HER |
| 1-1/4 | HER | BLDT | 25.0 | 1190 | REM | SP20 | WIN 209 | HER |
| 1-1/4 | HER | 2400 | 34.5 | 1240 | REM | RP20 | WIN 209 | HER |

# Appendix

**SHOT-CHOICE TABLE**

| Shot Size | Game |
|---|---|
| 9<br>8<br>7½ | Dove, quail, woodcock & like-sized birds |
| 7½<br>6<br>5 | Pheasant, grouse, ducks, woodchucks |
| 4<br>2<br>BB | Squirrel, turkey, fox, pass shooting, ducks & geese |

The following tables are courtesy of Hercules Incorporated.

## Reference Tables

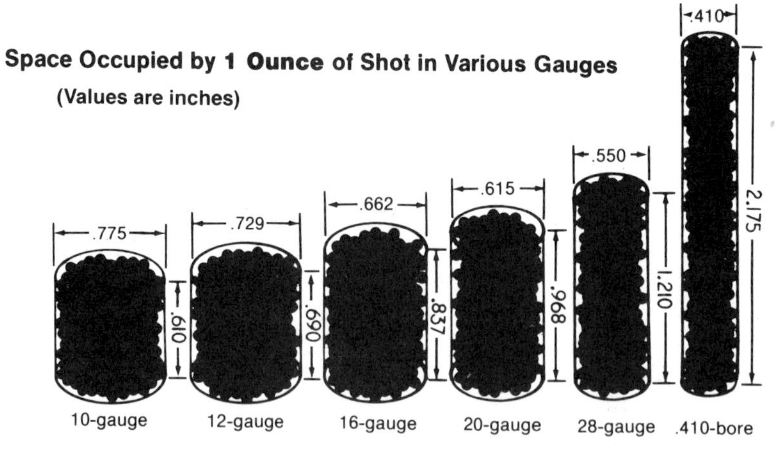


**Internal Diameter of the Barrel in Several Shotgun Gauges**

10-Gauge—0.775-Inch
12-Gauge—0.730-Inch
16-Gauge—0.670-Inch
20-Gauge—0.615-Inch
28-Gauge—0.550-Inch
.410 Bore—0.410-Inch

## Approximate Number of Pellets in Specific Weights of Lead Shot (Sizes 2 Through 9)

| Weight, oz | No. 2 | No. 4 | No. 5 | No. 6 | No. 7½ | No. 8 | No. 8½ | No. 9 |
|---|---|---|---|---|---|---|---|---|
| ½ | 45 | 67 | 85 | 112 | 175 | 205 | 242 | 292 |
| ¾ | 67 | 101 | 127 | 168 | 262 | 308 | 363 | 439 |
| ⅞ | 79 | 118 | 149 | 197 | 306 | 359 | 425 | 512 |
| 1 | 90 | 135 | 170 | 225 | 350 | 410 | 485 | 585 |
| 1⅛ | 101 | 152 | 191 | 253 | 393 | 461 | 545 | 658 |
| 1¼ | 112 | 169 | 213 | 281 | 437 | 513 | 605 | 731 |
| 1⅜ | 124 | 186 | 234 | 309 | 481 | 564 | 665 | 804 |
| 1½ | 135 | 202 | 255 | 337 | 525 | 615 | 730 | 877 |

## Number of Shells That Can Be Loaded With 1 Pound of Powder on Various Grains per Load

(The term grains is a measure of weight: 7,000 grains equals 1 pound)

| Grains/Load | Loads/Pound | Grains/Load | Loads/Pound | Grains/Load | Loads/Pound |
|---|---|---|---|---|---|
| 12 | 583 | 23 | 304 | 34 | 205 |
| 13 | 538 | 24 | 291 | 35 | 200 |
| 14 | 500 | 25 | 280 | 36 | 194 |
| 15 | 466 | 26 | 269 | 37 | 189 |
| 16 | 437 | 27 | 259 | 38 | 184 |
| 17 | 411 | 28 | 250 | 39 | 179 |
| 18 | 388 | 29 | 241 | 40 | 175 |
| 19 | 368 | 30 | 233 | 41 | 170 |
| 20 | 350 | 31 | 225 | 42 | 166 |
| 21 | 333 | 32 | 218 | 43 | 162 |
| 22 | 318 | 33 | 212 | | |

## Typical Percentage of Pellets in a 30-Inch Circle at 40 Yards (Pattern) for Various Choke Sizes (Choke is a Constriction at the Muzzle of a Shotgun Barrel)

Full Choke—70%  Improved Modified Choke—65 to 70%  Modified Choke—55%
Skeet—50 to 60%  Improved Cylinder—50%  True Cylinder—40%

# Index

References to reloading tables are printed in boldface type